Ellen Strauss

The Molecular Biology of the Positive Strand RNA Viruses

The Molecular Biology of the Positive Strand RNA Viruses

Editors

D. J. Rowlands

Wellcome Biotechnology Ltd, Pirbright

M. A. Mayo

Scottish Crop Research Institute, Dundee

B. W. J. Mahy

Animal Virus Research Institute, Pirbright

1987

ACADEMIC PRESS

Harcourt Brace Jovanovich, Publishers

London   Orlando   San Diego   New York
Austin   Boston   Sydney   Tokyo   Toronto

ACADEMIC PRESS INC. (LONDON) LTD.
24/28 Oval Road,
London NW1 7DX

United States Edition published by
ACADEMIC PRESS INC.
Orlando, Florida 32887

Copyright © 1987 by
ACADEMIC PRESS INC. (LONDON) LTD.

All rights reserved
No part of this publication may be reproduced or transmitted in any form or by any means, electronic or mechanical, including photocopy, recording, or any information storage and retrieval system, without permission in writing from the publisher.

British Library Cataloguing in Publication Data

Molecular biology of positive strand RNA viruses.—(FEMS Symposium).
1. Viruses, RNA 2. Molecular biology
I. Rowlands, D. J. II. Mayo, M. A.
III. Mahy, B. W. J. IV. Series
576'.64   QR458
ISBN 0-12-599930-5

Phototypeset by H Charlesworth & Co Ltd,
Huddersfield

Printed by The Alden Press, Oxford, London, Northampton.

Contributors

Almond, J. W. University of Reading, London Road, Reading RG1 5AQ, UK
Arnold, E. Department of Biological Sciences, Purdue University, W. Lafayette, IN 47907, USA
Blaas, D. Institut für Biochemie, der Universität Wien, Währinger Strasse 17, A-1090 Wien, Austria
Butler, P. J. G. Laboratory of Molecular Biology, Hills Road, Cambridge CB2 2QH, UK
Erickson, J. W. Department of Physical Biochemistry, AP-9A, D-47E, Abbott Labs, Abbott Park, North Chicago, IL 60064, USA
Evans, D. M. A. National Institute of Biological Standards and Control, Holly Hill, Hampstead, London NW3 6RB, UK
Frankenberger E. A. Department of Agronomy, Purdue University, W. Lafayette, IN 47907, USA
Griffith, J. P. Department of Biological Sciences, Purdue University, W. Lafayette, IN 47907, USA
Gruendler, P. Institut für Biochemie der Universität Wien, Währinger Strasse 17, A-1090 Wien, Austria
Haenni, A.-L. Institut Jacques Monod, CNRS and Université Paris VII, 2 Place Jussieu – Tour 43, 75251 Paris Cédex 05, France
Hahn, C. S. Division of Biology 156–29, California Institute of Technology, Pasadena, CA 91125, USA
Hecht, H.-J. FG Roentgenstrukturanalyse, Universität Wurzburg, Zentralblau Chemie, AM Hubland, D-8700 Wurzburg, Federal Republic of Germany
Horzinek, M. C. Institute of Virology, Veterinary Faculty, State University, Yalelaan 1, 3506 TD Utrecht, The Netherlands
Johnson, J. E. Department of Biological Sciences, Purdue University, W. Lafayette, IN 47907, USA
Kaesberg, P. Biophysics Laboratory & Biochemistry Department, University of Wisconsin, Madison, WI 53706, USA
Kamer, G. Department of Biological Sciences, Purdue University, W. Lafayette, IN 47907, USA
King, A. M. Q. A.F.R.C. Institute for Animal Disease Research, Pirbright Laboratory, Woking GU24 0NF, UK
King, L. A. Department of Biology, Oxford Polytechnic, Gypsy Lane, Headington, Oxford, UK
Kuechler, E. Institut für Biochemie der Universität Wien, Währinger Strasse 17, A-1090 Wien, Austria
Kuhn, R. J. Department of Microbiology, School of Medicine, State University of New York at Stony Brook, Stony Brook, NY 11794, USA
Luo, M. Department of Biological Sciences, Purdue University, W. Lafayette, IN 47907, USA
Magrath, D. National Institute of Biological Standards and Control, Holly Hill, Hampstead, London NW3 6RB, UK

Marsh, M. Institute of Cancer Research, Chester Beatty Laboratories, Fulham Road, London SW3 6JB, UK
Mayo, M. A. Scottish Crop Research Institute, Invergowrie, Dundee DD2 5DA, UK
McCahon, D. A.F.R.C. Institute for Animal Disease Research, Pirbright Laboratory, Woking GU24 0NF, UK
Minor, P. D. National Institute for Biological Standards and Control, Holly Hill, Hampstead, London NW3 6RB, UK
Moore, N. F. Chemical Defence Establishment, Porton Down, Salisbury, Wiltshire SP4 0JQ, UK
Morch, M.-D. Institut Jacques Monod, CNRS and Université Paris VII, 2 Place Jussieu – Tour 43, 75251 Paris Cédex 05, France
Mosser, A. G. Biophysics Laboratory, University of Wisconsin, 1525 Linden Drive, Madison, WI 53706, USA
Newman, J. W. I. A.F.R.C. Institute for Animal Disease Research, Pirbright Laboratory, Woking GU24 0NF, UK
Ortlepp, S. A. A. F. R. C. Institute for Animal Disease Research, Pirbright Laboratory, Woking GU24 0NF, UK
Palmenberg, A. C. Biophysics Laboratory of the Graduate School, University of Wisconsin, Madison, WI 53706, USA
Pullin, J. S. K. NERC Institute of Virology, Mansfield Road, Oxford OX1 3SR, UK
Rice, C. M. Division of Biology 156-29, California Institute of Technology, Pasadena, CA 91125, USA
Rossmann, M. G. Department of Biological Sciences, Purdue University, W. Lafayette, IN 47907, USA
Rueckert, R. R. Biophysics Laboratory, University of Wisconsin, 1525 Linden Drive, Madison, WI 53706, USA
Schild, G. C. National Institute of Biological Standards, Holly Hill, Hampstead, London NW3 6RB, UK
Sherry, B. Biophysics Laboratory, University of Wisconsin, Madison, WI 53706, USA
Siddell, S. Institute of Virology, Versbacher Strasse 7, D-8700 Würzburg, Federal Republic of Germany
Skern, T. Transgene SA, 11 rue de Molsheim, 67082 Strasbourg, France
Sommergruber, W. Institut für Biochemie der Universität Wien, Währinger Strasse 17, A-1090 Wien, Austria
Spaan, W. J. M. Institute of Virology, Veterinary Faculty, State University, Yalelaan 1, 3506 TD Utrecht, The Netherlands
Strauss, E. G. Division of Biology 156-29, California Institute of Technology, Pasadena, CA 91125, USA
Strauss, J. H. Division of Biology 156-29, California Institute of Technology, Pasadena, CA 91125, USA
van Berlo, M. F. Institute of Virology, Veterinary Faculty, State University, Yalelaan 1, 3506 TD Utrecht, The Netherlands
Vriend, G. Department of Biological Sciences, Purdue University, W. Lafayette, IN 47907, USA
Westrop, G. D. University of Glasgow, Institute of Virology, Church Street, Glasgow G11 5JR, UK
Wimmer, E. Department of Microbiology, School of Medicine, SUNY at Stony Brook, Stony Brook, NY 11794, USA

Preface

This book owes its origin to a meeting, held at Churchill College, Cambridge in June 1985, which was supported by the Society for General Microbiology and the Federation of European Microbiological Societies. The meeting aimed to bring together scientists working with positive-strand RNA viruses of both plants and animals. It was hoped that the occasion would result in mutual education for researchers working on different problems within the general field of positive-strand RNA viruses and would stimulate the cross-fertilisation of ideas.

Over many years the main thrusts of research into plant and animal viruses have been rather different, in part because of the restrictions imposed by the constraints of the culture systems available. The advent of *in vitro* tissue culture techniques meant that most animal viruses could be cultured in simple, well defined conditions ideal for the study of the molecular mechanisms involved in their replication by the use of radio-actively labelled precursors. No such convenient cell culture system was available to plant virologists and although some of the problems have been alleviated by the more recent development of plant cell protoplasts, these are by no means easy to use on a routine basis. However, many plant viruses are produced in very large amounts and because of the ready availability of virus particles much of the molecular biological research has been concentrated on determining the detailed structure of the particles and the forces directing their assembly and stability.

Another difference between viruses from plants and animals (or at least vertebrates) relates to their antigenic structure. This is incidental to the biology of plant viruses, although antigenic differences may be important in a diagnostic sense, but with vertebrate viruses antigenicity represents a crucial point of interaction between host and parasite. This means that study of the antigenic structures and characteristics of animal viruses is vital to understanding this interaction and, of course, has great relevence to the development of vaccines.

In recent years there has been a blurring of these boundaries of the types of study available to plant or animal virologists. The advent of molecular cloning and nucleotide sequence determination is probably the biggest single contribution to this change since it makes the determination of genome structure and organisation and the derivation of protein sequences a relatively trivial task. In so doing it has paved the way to comparisons between virus groups which hint at evolutionary relationships that could not be guessed from classification schemes based on morphology or host ranges. Developments in cell-free *in vitro* techniques for the analysis of specific aspects of virus replication, such as protein translation and processing are also helping to breakdown host-determined barriers in research. Finally, the application of X-ray crystallographic techniques to determine the structure of animal viruses (such as the structure of human rhinovirus reported at the meeting) will have enormous impact on the understanding of antigenic structure, receptor interactions, antibody-mediated neutralisation and many other important factors.

This was a particularly fortunate time to hold an interdisciplinary meeting. We are most grateful to all the contributors who have provided masterful reviews of their subjects as well as presenting up to date information and, in the spirit of the meeting, have highlighted the interdisciplinary implications of their work. Their presentation excited the participants at the meeting in Cambridge and are now available to a wider audience in this book.

We wish to thank the Society for General Microbiology, the Federation of European Microbiological Societies and the Wellcome Foundation who provided sponsorship for the meeting, and Penny Scott for invaluable assistance in correcting the proofs.

D. J. Rowlands

Contents

	Contributors	v
	Preface	vii
1.	Genome Organization, Translation and Processing in Picornaviruses A. C. Palmenberg	1
2.	The Replication of Picornaviruses R. J. Kuhn and E. Wimmer	17
3.	Investigation of the Molecular Basis of Attenuation in the Sabin Type 3 Vaccine Using Novel Recombinant Polioviruses Constructed from Infectious cDNA G. D. Westrop, D. M. A. Evans, P. D. Minor, D. Magrath, G. C. Schild and J. W. Almond	53
4.	Molecular Biological Aspects of Human Rhinovirus T. Skern, W. Sommergruber, P. Gruendler, D. Blaas and E. Kuechler	61
5.	Insect Picornaviruses N. F. Moore, L. A. King and J. S. K. Pullin	67
6.	The Genomes of Alphaviruses and Flaviviruses: Organization and Translation J. H. Strauss, E. G. Strauss, C. S. Hahn and C. M. Rice	75
7.	Replication of Equine Arteritis Virus (EAV): A Comparative Review M. F. van Berlo, W. J. M. Spaan and M. C. Horzinek	105
8.	The Organization and Expression of Coronavirus Genomes S. Siddell	117
9.	Genetic Recombination in RNA Viruses A. M. Q. King, S. A. Ortlepp, J. W. I. Newman and D. McCahon	129
10.	Organization of Plant Virus Genomes that Comprise a Single RNA Molecule M.-D. Morch and A.-L. Haenni	153
11.	A Comparison of the Translation Strategies used by Bipartite Genome, RNA Plant Viruses M. A. Mayo	177
12.	Organization of Bipartite Insect Virus Genomes: The Genome of Black Beetle Virus P. Kaesberg	207

13.	Organization of Tripartite Plant Virus Genomes: The Genome of Brome Mosaic Virus *P. Kaesberg*	219
14.	Molecular Architecture and Assembly of Tobacco Mosaic Virus Particles *P. J. G. Butler and M. A. Mayo*	237
15.	Structure of Picornavirus Coat Proteins and Their Antigenicity *P. D. Minor*	259
16.	Enveloped Virus Entry *M. Marsh*	281
17.	The Structure of a Human Common Cold Virus (Rhinovirus 14) and its Structural and Functional Relations to Other Picornaviruses and Plant Viruses *M. G. Rossmann, E. Arnold, J. W. Erickson, E. A. Frankenberger, J. P. Griffith, H.-J. Hecht, J. E. Johnson, G. Kamer, M. Luo, A. G. Mosser, R. R. Rueckert, B. Sherry and G. Vriend*	301

1. Genome Organization, Translation and Processing in Picornaviruses

Ann C. Palmenberg

Biophysics Laboratory of the Graduate School, The University of Wisconsin, Madison, Wisconsin 53706, U.S.A.

I. Classification of picornaviruses	1
II. Structure of the RNA genome	3
III. Polyprotein translation	5
IV. Organization of the viral polyprotein	7
V. Polyprotein homology	9
VI. Proteolytic processing	11
References	13

I. CLASSIFICATION OF PICORNAVIRUSES

Although physically among the smallest of positive strand RNA viruses, the picornaviruses are of major historic, economic and medical importance. The family contains a variety of highly virulent human and animal pathogens, including polio-, rhino- (common cold), hepatitis A, coxsackie-, Mengo- (murine encephalomyocarditis), swine vesicular disease, and foot-and-mouth disease viruses. In spite of the disparate afflictions caused by these agents, recent advances in molecular virology, nucleotide sequencing and x-ray crystallography have shown that all picornaviruses are remarkably similar in their particle structure and genome organization.

Classification of a virus as a picornavirus is based upon the structure of its virion and its overall biological character. The single-stranded viral RNA is encapsidated within a 60-subunit protein shell (20–30 nm diameter) having intrinsic 5 : 3 : 2 icosohedral symmetry. Each subunit (pro-

Table 1.1. Physical properties of picornaviruses*

	Enterovirus	Rhinovirus	Cardiovirus	Aphthovirus
Sedimentation coefficient	156	149	156	142–146
Buoyant density in CsCl	1.33–1.35	1.38–1.41[a]	1.34	1.43
pH 3	Stable	Labile	Stable	Labile
pH 5–6	Stable	Becoming labile	Stable[b]	Labile[c]
% RNA[d]	31	31	34	38
Average RNA length[e]	7445	7155	7832	8400
poly C tract	−	−	+	+

*Data from Rueckert (1985) and Rueckert and Wimmer (1984).
[a]Rhinovirus takes up CsCl, which changes the buoyant density.
[b]EMC is labile in 0.1 M Cl^- or Br^-.
[c]Becomes stable at high ionic strength.
[d]Calculated from sequence data, as sodium salt of the RNA.
[e]Heteropolymeric region, without 3' poly(A) tail.

tomer) is composed of four non-identical polypeptide chains that derive collectively from proteolytic cleavage of a large precursor molecule. Individual variation in capsid peptide size and sequence among the viruses is responsible for the diversity in picornaviral antigenicity and host-cell receptor interaction.

Electron microscopy cannot distinguish particle surface differences among the viruses. However, physical properties such as pH lability, buoyant density and thermostability are commonly used to subdivide picornaviruses into four groups, or genera (Table 1.1). (For reviews of structure and classification, see Rossmann *et al.*, 1985, this volume Ch. 17; Rueckert, 1976, 1985).

Enteroviruses are typified by human poliovirus (three serotypes), but also include murine poliovirus (Theiler's virus), coxsackie viruses A and B, swine vesicular disease virus, human echoviruses (enteric cytopathic human orphan), hepatitis A virus, and other human and animal enteroviruses. All enterovirus particles are stable at a wide range of pH values. In the absence of $MgCl_2$, enterovirus capsid structures can be denatured by heat (50°C, 1 h).

Rhinoviruses are a major causative agent of the common cold in humans (89 identified serotypes) and cattle (two serotypes). These viruses are acid-labile below pH 6, and unlike enteroviruses, can be denatured by heat in the presence of $MgCl_2$. Despite these physical differences, homologous nucleotide and protein sequences suggest that enteroviruses and rhinoviruses are actually quite closely related. Equine rhinoviruses are unlike those of humans and cattle and are usually classified within a separate, or "unassigned" subdivision.

Aphthoviruses or foot-and-mouth disease viruses (FMDV, 7 serotypes) infect many types of cloven-hoofed animals (pigs, sheep, goats, cattle). Lack of effective disease control has created devastating agricultural problems in many areas of the world. Aphthoviruses are acid-labile; the capsid structures dissociate at pH <7. They have the largest RNA genome, the smallest protomer structure and, correspondingly, the highest buoyant density (1.42 g ml^{-1}) of any picornavirus.

Cardioviruses such as encephalomyocarditis virus (EMC), Mengovirus, Maus Elberfeld (ME) and Columbia SK are primarily murine in host range, though humans and swine are sometimes also susceptible. The cardioviruses are serologically indistinguishable from one another, but sequence studies indicate individual features that could influence receptor binding and host-cell specificity. Cardioviruses exibit biphasic pH stability and are not thermolabile (when compared to enteroviruses or rhinoviruses). Genome structure and nucleotide comparisons suggest that cardioviruses may represent an evolutionary "intermediate" group between the entero-rhinoviruses, and the aphthoviruses.

II. STRUCTURE OF THE RNA GENOME

Complete nucleotide sequences are now available for many picornaviral RNAs (see Table 1.2). The cardioviruses and aphthoviruses are distinguished by the presence of a 5'-proximal polycytidylate (poly(C)) tract, whose length (50–200 bases) and exact location relative to the 5' end of the genome (150–400 bases) vary with different isolates of virus (Black *et al.*, 1979). Artificial deletion of the poly(C) region abolishes infectivity of these RNAs, but does not destroy coding capacity or efficiency as mRNA (Rowlands *et al.*, 1978; Sangar *et al.*, 1980; Costa Giomi *et al.*, 1984). Computer-generated models predict that the poly(C) tract may play a role in stabilizing certain RNA secondary structures (Palmenberg, unpublished observation), but since rhinovirus and enterovirus genomes do not have a 5'-proximal poly(C) and yet function as effective, infective pathogens, the biological role of this region remains unclear.

The 3' ends of all picornaviral RNAs are polyadenylated, which is characteristic of most eukaryotic mRNAs (Ahlquist and Kaesberg, 1979). However, the 5' ends are not capped in the usual manner with 5'–5' triphosphate linkages. Instead, the viruses have a small (20–24 amino acids) virus-coded, genome-linked protein (VPg) attached by a tyrosine-O^4-phosphodiester bond to the 5' pUp of the RNA (Nomoto *et al.*, 1976; Rothberg *et al.*, 1980). VPg sequences are rich in basic, hydrophilic amino acids, and have only one tyrosine residue (the attachment site) at position 3 from the amino end of the peptide. The specific function of a VPg linkage to RNA is unknown, although some lines of evidence suggest the protein may serve a role in the initiation of positive and negative strand RNA synthesis (Ambros and Baltimore, 1980; Nomoto *et al.*, 1976; Pettersson *et al.*, 1978a; Rothberg *et al.*, 1980; this volume, Chapter 2).

The length of genomic RNAs varies from 7102 bases (rhinovirus type 2)

Table 1.2. Properties of the RNAs

	Genome length bases	5' Non-coding bases	3' Non-coding bases	Poly-protein codons	AUGs before translation start site codons	Reference
Polio 1 Mahoney	7440	742	71	2209	8	[20, 31]
Polio 1 Sabin	7441	742	72	2209	8	[47]
Polio 2 Sabin	7440	747	72	2207	6	[47]
Polio 3 Sabin	7432	742	72	2206	7	[47]
Coxsackie B3	(~7400)[a]	738	101	(~2200)[a]	6	[42, 49]
Hepatitis A	7478	733	64	2227	10	[23]
Rhino 14	7208	624	47	2179	13	[8, 43]
Rhino 2	7102	610	42	2150	11	[41]
EMC	7835	833	126	2292	10	[29, 52]
FMDV A10	8280	1188	96	2332	8	[6, 9[b]]
FMDV A12	(~8400)[a]	(~1200)[a]	96	2332	>6[a]	[32]
FMDV 01K	(~8450)[a]	(~1300)[a]	92	2332	>8[a]	[15]

[a]Based on partial sequence data.
[b]Also Rowlands, personal communication.

Table 1.3. Genome base composition (%)*

	A	C	G	U	Reference
Polio 1 Mahoney	29.6	23.3	23.0	24.1	[20, 31]
Polio 1 Sabin	29.7	23.2	23.0	24.1	[47]
Polio 2 Sabin	29.2	23.1	23.3	24.4	[47]
Polio 3 Sabin	28.9	23.4	23.5	24.2	[43]
Coxsackie B3[a]	28.5	23.3	24.6	23.6	[40, 45]
Hepatitis A	29.2	16.2	22.0	32.6	[23]
Rhino 14	32.1	20.1	20.4	27.4	[8,43]
Rhino 2	32.7	19.0	20.0	28.3	[41]
EMC	25.7	25.9	23.6	24.8	[29, 52]
FMDV A10[b]	25.7	27.6	25.2	21.5	[6, 9]
FMDV A12[c]	25.4	27.9	25.5	21.2	[32]
FMDV 01K[d]	25.3	28.1	25.5	21.1	[15]

*All values exclude 3' poly(A) tail.
[a]Based on partial sequence of 5366 nucleotides [40, 45].
[b]Based on 3' partial sequence of 7322 nucleotides [6, 9].
[c]Based on 3' partial sequence of 7712 nucleotides [32].
[d]Based on 3' partial sequence of 7815 nucleotides [15].

to 8450 bases (approximate size of FMDV type O1K), not including the 3' polyadenylate tail of 50–150 bases (Table 1.2). Each encodes a single, giant peptide called the "polyprotein", which represents 85–90% of the theoretical coding capacity of the genome. The remainder of the bases are distributed between the 5' (610–1300 bases) and 3' (42–126 bases) noncoding regions.

Nucleotide compositions are different and generally characteristic for each genus (Table 1.3). Rhinoviruses (and hepatitis A virus) have a distinct bias toward high $A+U$ content ($>60\%$). The remaining enteroviruses ($A+U > 52\%$), cardioviruses ($A+U \sim 50\%$), and aphthoviruses ($A+U < 47\%$) have progressively lower $A+U$ content. The unusual skewing of base compositions is not unduly affected by the presence or absence of poly(C) tracts, since these regions do not contribute a significant numerical weight to the calculations. The high $A+U$ content of the rhino- and hepatitis viruses is distributed throughout their genomes, and is also reflected by the codon preference patterns used during translation (see Fig. 1.1).

III. POLYPROTEIN TRANSLATION

Virion RNAs are exceedingly good templates for translation, both in infected cells, and in cell-free extracts. Reticulocyte lysates programmed with EMC or FMDV routinely synthesize an average of 25–30 polyproteins from each RNA molecule (Grubman and Baxt, 1982; Shih et al., 1979). Translation in vitro of poliovirus RNA is somewhat less efficient, but this property may be dependent upon the activity of initiation factors within a given lysate (Brown and Ehrenfeld, 1979). Pronase digestion, to remove VPg, does not affect mRNA activity (Ambros and Baltimore, 1980).

Except for the aphthoviruses, translation always proceeds from a unique start site on each RNA. With FMDV, two in-phase AUG codons (separated by 84 bases) initiate with equal frequency (Beck et al., 1983). Very little is known about mechanisms by which ribosomes select or initiate at the correct translational start sites. Scanning hypotheses predict that 40S subunits should enter at the 5' end of the RNA, and move in a 3' direction until reaching appropriate AUG sequences (Kozak, 1981; Kozak, 1983). However, structural features of picornavirus RNAs may be inconsistent with this type of model. The 5' VPg protein, poly(C) tracts, numerous (6–13) non-coding AUG triplets and extensive heteropolymeric sequences (600–1300 bases) must be passed over before scanning ribosomes could reach the beginning of the long open reading frame. Correct initiation in these viruses may therefore be aided by use of RNA secondary structure, special sequences, or other features that enhance ribosome positioning and recognition.

Elongation of the nascent polypeptide proceeds at an uneven rate. Cell-free experiments have shown that ribosomes slow down or pause for several minutes as they reach the middle of aphthoviral and cardioviral RNAs (Shih et al., 1979; Svitkin and Agol, 1983). The translational barrier

Polyprotein Codon Selection (Number used per 1000 Codons)

Codon	FMD;EMC;Polio;Rhino	AA	Codon	FMD;EMC;Polio;Rhino	AA	Codon	FMD;EMC;Polio;Rhino	AA	Codon	FMD;EMC;Polio;Rhino	AA
UUU	20;28;18;23	Phe	UCU	6;20; 9;13	Ser	UAU	4;16;18;23	Tyr	UGU	7; 6;10;15	Cys
UUC	27;28;20;16	Phe	UCC	15;17;15; 8	Ser	UAC	33;17;26;16	Tyr	UGC	8; 7; 9; 6	Cys
UUA	1; 6;11;19	Leu	UCA	9;16;22;29	Ser	UAA	0; 0; 0; 0	End	UGA	0; 0; 0; 0	End
UUG	18;20;16;20	Leu	UCG	10; 6; 5; 1	Ser	UAG	0; 0; 0; 0	End	UGG	11;14;13;11	Trp
CUU	21;14;10;13	Leu	CCU	15;16;14;15	Pro	CAU	2;11; 9;14	His	CGU	7; 7; 4; 3	Arg
CUC	23;14;12; 7	Leu	CCC	16;18; 9;14	Pro	CAC	24; 7;15;11	His	CGC	12; 5; 3; 1	Arg
CUA	3;12;15;16	Leu	CCA	12;22;25;28	Pro	CAA	14;18;17;24	Gln	CGA	2; 2; 1; 2	Arg
CUG	17;20;15;13	Leu	CCG	11; 7; 6; 2	Pro	CAG	21;26;19;13	Gln	CGG	5; 6; 3; 0	Arg
AUU	19;22;24;28	Ile	ACU	16;25;23;25	Thr	AAU	6;26;22;33	Asn	AGU	6; 7; 9;15	Ser
AUC	24; 8;21;14	Ile	ACC	27;26;24;13	Thr	AAC	37;19;31;21	Asn	AGC	10; 5; 9; 6	Ser
AUA	2;10;16;26	Ile	ACA	18;23;22;37	Thr	AAA	27;19;29;39	Lys	AGA	12;22;22;20	Arg
AUG	25;22;30;24	Met	ACG	11; 5; 7; 4	Thr	AAG	36;31;28;23	Lys	AGG	6; 6;11;10	Arg
GUU	18;17;10;24	Val	GCU	18;24;22;14	Ala	GAU	18;27;24;41	Asp	GGU	15;15;17;26	Gly
GUC	20;19;13;14	Val	GCC	30;29;16;10	Ala	GAC	47;20;28;12	Asp	GGC	22;14;11; 8	Gly
GUA	6;14;13;16	Val	GCA	24;15;27;26	Ala	GAA	16;25;29;36	Glu	GGA	15;17;22;21	Gly
GUG	32;28;26;16	Val	GCG	9;11; 8; 1	Ala	GAG	37;30;22;11	Glu	GGG	15;12;14;12	Gly

Sum by Base Position (%)		Sum by Base Position (%)		Sum by Base Position (%)		Sum by Base Position (%)	
Uxx	17;20;19;20	Cxx	21;21;18;18	Axx	28;28;33;33	Gxx	34;31;30;29
xUx	28;28;27;29	xCx	25;28;25;24	xAx	32;29;32;32	xGx	15;15;16;15
xxU	20;28;24;33	xxC	38;25;27;17	xxA	16;22;27;34	xxG	26;25;22;16

Figure 1.1. Polyprotein codon selection. Codons within the long open reading frame of FMDV A10, EMC, polio 1 Sabin and rhino 14 were summed, and their frequency tabulated as use per 1000 codons (see references in Table 1.2). The bottom frame shows the total percentage use for each base within a codon.

can be relieved by addition of elongation factor eEF-2 and high concentrations of tRNA, but the impediment that causes the pause has yet to be determined. Examination of RNA sequences has shown that rare codons are not uniquely distributed within this region and cannot be solely responsible for the translational delay.

All picornavirus polyproteins have similar amino acid compositions, but codon selection, especially in the third position of the triplet, reflects the nucleotide composition bias that is characteristic of the genus (Fig. 1.1). Most rhinovirus codons end in A or U (xxA + xxU = 67%) rather than G or C (xxG + xxC = 33%). The preference is reversed for the aphthoviruses (xxA + xxU = 36%, xxG + xxC = 64%). Cardioviruses and enteroviruses (except hepatitis A, which is like rhinovirus) have no obvious discrimination in the third codon position. The factors that influence codon and nucleotide usage are subject to speculation. Presumably, host and tissue tropism must be important, although it is interesting that rhinovirus and poliovirus, both human pathogens, have somewhat different coding patterns. These trends may be useful in categorizing specific viruses or in defining evolutionary relationships.

IV. ORGANIZATION OF THE VIRAL POLYPROTEIN

Mature viral proteins are derived by progressive, post-translational cleavage of the polyprotein. The full-length precursor is only rarely observed experimentally, because the first processing events usually occur while the peptides are still nascent on ribosomes. Individual molecular weights can vary somewhat for different isolates, but the overall proteolytic cascade scheme is very similar along the viruses (Fig. 1.2).

To simplify homologue identification, a uniform nomenclature system, designated L-4-3-4, was adopted in 1983 by the European Study Group on the Molecular Biology of Picornaviruses (Rueckert and Wimmer, 1984). Accordingly, mature proteins and their precursors are subdivided into four groups (L, P1, P2, P3) on the basis of structure, enzymatic function and position of the primary cleavages

The leader or "L" proteins are present only in cardioviruses and aphthoviruses. The EMC and Mengo leaders have molecular weights of about 8kD. FMD viruses have two nested L peptides (mol. wts 16 kD and 23 kD), which share common carboxyl termini, but have different, in-phase, translational start sites (Beck et al., 1983; Carroll et al., 1984; Forss et al., 1984; Robertson et al., 1985). A proteolytic activity has been ascribed to the leader peptide (Strebel and Beck, 1980), but the reason why the protein invariably occurs in two forms has not been determined.

The four P1 polypeptides are the capsid structural proteins. They are commonly called VP1, VP2, VP3 and VP4 (1D, 1B, 1C, 1A), in order of descending molecular weight on polyacrylamide gels (polio: 33, 30, 26, 7 kD). Protein VPO (1AB), the uncleaved precursor of VP4 + VP2, can also be detected at trace levels, in virions. Cell-free processing experiments

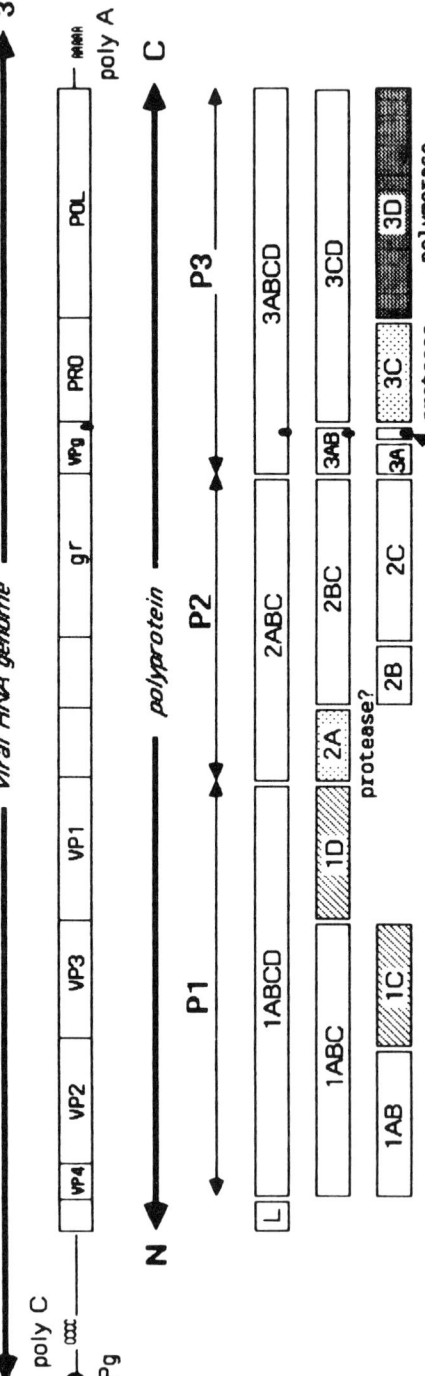

Figure 1.2. Structure of the picornaviral genome. The poly C tract within the 5' non-coding region and the leader protein, L, are found in the cardioviruses and aphthoviruses but not in enteroviruses or rhinoviruses. The polyprotein of FMDV contains three 3B (VPg) segments, while that of the other viruses contain only one. g^r represents the guanidine resistance marker, a genetic locus affecting the action of a drug thought to block initiation of RNA synthesis. Protein nomenclature is that adopted in 1983 by the European Study Group on the Molecular Biology of Picornaviruses.

suggest that capsid polypeptides that derive from a common precursor molecule stay together as a protomer unit throughout particle morphogenesis (Palmenberg, 1982).

The middle portion of the polyprotein yields polypeptides 2A, 2B and 2C (polio: 12, 14 and 36 kD). FMDV genomes have very small or deleted 2A sequences compared to the other viruses. While not all the P2 polypeptides have been assigned definitive biological roles, recent experiments have shown that poliovirus 2A protein is proteolytically involved in the initial step of polyprotein processing (Toyoda et al., 1986). Protein 2C is the probable genetic locus of the guanidine resistance marker, a compound that affects the initiation of RNA synthesis (Anderson-Sillman et al., 1984). However, 2C is not a polymerase, and its contribution to the replication cycle remains unclear.

The P3 polypeptides, 3A, 3B, 3C and 3D (polio: 9, 2, 20 and 51 kD) are more closely associated with genome replication. Purified preparations of 3D can catalyse elongation of nascent RNA chains in primer-dependent reactions, an activity that identifies this enzyme as the central element of viral polymerase complexes (Flanegan and Baltimore, 1977, 1979; Van Dyke and Flanegan, 1980; Van Dyke et al., 1982). Protein 3B is VPg, the peptide attached to the 5' end of the genome (Pallansch et al., 1980). Aphthoviruses have three tandemly-linked different VPg sequences at this position of the polyprotein, making the FMDV P3 segment somewhat longer than in other viruses (Forss and Schaller, 1982). Initiation of positive and negative strand RNA synthesis may require VPg, either as free protein, or as part of the donor polypeptide, 3AB. Protein 3C is a virus-specific proteolytic enzyme responsible for many of the post-translational cleavage events within the polyprotein (see below) (Gorbalenya et al., 1979; Palmenberg et al., 1979; Svitkin et al., 1979).

V. POLYPROTEIN HOMOLOGY

The dot matrix comparisons in Fig. 1.3 graphically depict the homologous relationships among representative polyprotein sequences. Points along the diagonal (left to right) represent regions where five out of nine amino acids match in orientation and identity. This particular stringency shows the general distribution of sequence similarities, but should be considered only a minimum estimate of the overall amino acid homology. Although these plots were calculated for polio 1 Sabin, EMC, rhino 14 and FMDV A10 viruses, the patterns are typical for the genera, and do not change substantially when other viruses are compared (e.g. polio 2 Sabin, polio 3 Sabin, coxsackie B3, Mengo, FMDV A12, FMDV O1K or rhino 2).

As can be seen, the carboxyl end regions of all polyproteins (protein 3D-region, upper right of each plot) share strong amino acid homology, as does a segment from the middle region, representing protein 2C. Since these peptides are involved in RNA replication, a vital biological function, conservation (or preservation) among the viruses is not unexpected.

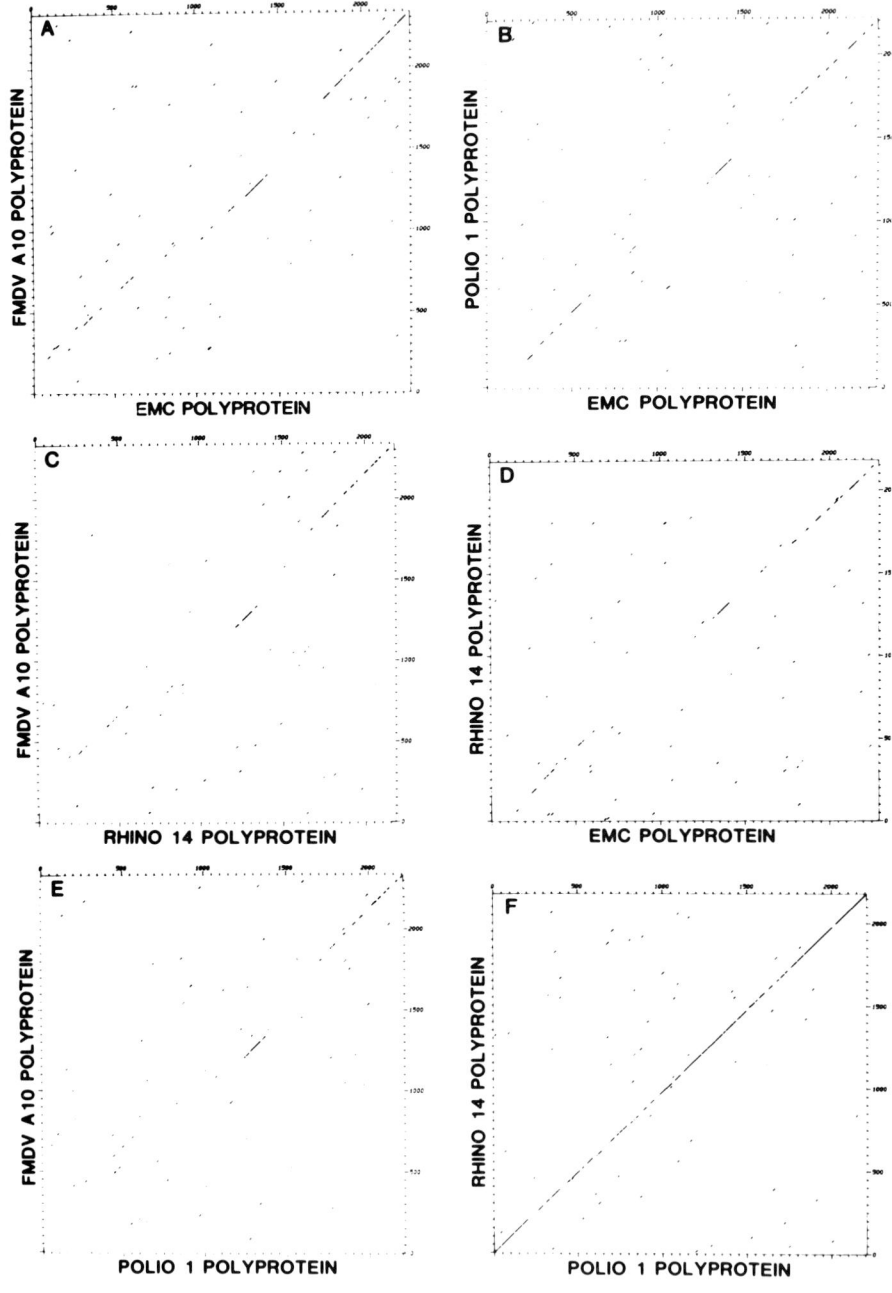

Comparison of P1 peptides (lower left of each plot) shows a lower degree of similarity. The variation in capsid sequences is responsible for specific differences in virion stability, antigenicity and host-cell binding properties.

Clearly, the most closely related polyproteins, overall, are those of poliovirus and rhinovirus (plot F), which share extensive common segments throughout their lengths. Polio and FMDV, or rhino and FMDV (plots C and E) have many fewer dots on the diagonal, indicating their proteins have much less homology. EMC, when compared to any other sequence (plots A, B, and D), gives scattered dots of intermediate density and distribution.

When taken together with data for base composition, genome organization and codon frequency, the protein dot plots strongly suggest a linear developmental relationship among picornaviral sequences. The comparisons indicate that rhinoviruses and enteroviruses can (almost) be considered members of the same subgroup. The cardioviruses have genome structures, compositions and sequence arrangements that place them intermediate between the aphtho- and rhino-enterovirus groups. The validity of a linear correlation among the genera will be determined as more viruses are analysed.

VI. PROTEOLYTIC PROCESSING

The first proteolytic cleavage within the polyprotein occurs while the chain is still nascent on a ribosome. The processing site for poliovirus is between the P1/P2 regions at a Tyr-Gly dipeptide pair (Kitimura et al., 1981; Racaniello and Baltimore, 1981; Toyoda et al., 1984). In EMC, the initial cleavage is between peptides 2A/2B, a Gln-Gly sequence (Palmenberg et al., 1984). These events have been attributed to catalysis by a hypothetical "host" protease, whose origin and specificity were never accurately characterized. However, recent genetic cloning experiments with poliovirus present compelling evidence that these sites may actually be cleaved by viral peptide 2A from within the P2 region (Toyoda et al., 1986). The potential autocatalytic nature of this cleavage is presently being investigated.

Considerable effort has also been directed towards characterization of the protease(s) responsible for subsequent, post-translational polyprotein cleavages. Attention has focused primarily on viral peptide 3C, because partially purified preparations of 3C from EMC virus can be shown to catalyse P1 region processing of 1AB/1C and 1C/1D (Gorbalenya et al.,

---

Figure 1.3. Dot matrix comparisons of polyproteins. Complete polyprotein sequences of FMDV A10, EMC, polio 1 Sabin and rhino 14 were analysed in all possible pairwise combinations by the dot matrix comparison program COMPARE (Devereux et al., 1984, Genetics Computer Group, University of Wisconsin). The indicated amino acid sequences are represented horizontally and vertically. Dots are placed within the matrix wherever five out of nine amino acids from the two sequences match exactly in identity and orientation.

Table 1.4. Proteolytic cleavage sequences within the polyproteins. Cleavage occurs between the indicated dipeptide pairs

	L/1A	1A/1B	1B/1C	1C/1D	1D/2A	2A/2B	2B/2C	2C/3A	3A/3B	3B/3C	3C/3D
Polio (all types)	—[d]	NS	QG	QG	YG	QG	QG	QG	QG	QG	QG
Coxsackie B3	—[d]	NS	QG	QG	TI[a]	QG	?[d]	?[d]	?[d]	?[d]	QG
Hepatitis A	—[d]	AD	QM	QV	QA	QG	QM[a]/QS[b]	QG[a]/QS[b]	HF[a]/EG[b]	QV[a]/ES[b]	QA[a]/ES[b]
Rhino 14	—[d]	NS	QG	EG	YG	QG	QA	QG	QG	QG	QG
Rhino 2	—[d]	QS	QG	QN	YV	QG	ES	QG	QG	QG	QG
EMC	QG	AD	QS	QG	ES	QG	QS	QG	QG	QG	QG
FMDV A10	GQ[a]/KG[b]	AD	EG	QT	LN	ES[b]	QL	QI	EG[c]	ES	EG
FMDV A12	GQ[a]/KG[b]	AD	VG	QT	LN	RP[a]/ES[b]	QL	QI	EG[c]	ES	EG
FMDV 01K	GN[a]/KG[b]	AD	EG	ET	LN	GP[a]/ES[b]	QL	QI	EG[c]	ES	EG

[a] Considered tentative, based on published sequence.
[b] Possible site based on homologous protein alignment, suggested by author.
[c] FMDV cleavages between tandomly linked 3B proteins all occur between identical EG dipeptides.
[d] (—) indicates no analogous cleavage site exists within that polyprotein; (?) indicates that the protein sequence is not known.

1979; Palmenberg et al., 1979; Svitkin et al., 1979). Dilution experiments with reticulocyte extracts have also demonstrated that 3C sequences are capable of monomolecular, autocatalytic reactions within the P3 region (3AB/3C and 3C/3D) (Palmenberg and Reuckert, 1982), and monoclonal antibodies directed against poliovirus 3C peptide inhibit cell-free cleavage at all major sites except P1/P2 (the 2A cleavage) and 1A/1B (the final maturation cleavage, see below) (Hanecak et al., 1982). Accordingly, it has been proposed that monomolecular, self-cleavage reactions within the P3 region (3AB/3CD and 3C/3D) are the probable mechanism by which the 3C peptide is released initially from the polyprotein (Palmenberg and Reuckert, 1982). The protease, as free enzyme, then catalyses subsequent P1, P3 (and P2?) processing reactions (3A/3B, 1AB/1C, 1C/1D etc.).

Protease 3C-directed cleavages appear to be completely virus-specific. That is, aside from the processing sites within the polyprotein itself, the enzyme has no other known natural substrates. Proteolytic cross-reactivity between different viruses has never been rigorously tested, but preliminary mixing experiments with poliovirus and EMC or FMDV and EMC have failed to detect processing (Palmenberg, unpublished observation). The specificity is somewhat surprising in view of the similarity among cleavage sequences (Table 1.4). Most 3C-directed cleavages occur between glutamine (or glutamate) and glycine (or serine) dipeptide pairs. The surrounding sequences (not shown) have a high frequency of helix-breaking residues, such as proline and threonine, which leads to speculation that protein secondary structure probably plays a role in cleavage site identification.

The last maturation cleavage within a picornaviral polyprotein occurs between capsid peptides 1A/1B. This event is observable only during the final stages of virion morphogenesis, and is presumably dependent upon RNA encapsidation (Reuckert, 1976, 1985). Within a particle, all 60 subunits seem to process simultaneously (or at least very rapidly), because partially cleaved virions are only rarely observed. Translation experiments *in vitro* with EMC and crystallographic data from rhinovirus 14 both point tantalizingly toward a self-cleavage reaction within the 1AB protein (Rossmann et al., 1985, this volume Chapter 17; Palmenberg, unpublished observations). If this possibility is substantiated, then every proteolytic step within the picornaviral protein cascade can be attributed to viral proteases.

REFERENCES

1. Ahlquist, P. and P. Kaesberg (1979). *Nucl. Acids Res.* **7**, 1195–1204.
2. Ambros, V. and D. Baltimore (1980). *J. Biol. Chem.* **255**, 6739–6744.
3. Anderson-Sillman, K., S. Bartal and D. R. Tershak (1984). *J. Virol.* **50**, 922–928.
4. Beck, E., S. Forss, K. Strebel, R. Cattaneo and G. Feil (1983). *Nucl. Acids Res.* **11**, 7873–7885.
5. Black, D. N., P. Stephenson, D. J. Rowlands and F. Brown (1979). *Nucl. Acids Res.* **6**, 2381–2390.

6. Boothroyd, J., T. Harris, D. Rowlands and P. Lowe (1982). *Gene* **17**, 153–161.
7. Brown, B. A. and E. Ehrenfeld (1979). *Virology* **97**, 396–405.
8. Callahan, P., S. Mizutani and C. Colonno (1985). *Proc. Natl Acad. Sci. USA* **82**, 732–736.
9. Carroll, A., D. Rowlands and B. E. Clarke (1984). *Nucl. Acids Res.* **12**, 2461–2472.
10. Costa Giomi, M. P., I. E. Bergmann, E. A. Scodeller, P. Auge de Mello, I. Gomez and J. L. La Torre (1984). *J. Virol.* **51**, 799–805.
11. Devereux, J., P. Haeberli and O. Smithies (1984). *Nucl. Acids Res.* **12**, 378–395.
12. Flanegan, J. B. and D. Baltimore (1977). *Proc. Natl Acad. Sci. USA* **74**, 2677–2680.
13. Flanegan, J. B. and D. Baltimore (1979). *J. Virol.* **29**, 352–360.
14. Forss, S. and H. Schaller (1982). *Nucl. Acids Res.* **10**, 6441–6450.
15. Forss, S., K. Strebel, E. Beck and H. Schaller (1984). *Nucl. Acids Res.* **12**, 6587–6601.
16. Gorbalenya, A. E., Y. V. Svitkin, Y. A. Kazachkov and V. I. Agol (1979). *FEBS Lett.* **108**, 1–5.
17. Grubman, M. J. and B. Baxt (1982). *Virology* **116**, 19–30.
18. Hanecak, R., B. L. Semler, C. W. Anderson and E. Wimmer (1982). *Proc. Natl Acad. Sci. USA* **79**, 3973–3977.
19. Harris, T. J. R. and F. Brown (1976). *J. Gen. Virol.* **33**, 493–501.
20. Kitimura, N., B. Semler, P. G. Rothberg, G. R. Larsen, C. J. Adler, A. J. Dorner, E. A. Emini, R. Hanecak, J. J. Lee, S. van der Werf, C. W. Anderson and E. Wimmer (1981). *Nature (London)* **291**, 547–553.
21. Kozak, M. (1981). *Nucl. Acids Res.* **9**, 5237–5252.
22. Kozak, M. (1983). *Microbiol. Rev.* **47**, 1–49.
23. Najarian, R., D. Caput, W. Gee, S. Potter, A. Renard, J. Merryweather, G. Van Nest and D. Dina (1985). *Proc. Natl Acad. Sci. USA* **82**, 2627–2631.
24. Nomoto, A., Y. F. Lee and E. Wimmer (1976). *Proc. Natl Acad. Sci. USA* **73**, 375–380.
25. Pallansch, M. A., O. M. Kew, A. C. Palmenberg, F. Golini, E. Wimmer and R. R. Rueckert (1980). *J. Virol.* **35**, 414–419.
26. Palmenberg, A. C. (1982). *J. Virol.* **44**, 900–906.
27. Palmenberg, A. C. and R. R. Rueckert (1982). *J. Virol.* **41**, 244–249.
28. Palmenberg, A. C., M. A. Pallansch and R. R. Rueckert (1979). *J. Virol.* **32**, 770–778.
29. Palmenberg, A. C., E. M. Kirby, M. J. Janda, N. L. Drake, G. M. Duke, K. F. Potratz and M. S. Collett (1984). *Nucl. Acids Res.* **12**, 2969–2985.
30. Pettersson, R. F., V. Ambros and D. Baltimore (1978). *J. Virol.* **27**, 357–365.
31. Racaniello, V. R. and D. Baltimore (1981). *Proc. Natl Acad. Sci. USA* **78**, 4887–4891.
32. Robertson, B. H., M. J. Grubman, G. N. Wenddell, D. M. Moore, J. D. Welsh, T. Fischer, D. J. Dowbenko, D. G. Yanssura, B. Small and D. G. Kleid (1985). *J. Virol.* **54**, 651–660.
33. Rossmann, M. J., E. Arnold, J. W. Erickson, E. A. Frankenberger, J. P. Griffith, H.-J. Hecht, J. E. Johnson, G. Kamer, M. Luo, A. G. Mosser, R. R. Rueckert, B. Sherry and G. Vriend (1985). *Nature (London)*, **317**, 145–153.
34. Rothberg, P. G., T. J. R. Harris, A. Nomoto and E. Wimmer (1980). *Proc. Natl Acad. Sci. USA* **75**, 4868–4872.

35. Rowlands, D. J., T. J. R. Harris and F. Brown (1978). *J. Virol.* **26**, 335–343.
36. Rueckert, R. R. (1976). *In* "Comprehensive Virology", (Eds H. Fraenkel-Conrat, R. R. Wagner) Vol. 6, pp. 131–213. Plenum, New York.
37. Rueckert, R. R. (1985). *In* "Virology", (Eds B. Fields, D. M. Knipe, R. M. Chanock, J. L. Melnick, B. Roizman, R. E. Shope) pp. 705–738. Raven Press, New York.
38. Rueckert, R. R. and E. Wimmer (1984). *J. Virol.* **50**, 957–959.
39. Sangar, D. V., D. N. Black, D. J. Rowlands, T. J. R. Harris and F. Brown (1980). *J. Virol.* **33**, 59–68.
40. Shih, D. S., C. T. Shih, D. Zimmern, R. R. Rueckert and P. Kaesberg (1979). *J. Virol.* **30**, 472–480.
41. Skern, T., W. Sommergruber, D. Blass, P. Gruendler, F. Fraundorfer, C. Pieler, I. Fogy and E. Kuechler (1985). *Nucl. Acids Res.* **13**, 2111–2126.
42. Stalhandske, P., M. Lindberg and U. Pettersson (1984). *J. Virol.* **51**, 742–746.
43. Stanway, G., P. Hughes, R. Mountford, P. Minor and J. Almond (1984). *Nucl. Acids Res.* **12**, 7859–7875.
44. Strebel, K. and E. Beck (1986). *J. Virol.* **58**, 893–899.
45. Svitkin, Y. V. and V. I. Agol (1983). *Eur. J. Biochem.* **113**, 145–154.
46. Svitkin, Y. V., A. E. Gorbalenya, Y. A. Kazachkov and V. I. Agol (1979). *FEBS Letts.* **108**, 6–8.
47. Toyoda, H., M. Kohara, Y. Kataoka, T. Suganuma, T. Omata, N. Imura and A. Nomoto (1984). *J. Mol. Biol.* **174**, 561–585.
48. Toyoda, H., M. J. H. Nicklin, M. G. Murray, C. W. Anderson, J. J. Dunn, F. W. Studier and E. Wimmer (1986). *Cell* **45**, 761–770.
49. Tracy, S., H.-L. Liu and N. M. Chapmen (1985). *Virus Res.* **3**, 263–270.
50. Van Dyke, T. A. and J. B. Flanegan (1980). *J. Virol.* **35**, 732–740.
51. Van Dyke, T. A., R. J. Rickles and J. B. Flanegan (1982). *J. Biol. Chem.* **257**, 4610–4617.
52. Vartapetian, A. B., A. S. Mankin, E. V. Shripkin, K. M. Chumakov, V. D. Smirnov and A. A. Bogdanov (1983). *Gene* **26**, 189–195.

2. The Replication of Picornaviruses

Richard J. Kuhn and Eckard Wimmer

Department of Microbiology, School of Medicine, State University of New York at Stony Brook, Stony Brook, New York 11794, U.S.A

I. Introduction	17
II. Translation and processing of the polyprotein	20
III. RNA replication	23
A. Viral structures associated with the replication complex	23
1. RNA structures involved in replication	23
2. Viral proteins involved in replication	28
3. Cellular components implicated in picornavirus RNA replication	31
B. Mechanism of viral replication	33
1. RNA synthesis *in vivo*	33
2. Studies of RNA replication *in vitro*	34
3. Models of picornaviral RNA replication	40
IV. Conclusions	44
References	44

I. INTRODUCTION

The picornaviruses belong to one of the largest and most widely characterized families of RNA-containing animal viruses. The family is divided into four genera: the cardioviruses (e.g. encephalomyocarditis virus, mengovirus), the aphthoviruses (foot-and-mouth disease virus of ungulates), the rhinoviruses (whose members cause the common cold), and the enteroviruses (e.g. poliovirus, coxsackie virus, echovirus, hepatitis A virus). These viruses are both medically important and economically significant. For example, the human enteroviruses are a major cause of diseases of children, and foot-and-mouth disease virus (FMDV) is the single most important pathogen of livestock.

The picornavirion is a non-enveloped, icosahedral particle consisting of 60 copies each of four virus-specific coat polypeptides (VP1, VP2, VP3, and VP4) (Rossmann et al., 1985; Hogle et al., 1985). This protein shell encapsidates a single-stranded messenger-sense genomic RNA with an average chain length of 7500 nucleotides. The 3' terminus of the RNA is polyadenylated (Yogo and Wimmer, 1975) and the 5' terminus is covalently linked to a small polypeptide termed VPg (**g**enome-linked **V**iral **P**rotein; Lee et al., 1977). Poliovirus type 1 (Mahoney) was the first picornaviral genome to be completely sequenced (Kitamura et al., 1981; Racaniello and Baltimore, 1981). Its primary structure and gene organization are bizarre when compared to genomes of other RNA viruses (Fig. 2.1). Subsequent sequencing of other picornaviral genomes has greatly influenced our understanding not only of the individual viruses but of all picornaviruses, and has given us a new approach to understanding RNA viruses in general (Stanway et al., 1983, 1984; Carroll et al., 1984; Palmenberg et al., 1984; Toyoda et al., 1984). However, the significance of most of the unusual structural features of picornaviral RNAs for genome replication and gene expression remain to be explained.

Whereas all picornaviral RNAs possess a 5'-terminal VPg, an oligopeptide of molecular weight approximately 2200 daltons, and are 3' polyadenylated, the RNAs of cardioviruses and aphthoviruses also contain a segment of poly(C) (50–200 bases in length) located within the 5' non-coding region (Porter et al., 1974; for references, see Palmenberg et al., 1984). The non-coding region at the 5' end of picornaviral RNAs is unusually long: 743 nucleotides for poliovirus, 1400 nucleotides for FMDV. Neither the function of VPg nor the role of the poly(A) tail and the internal poly(C) segment are known, although there are many hypotheses implicating these features in replication, as will be discussed below. There is also no clue as to why these genomes have retained their unusually long 5'-terminal non-coding region through the course of evolution. Attempts to modify or shorten the non-coding region of poliovirus have usually led to non-viable or conditionally lethal mutants.

The infectious cycle is initiated by the attachment of the virus to a host-cell receptor. No information concerning the precise molecular structure of any receptor for picornaviruses has been obtained. However, this field of interest is rapidly moving, and the genes coding for rhinovirus and poliovirus receptors are currently being cloned. Cromwell and his colleagues have isolated a membrane protein that is likely to be the receptor for coxsackie B3 virus (Mapoles et al., 1985). The isolation of monoclonal antibodies that are presumably specific for the coxsackie B3 virus, rhinovirus and poliovirus receptors have been reported (Campbell and Cords, 1983; Colonno et al., 1986; Minor et al., 1984).

The mechanisms of uptake and uncoating are still obscure, although it has been proposed that poliovirus enters the cells via the endosomal pathway (Madshus et al., 1984). Uncoating is followed by translation of the single, large open reading frame of the RNA into a polyprotein

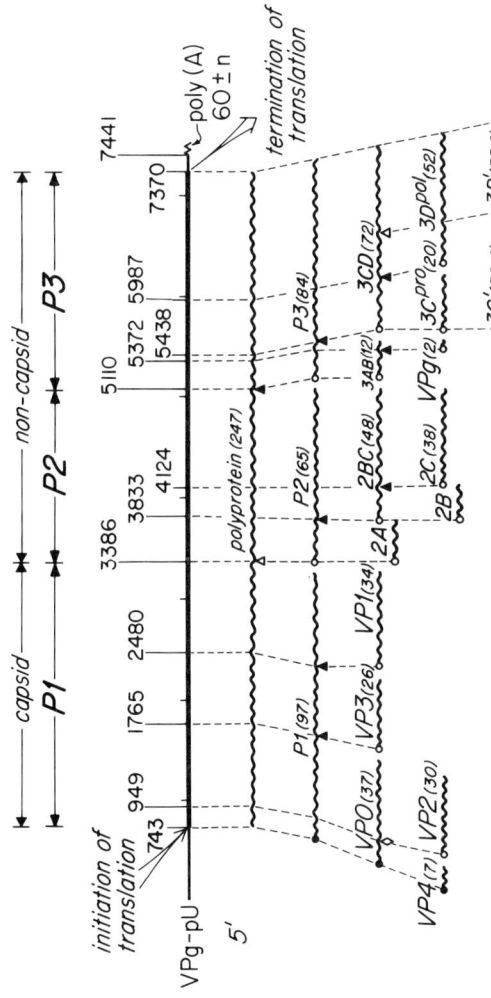

Figure 2.1. Gene organization of poliovirus and map of proteolytic processing. Virion RNA, terminated at the 5' end with the genome-linked protein VPg and at the 3' end with poly(A), is shown as a solid line, the translated region being more pronounced than the non-coding regions. The numbers above the virion RNA refer to the first nucleotide of the codon specifying the N-terminal amino acid for the viral specific proteins. The coding region has been divided into three regions (P1, P2, P3), corresponding to rapid cleavages of the polyprotein. The newly adopted nomenclature of polypeptides is according to Rueckert and Wimmer (1984). Numbers in parentheses are calculated molecular weights in kilodaltons. Open circles indicate that the terminal amino acids have been determined by sequence analysis. Solid circles indicate that the N-termini are known to be blocked. Solid triangles: Gln-Gly pairs that are cleaved during proteolytic processing of a polypeptide by the virus-coded proteinase 3C. Open triangles: Tyr-Gly pairs cleaved by viral proteinase 2A. Open diamond: Asn-Ser pair cleaved only during morphogenesis. Polypeptides 3C' and 3D' are products of an alternative cleavage, the biological significance of which is unknown.

The genome structure and processing map of poliovirus is probably applicable to all enteroviruses and rhinoviruses, with the exception of hepatitis A virus (human enterovirus 72) that has a truncated VP4. Cardioviruses and aphthoviruses have a segment of poly(C) inserted into their 5' non-coding region. Moreover, their polyprotein begins with a leader polypeptide ("L") that precedes the P1 region. Finally, aphthoviruses have most of their polypeptide 2A within the polyprotein deleted. For further details, see text.

that is cleaved proteolytically by two virus-encoded proteinases. Once the replication proteins have become available, the system begins to replicate the incoming RNA. Translation and replication then occur in parallel. It has been shown that picornaviruses can replicate in the cytoplasm of enucleated cells, an observation suggesting that nuclear functions of the cell are not required during the replication cycle (Pollack and Goldman, 1973; Detjen et al., 1978). Little is known about the precise pathway of morphogenesis (Putnak and Phillips, 1981) and this subject will not be dealt with in this paper. Instead, we will very briefly review newer results in proteolytic processing of poliovirus, and then review in some detail the current knowledge and confusion concerning RNA replication.

II. TRANSLATION AND PROCESSING OF THE POLYPROTEIN

After penetration and uncoating of the virion, the viral RNA (vRNA) uses the machinery of cellular protein synthesis to express its genetic information. Although the early events in the virus replication cycle are unclear, it is safe to assume that the incoming viral RNA is first translated. As will be discussed later, picornavirus mRNA differs from vRNA by the absence of VPg (Nomoto et al., 1977b). Whether or not VPg is cleaved from the infecting vRNA, prior to or during the initial virus-specific protein synthesis, is unknown; however, VPg-linked RNA can form an initiation complex of translation *in vitro* (Golini et al., 1980).

The mechanism by which picornaviruses initiate translation of their genomes in infected cells has not been completely solved. Shortly after infection there is a rapid decline in host-cell protein synthesis, concurrent with a steady increase in virus-specific protein synthesis. Picornavirus mRNA, quite unlike cellular mRNA, lacks the $^{7m}G(5')ppp(5')N$ capping group (Nomoto et al., 1976); instead of a capping group the viral mRNA has, remarkably, a pU terminus (Nomoto et al., 1977b; Pettersson et al., 1977). Available evidence strongly suggests that at least in the case of polio and rhinoviruses, host protein synthesis is blocked at the level of cap recognition (Sonenberg et al., 1981; Hansen and Ehrenfeld, 1981). Support of this theory comes from work identifying a defective component of the cellular translational machinery in poliovirus-infected cells (Tahara et al., 1981; Grifo et al., 1983; Edery et al., 1983; Etchison et al., 1982). The defect lies in the cap-binding protein (CBP) complex, which normally consists of three polypeptides with molecular weights of 24 kD (eIF-4E), 46 kD (eIF4A), and 220 kD (p220). In rhinovirus- and in poliovirus-infected HeLa cells, p220 is degraded (Etchison and Fout, 1985). Neither viral proteinases 3C or 2A is directly involved in the degradation of p220 (Lloyd et al., 1985, 1986; Lee et al., 1985). Cardioviruses, on the other hand, appear to shut off host cell protein synthesis by a different mechanism. Jen and Thach (1982) have shown that accumulating EMC mRNA out-competes cellular mRNA for one or more of the factors involved in initiating protein synthesis. Moreover, p220 remains intact in EMCV-infected cells (Mosenkis et al., 1985).

The sequencing of the poliovirus genome and the mapping of viral-specific polypeptides identified an open reading frame of 2207 consecutive triplets, spanning most (89%) of the viral genome (Kitamura et al., 1981; Semler et al., 1981a; Racaniello and Baltimore, 1981). It confirmed earlier suggestions that all viral proteins are derived from a single precursor molecule that is cleaved by proteinase(s) to generate the viral gene products (Jacobson and Baltimore, 1968; Summers and Maizel, 1968) and this strategy is followed by all picornaviruses. The AUG codon that initiates synthesis of the poliovirus type 1 polyprotein is located 743 nucleotides downstream from the 5' end of the RNA and is preceded by eight other AUG codons (Dorner et al., 1982).

There is no evidence that the AUG codons upstream from nt743 serve to initiate short viral polypeptides, the largest having the potential of being a protein of 7 kD, or might serve some other function. Not even in the RNAs of the three poliovirus serotypes are all of these AUGs or the coding sequences following them conserved (Toyoda et al., 1984).

It is not clear whether initiation at the AUG at nt743 follows the "scanning" of the small ribosomal subunit along the non-coding region (Kozak, 1983), or is the result of "internal" initiation. *In vitro*, poliovirus RNA can initiate protein synthesis several thousand nucleotides downstream from the 5' end; this phenomenon, however, is observed only when translation occurs in the reticulocyte lysate and is abolished when unknown HeLa cell factor(s) are added or translation is carried out in HeLa or ascites cell extracts (Dorner et al., 1984; Svitkin et al., 1985). Nevertheless, direct binding of a ribosome to the proper initiation codon of the polyprotein of picornaviruses cannot be excluded at present.

The polyprotein can be divided into three regions according to function and processing: P1 corresponds to the region of capsid polypeptides and their precursors; P2 corresponds to the central portion of the genome encoding non-structural proteins; and P3 corresponds to a group of non-structural polypeptides known to be involved in replication (Fig. 2.1). The polyprotein is processed into functional proteins by proteolytic cleavages (see the recent reviews by Nicklin et al., 1986; Toyoda et al., 1986a and Chapter 1 of this volume). Cleavages were found to occur between glutamine-glycine (Q-G) amino acid pairs at nine processing sites within the polyprotein, between tyrosine-glycine (Y-G) at two sites, and between asparagine-serine (N-S) at one site (Kitamura et al., 1981; Semler et al., 1981a,b; Emini et al., 1982; Larsen et al., 1982; Pallansch et al., 1984). All Q-G cleavages are carried out by the virus-encoded protease $3C^{pro}$, a polypeptide of 20 kD (Hanecak et al., 1982). Since deproteinized vRNA is infectious (Nomoto et al., 1977b) and because $3C^{pro}$ itself is a product of Q-G cleavage, it appears likely that $3C^{pro}$ is generated in the initial stages of translation by an intramolecular cleavage of the polyprotein. This mechanism has been suggested by Palmenberg and Rueckert (1982) and is based on their studies with EMC virus. Briefly, EMC RNA was translated in a rabbit reticulocyte lysate and the rate of cleavage of a $3C^{pro}$ precursor under conditions of increasing dilution was measured. It was found that cleavage of 3CD of EMC was dilution-independent. Hanecak et al. (1984) have subsequently constructed an expression plasmid that contains a cloned segment of the poliovirus genome corresponding to a segment slightly larger than the $3C^{pro}$ coding region. Upon induction, three virus-specific polypeptides are produced that can be immunoprecipitated with anti-$3C^{pro}$. The two smaller polypeptides are the result of proteolytic processing at Q-G amino acid pairs, the smallest polypeptide being authentic $3C^{pro}$. Insertion of a DNA linker into the $3C^{pro}$ coding region resulted in loss of this cleavage activity. The kinetics of the cleavage reaction suggest intramolecular cleavage (Hanecak et al., 1984). Recently an E. coli expression plasmid containing the P1 region plus part of the P2 region of poliovirus type 3 has been constructed. Expression of the poliovirus-specific sequences yields a protein that is rapidly processed at exactly that Y-G site which severs P1 from P2 (see Fig. 2.1; Toyoda et al., 1986a,b). Linker-insertion mutagenesis of four amino acids or deletion mutagenesis of nine amino acids in the 2A coding region eliminates this activity, an observation suggesting that a Y-G proteinase activity maps in polio protein 2A. Experiments using translations in vitro indicated that antibodies prepared against purified $2A^{pro}$ inhibit cleavage at a Y-G cleavage site within polypeptide 3CD (see Fig. 2.1). On the other hand, anti-$2A^{pro}$ antibodies did not prevent the Y-G cleavage at an N-terminal of $2A^{pro}$. The failure to inhibit the Y-G cut between P1 and P2 with anti-$2A^{pro}$ antibodies in the translation in vitro has been interpreted to mean that this cleavage is too rapid and possibly intramolecular. Cleavage to yield 3C' and 3D' (Fig. 2.1) definitely occurs intermolecularly and can therefore be prevented with specific antisera (Toyoda et al., 1986b).

Processing of VPO at the N-S amino acid pair occurs only during morphogenesis (Rueckert, 1976). It has been suggested that this cleavage is also a triggered autocatalytic process (Rossmann et al., 1985).

Whereas Q-G or E-G amino acid pairs are strong recognition signals for the picornavirus $3C^{pro}$, not all of these signals in the polyprotein are recognized. For example, the polyprotein of poliovirus contains 13 Q-G pairs of which only 8–9 are cleaved. The additional determinants recognized by $3C^{pro}$ are unknown, but may be an amino acid in the -4 position of the cleavage site and specific higher-order structures of the polypeptide chain (for a discussion, see Nicklin et al., 1986). Some Q-G pairs are cleaved much more efficiently than others. For example, the Q-G pairs flanking 2C are cleaved so fast that it was suggested that 2C may be synthesized from an extra cistronic region of the genome (for references, see Kitamura et al., 1981). A Q-G pair within $3C^{pro}$, on the other hand, is cleaved only rarely and leads to the production of yet another cleavage product of 3CD, termed P3-4a (Semler et al., 1983; note that the cleavage leading to P3-4a is not shown in Fig. 2.1).

Finally, EMCV codes for a polypeptide corresponding to poliovirus 2A^{pro}, but there is no homology between the two proteins. It is possible, therefore, that 2A of EMCV is not a proteinase. The proteolytic activity responsible for cleaving the EMCV polyprotein between P1–2A and 2B-2C-P3 may be catalysed by EMCV 2B, a mechanism favoured by Jackson (1987). Moreover, there is no 2A in FMDV but the leader polypeptide L (Rueckert and Wimmer, 1984 and references therein), a protein mapping ahead of P1, is a protease (Strebel and Beck, 1986). EMCV also codes for an L protein although it is much shorter than the L of FMDV. As seen in Fig. 2.1, the poliovirus polyprotein starts directly with the capsid protein region P1, a situation true for all enteroviruses and rhinoviruses whose genomes have been sequenced.

III. RNA REPLICATION

A. Viral Structures Associated with the Replication Complex

1. RNA Structures Involved in Replication
The RNA structures induced by picornavirus infection have been the most accessible and well-defined elements in the viral replication cycle. Three virus-specific RNAs, schematically shown in Fig. 2.2 for poliovirus, can be found in the infected cell: (i) single-stranded RNA (ssRNA) that is either virion RNA or mRNA (Zimmerman *et al.*, 1963); (ii) double-stranded or replicative form RNA (RF; Montagnier and Sanders, 1963); and (iii) replicative intermediate (RI), a molecule that is partially double-stranded, partially single-stranded RNA and is heterogeneous in size (Baltimore and Girard, 1966). The properties of these structures will be briefly discussed below.

Following the discovery that picornaviruses contain a single-stranded RNA genome that itself is infectious (see e.g. Alexander *et al.*, 1958) it was shown that ssRNA was also the predominant RNA molecule synthesized in infected cells. Sucrose density gradient analysis of newly synthesized RNA identified only one major class of RNA; this RNA sedimented at 35S, as does purified poliovirus RNA (Zimmerman *et al.*, 1963). An analysis of the base ratio of the RNA synthesized in the infected cells matched that of viral RNA. An additional minor species sedimenting in sucrose gradients at 20S was also observed. When 20S and 35S RNA were treated with ribonuclease, 90% of the 35S RNA was digested, whereas the majority of the 20S RNA remained intact. This, together with the known differences in the effects of salt concentration on the sedimentation rates of ssRNA and dsRNA (Montagnier and Sanders, 1963; Baltimore, 1964), identified the 35S RNA as single-stranded and the 20S RNA as double-stranded (Baltimore, 1964, and references therein). All ssRNA isolated from infected HeLa cells or from virions has plus-strand polarity, that is, it has the same polarity as messenger RNA. This conclusion was based on translation experiments in *E. coli* cell-free extracts in which polypeptides

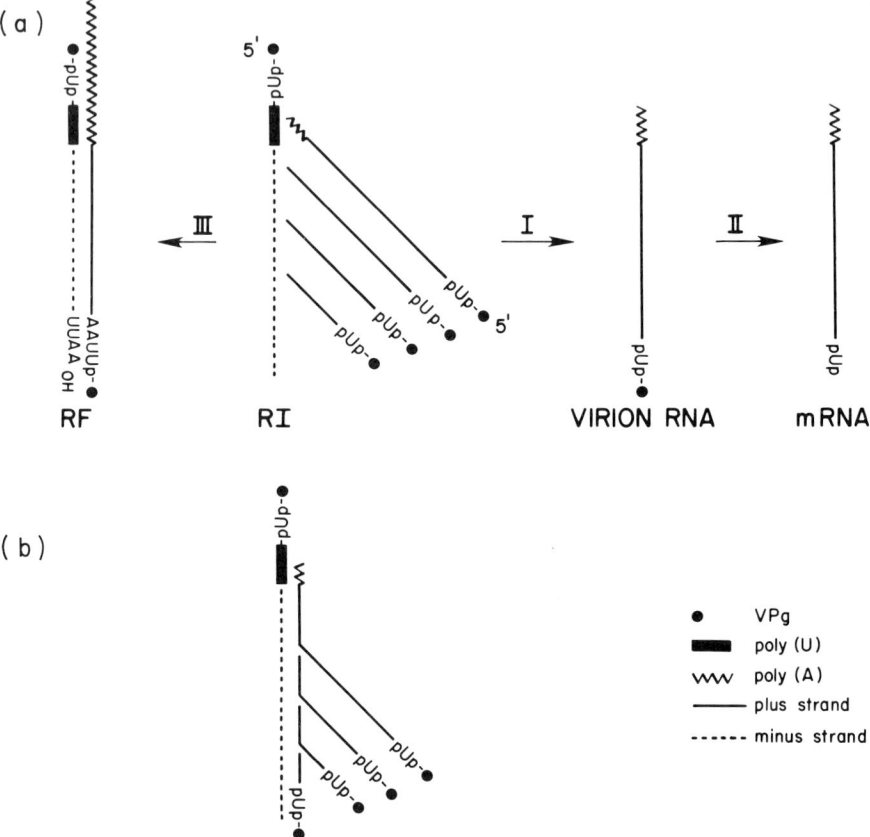

Figure 2.2. Intracellular structures of poliovirus RNA. The RNAs are represented with the known sequence structures at their termini. The replicative intermediate (RI) is shown in two possible forms: (a) the "open" form (double-stranded regions are supposedly very short) and (b) the "closed" form. Form B requires that strand replacement occurs during replication, a mechanism strongly favoured by data of Lundquist and Maizel (1978) and Yogo and Wimmer (1975), although several investigations support also form A (see text). Arrows indicate the possible relationship between the structures. Various forms of the replicative form (RF) that occur in small quantities in the infected cell (Senkovich et al., 1980) have been omitted. (Reproduced with permission from Kitamura et al., 1980.)

were produced having properties similar to those of virus-specific polypeptides isolated from infected HeLa cells (Rekosh et al., 1969). Moreover, when poliovirus-induced messenger RNA was isolated from polyribosomes of infected cells and compared with viral RNA, it was found that they have identical size and base composition (Penman et al., 1964; Summers and Levintow, 1965; Summers et al., 1967).

As was mentioned previously, the 5' end of viral RNA is blocked by the genome-linked protein VPg. This small protein is covalently attached to the polynucleotide chain via a phosphodiester bond between the 5'-terminal uridine residue and the O^4 of a tyrosine residue (Rothberg et al., 1978; for a review, see Wimmer, 1979). All viral-induced RNAs found in the infected cell, with the exception of messenger RNA found in polyribosomes, contain VPg at the 5' end (Nomoto et al., 1977a,b). The removal of VPg from the 5' end of messenger RNA is most likely the result of its post-transcriptional cleavage from the newly made RNA. The release of VPg is accomplished by a cellular protein, termed the unlinking enzyme, that leaves a pU at the 5' end of the RNA (Ambros et al., 1978). VPg released from RNA is rapidly degraded by cellular enzymes (Dorner et al., 1981; Sangar et al., 1981). It is not understood why VPg is cleaved from those plus strands destined to become messenger RNA. One hypothesis suggests that VPg might serve as a recognition signal for encapsidation of RNA into virions (Nomoto et al., 1977b). Since the nucleotide sequences of mRNA and virion RNA are identical, an encapsidation signal consisting solely of a unique nucleotide sequence might lead to encapsidation of mRNA and hence would cause an inhibition of translation. If VPg were an additional signal in morphogenesis, the structure of the 5' ends of newly made RNA would allow the capsid proteins to distinguish between virion RNA and mRNA and thus ensure the separation between encapsidation and translation. An alternative explanation for the removal of VPg is that VPg interferes with the formation of an initiation complex of translation. However, as was mentioned before, VPg-linked RNA can form a translation initiation complex, at least *in vitro* (Golini et al., 1980), but this conclusion may have to be modified in view of the "internal" initiation of translation of poliovirus RNA in the reticulocyte extract (Dorner et al., 1984).

The 3' end of the genomic RNA of poliovirus contains a poly(A) tract that is heterogenous in size (Yogo and Wimmer, 1972; Miller and Plagemann, 1972; Armstrong et al., 1972) with an average chain length of 60 nucleotides (Alquist and Kaesberg, 1979). The presence of this homopolymeric tract in picornaviral RNA may not be surprising, since most eukaryotic messenger RNAs are polyadenylated. Cellular mRNA is polyadenylated post-transcriptionally by nuclear adenylate transferases. Picornaviruses, however, replicate in the cytoplasm with no apparent need for nuclear functions. The mechanism of poly(A) synthesis of picornaviral RNA is therefore different: the homopolymeric tract is genetically encoded, that is, transcribed from a poly(U) tract at the 5' end of minus RNA strands (Dorsch-Hasler et al., 1975). The mean size of the poly(A) tract varies for RNAs of different picornavirus groups but is usually between 40 and 120 A residues. However, the poly(A) tract of polyribosomal mRNA isolated late from poliovirus-infected cells may be 5–8 times longer than that of virion RNA (Spector and Baltimore, 1975a). The function of the picornaviral poly(A) is not known. A reduction in size of the poly(A) by RNAase H treatment after hybridization with oligo (dT) has been reported to abolish the infectivity of the virion RNA (Spector and Baltimore, 1974). More recently a role of poly(A) in the initiation of minus strand synthesis has been proposed (see below).

Genomic RNAs of aphthoviruses and cardioviruses were shown to contain a large T1-ribonuclease resistant region consisting of a poly(C) tract 100–300 nucleotides in length (Porter et al., 1974; Brown et al., 1974). In the case of EMCV, this poly(C) tract is positioned 149 nucleotides downstream from the 5′ end, in FMDV this distance is 362–367 nucleotides (Newton et al., 1985). It was later shown that cleavage of FMDV RNA at the poly(C) tract had no effect on proteins synthesized in vitro by a translation system (Sangar et al., 1980). An interesting observation is that more virulent strains of cardioviruses or aphthoviruses have longer poly(C) tracts than laboratory strains (Fellner, 1979; Harris and Brown, 1977). The function of the poly(C) tract is unknown. Some investigators have suggested it plays a role in RNA replication, perhaps as a polymerase binding site, others have suggested it might function as a nucleation site in assembly. However, if poly(C) plays an important role in replication, the need for this function is obviously absent for enteroviruses and rhinoviruses.

The 20S dsRNA, first identified by Montagnier and Sanders (1963) in EMCV-infected cells contains a complete copy of the genomic plus-stranded RNA that is hydrogen-bonded to a full-length complementary strand (Baltimore, 1966). However, this molecule, referred to as replicative form (RF), proved much more complex in structure than originally anticipated: one end is a poly(A)/poly(U) homopolymeric duplex (Yogo and Wimmer, 1973, 1975; Yogo et al., 1974), the other end a heteropolymeric duplex. Both 5′ ends carry a VPg (Wu et al., 1978); in the case of the minus strands, the VPg is linked to poly(U) (Nomoto et al., 1977a). Interestingly, the 3′-terminal poly(A) tract in RF is always longer than the 5′-terminal poly(U) of minus strands and thus some 100 adenylate residues protrudes as a single-stranded tail (Larsen et al., 1980; see Fig. 2.2). Larsen et al. (1980) reported that the heteropolymeric end of RF RNA is a perfect blunt-ended duplex; Richards and Ehrenfeld (1980), on the other hand, found that the 3′ end of the minus strand carries several extra A residues in a small fraction of RF molecules. Also of particular interest are the reports that the strands of some of the RF molecules isolated from EMCV- or mengovirus-infected cells are somehow covalently linked; they may form linear dimers or apparent circular dimers (Romanova and Agol, 1979; Robberson et al., 1982). Moreover, Richards et al. (1979) found a population of poliovirus RF that seemed to lack VPg at one end. Finally, Young et al. (1985) reported that a fraction of poliovirus RF is apparently covalently linked at one end through a hairpin, thereby migrating in denaturing gels as a single-stranded RNA molecule twice the length of genomic RNA. These observations are significant for two reasons. First, the unusual structures of RF clearly indicate that the existence of these molecules is not the result of an artifact of rapid hybridization between plus and minus strands occurring during isolation. Second, some of the structures have been seen as intermediates in a unique pathway of RNA replication (snap-back model leading to self-initiation of RNA synthesis; see below).

Picornavirus RF is infectious with a specific infectivity 30-fold higher than ssRNA (Bishop and Koch, 1967). The higher infectivity of the dsRNA may reflect its decreased sensitivity to ribonuclease or an increased ability of the ds nucleic acid to penetrate into the cell. Pretreatment of the cells with DEAE-dextran prior to transfection results in equal specific infectivity of RF and ssRNA. Treatment of cells with actinomycin D decreases their sensitivity to infection with dsRNA but has the opposite effect on ssRNA (Koch et al., 1967). Moreover, it was found that the infectivity of the dsRNA is dependent on a nuclear function since enucleated HeLa cells cannot be transfected with poliovirus RF (Detjen et al., 1978). How RF molecules initiate an infectious cycle is not known, but it appears that its strands must be separated. The following observations support this: (i) dsRNA is not an active message in a translation in vitro (Baltimore, 1968); (ii) isolation of minus strands from dsRNA molecules and subsequent transfection indicates that minus strands are not infectious (Roy and Bishop, 1970); and (iii) when a genetic hybrid RF molecule was used to infect cells, the resulting virus had a phenotype that corresponded to a genotype of the positive strand (Best et al., 1972), and RF molecules in which one of the strands has been inactivated are infectious only in those molecules in which the plus strand is intact (Chumakov and Agol, 1976). The involvement of cellular factors in transfection with picornavirus RF may give a clue to processes occurring during normal infection. Proteins that unwind nucleic acids are known to be required for DNA replication and may also be required for the replication of picornaviral RNA. Synthesis of RNA at some time in the replication cycle involves the displacement of a strand in a hydrogen-bonded, double-stranded region. Whether this is accomplished by the viral polymerase $3D^{pol}$, another viral product, or a cellular protein has yet to be determined.

A third structure identified in picornavirus-infected cells is the replicative intermediate (RI). It occurs only in small amounts in infected cells, but it can be easily identified after a pulse with a suitable radiolabelled precursor. RI molecules are partially ribonuclease-resistant. They sediment heterogeneously in a sucrose gradient between 20S and 70S (Baltimore and Girard, 1966). When isolated from infected cells, the RI consists of a full-length minus-sense template strand with 6–8 nascent plus-sense daughter strands (Baltimore, 1969) (see Fig. 2.2). The opposite polarity — full-length plus-strand template and nascent minus strands — has never been found (Bishop et al., 1969). Whether or not RI possesses a fully double-stranded backbone or is predominantly single-stranded in vivo (two structures shown in Fig. 2.2) is not known. Various publications support one or the other structure (see papers by Nilsen et al., 1981, and Richards et al., 1984, and discussions and references therein).

It should be noted that all nascent plus strands of RI molecules are linked to VPg (Nomoto et al., 1977a; Flanegan et al., 1977; Pettersson et al., 1978). The VPg molecule is bound also to the 5'-terminal poly(U) of minus strands in RI (Yogo and Wimmer, 1975; Nomoto et al., 1977a). Nascent RNA strands with a 5'-terminal pppUp have never been found in spite of labelling of these molecules with ^{32}P to very high specific radioactivity *in vivo* (Nomoto et al., 1977a). This result suggested that RNA synthesis may not be initiated *de novo* (which would yield 5'-terminal pppNp termini) but may be a primer-dependent event (see Wimmer, 1982).

A fourth RNA molecule distinct from all other viral RNAs is mRNA. It is identical to virion RNA except that it lacks VPg. It is generally agreed that *all* newly made RNA is VPg-linked and that the mRNA is a product of processing (see Fig. 2.2).

2. Viral Proteins Involved in Replication

A poliovirus-induced RNA polymerase was originally found by Baltimore and co-workers in infected cells; its appearance was correlated with the time course of infection, an observation suggesting that the enzyme was virus-encoded (Baltimore et al., 1963). It was subsequently shown that the virus-induced polymerase sedimented in a membrane-associated complex of 60–250S (Baltimore, 1964; Girard et al., 1967; Ehrenfeld et al., 1970). Neither puromycin nor EDTA changed the sedimentation rate of the complex and radiolabelled ribosomes did not associate with it. It was thus concluded that the replication complex was free of ribosomes. Attempts to isolate polymerase in an active form were originally unsuccessful owing to the lack of an assay system that was dependent upon the addition of exogenous RNA. Partial purification of the replication complex by precipitation in 2 M LiCl and gel filtration resulted in the identification of a protein of molecular weight 58 kD that was shown to be the virus-specific polypeptide 3D (Lundquist et al., 1974). This result implied that RNA replication may be catalysed by 3D. Whether host-cell factor(s) or smaller viral polypeptides were also involved remained an open question.

Direct identification of a poliovirus-encoded RNA polymerase was accomplished through an assay that was based upon the realization that the 3'-terminal poly(A) is genetically coded (Dorsch-Hasler et al., 1975; Spector and Baltimore, 1975b). Accordingly, the first step of replication would be the transcription of the 3'-terminal poly(A) tract of vRNA into the 5'-terminal poly(U) of minus strands. Initial attempts to isolate a soluble poly(A)-dependent "poly(U) polymerase" failed. However, addition of oligo(U) to the poly(A) containing assay mixture as a possible primer for RNA synthesis resulted in the strong stimulation of the expected polymerase activity (Flanegan and Baltimore, 1977). This led to the discovery that the poliovirus-encoded RNA polymerase is dependent not only on template but also on primer. The activity copurified exclusively with viral polypeptide 3D^{pol} (formerly P63, P3-4b, NCVP4,

52 kD). On the other hand polypeptide 3CD (formerly NCVP2, P3–2, 72 kD), was found to be inactive. $3D^{pol}$ was subsequently shown capable of transcribing vRNA as long as an oligo(U) primer was added (Dasgupta et al., 1979). Although purification of the soluble, membrane-free polymerase yields an active protein, attempts to purify a membrane-bound form of the polymerase with poly(U) polymerase activity have been unsuccessful (Takeda, Takegami, and Wimmer, unpublished results).

A second virus-encoded protein involved in RNA replication is the genome-linked protein VPg, a small (20–24 amino acids, depending upon group and serotype) basic oligopeptide, with an invariant tyrosine residue at position 3 (Burroughs and Brown, 1978; Hruby and Roberts, 1978; Golini et al., 1978; Sangar et al., 1977; Wimmer, 1979; Adler et al., 1983; for a review, see Wimmer, 1982). A high energy ($-9.6 \text{ kcal mol}^{-1}$) phosphodiester bond links the phenolic hydroxyl group of the tyrosine to the 5' phosphoric group of the terminal uridylic acid residue (Ambros and Baltimore, 1978; Rothberg et al., 1978; Vartapetian et al., 1980). As discussed before, VPg is found at the 5' end of all newly synthesized viral RNAs but is absent from mRNA. The existence of a genome-linked protein is not unique to picornaviruses. It has been found in other animal RNA viruses (Revet and Delain, 1982), in DNA bacteriophages (Salas et al., 1978; Yehle, 1978; Ito, 1978; Harding et al., 1978) and animal DNA viruses (Rekosh et al., 1977; Keegstra et al., 1977; Gerlich and Robinson, 1980; Ganem et al., 1982; Molnar-Kimber et al., 1983), and plant RNA viruses (Stanley et al., 1978; Daubert et al., 1978; Ghosh et al., 1979; Mayo et al., 1982; Zabel et al., 1984).

VPg is very difficult to detect in extracts of HeLa cells infected with poliovirus although in the course of the infection more than 6×10^6 VPg molecules should be synthesized. VPg escapes detection because it is very small and acid-soluble, properties that make it elute easily from polyacrylamide gels (Crawford and Baltimore, 1983). Moreover, VPg is rapidly degraded in cell-free extracts (Dorner et al., 1981). We currently believe that VPg occurs predominantly in the form of one of its precursor polypeptides, the most prominent precursor being the membrane-associated protein 3AB (Takegami et al., 1983a; Semler et al., 1982). For example, membrane complexes of poliovirus-infected HeLa cells that contain an active RNA synthesizing complex have been found to be rich in this viral polypeptide 3AB (Takegami et al., 1983a). Two other membrane-associated viral polypeptides, 2ABC-3AB (formerly 3b/9) and 2C-3AB (formerly X/9), that share primary sequence with 3AB have also been found in these replication complexes. Their function, if any, is obscure. The availability of a membrane anchor protein such as 3AB may not only serve to direct other viral polypeptides to the site of RNA synthesis (see below); it may also deliver the hydrophilic protein VPg to the 5' ends of nascent RNA strands in a hydrophobic environment (Semler et al., 1982; Takegami et al., 1983a,b).

FMDV is unique among picornaviruses in that the VPg attached to its

RNA can be separated into two fractions differing in charge, tryptic peptide maps, and amino acid composition (King et al., 1980). Subsequent nucleotide sequence analysis of a cloned cDNA from FMDV identified three tandemly arranged VPg genes (Forss and Schaller, 1982). Since all three VPgs are utilized *in vivo* in an equimolar ratio, their position relative to the precursor protein does not seem to affect transfer of each VPg to the 5' end of the viral RNA.

Attempts to purify the VPg precursor polypeptide 3AB from infected HeLa cells were frustrated owing to an apparent high affinity between viral polypeptides 2C and 3AB (Kuhn and Wimmer, unpublished results). These polypeptides copurify through ion-exchange chromatography, gel-filtration, and antibody-affinity chromatography, even in the presence of NP-40 and urea. Although it copurifies with the membrane-bound replication complex (Butterworth et al., 1976; Traub et al., 1976), 2C is rapidly degraded with trypsin in the absence of detergents (Takegami et al., 1983a). Since 2C lacks a hydrophobic region, it may be associated with membranes only by its affinity to 3AB. This tight association of both 2C and 3AB with membranes is further suggested by the observation that high-ionic-strength buffer containing EDTA or 4 M urea cannot dislodge them (Tershak, 1984). Aside from its association with the membrane-bound replication complex, 2C has been directly implicated in RNA replication.

Guanidine, at concentrations (0.5–2 mM) that have no detectable effects on cell growth, selectively blocks the growth of several togaviruses (Friedman, 1970), several plant viruses (Dawson, 1975; Varma, 1968), and many picornaviruses including poliovirus and FMDV (Crowther and Melnick, 1961; Loddo et al., 1962; Pringle, 1964; Rightsel et al., 1961). Although guanidine is known to interfere with a number of viral processes, the major block appears to be in the synthesis of viral RNA (Caliguiri and Tamm, 1968; Tershak, 1982). The blockage is thought to occur at the level of initiation, as will be discussed later (Caliguiri and Tamm, 1968; Tershak, 1982). Mutants capable of growing in the presence of guanidine were isolated from both FMDV and poliovirus. Analysis of virus-induced polypeptides in guanidine-resistant FMDV mutants showed that polypeptide 2C (formerly P34 in FMDV) was altered in five out of ten mutants, an observation suggesting that 2C may be the target of the antiviral action of guanidine (Saunders and King, 1982). Some of the guanidine-resistant poliovirus mutants were also shown to express an altered 2C polypeptide (Anderson-Sillman et al., 1984) and recombinants between appropriate poliovirus strains unambiguously mapped the g^r mutation(s) downstream from polypeptide $2A^{pro}$ (Emini et al., 1984). More recently, Pincus et al. (1986, 1987) have sequenced polio type 1 genomic RNA isolated from: (i) six mutants resistant to 2 mM guanidine; (ii) one mutant resistant to 0.5 mM guanidine; and (iii) two mutants whose growth is dependent upon the addition of guanidine. The mutants resistant to 2 mM guanidine all contain an amino acid substitution in 2C at the same position. Interestingly, resistance to this "high" concentration required two nucleotide substitutions in the same codon that resulted in an exchange of either Asn→Gly or Asn→Ala. The other mutants each contain one or two amino acid changes, also within the 2C coding region. All mutations occurred in a region of 2C that is highly conserved among picornaviruses and is therefore thought to be the reactive site of 2C (Pincus et al., 1986, 1987; Argos et al., 1984). Moreover, a cDNA segment containing the g^r mutation was cloned into the wild-type background of an infectious poliovirus cDNA clone (Semler et al., 1984), and the virus isolated from transfected cells was found to express the g^r phenotype (Pincus and Wimmer, unpublished results). These data show that guanidine resistance can be mapped to the 2C region of the genome, implicating 2C, or a precursor thereof, as a protein involved in RNA synthesis, possibly in initiation.

The viral proteinase $3C^{pol}$ has also been identified as a component of the membrane-bound replication complex (Caliguiri and Mosser, 1971; Takegami et al., 1983a). Processing of the viral polyprotein into the individual polypeptides can be viewed as a regulation step in viral replication. Whether the proteinase regulates the release of VPg from its precursor and, in doing so, limits the rate of initiation of RNA synthesis is not known. However, it is quite clear that VPg is the only polypeptide linked to viral RNA (3AB, for example, is never found attached even to nascent strands), and therefore cleavage of a VPg precursor may directly affect the synthesis of RNA molecules.

In addition to these non-structural proteins, the capsid proteins VP0, VP1, and VP3 can also be found associated with the replication complex (Wright and Cooper, 1976; Loesch and Arlinghaus, 1975; Caliguiri and Mosser, 1971; Takegami et al., 1983a). It is not known what role, if any, these proteins play in the replication complex. Although virus maturation is thought to occur on membranes, complete virions are always found free in the cytoplasm.

Extensive comparisons of amino acid sequences were carried out recently between viral polypeptides of polio, EMCV, FMDV and cowpea mosaic virus B-RNA, a comovirus (see Chapter 12) (Argos et al., 1984; Franssen et al., 1984). Statistically significant homology was found in polypeptides 2C, VPg, $3C^{pro}$, and $3D^{pol}$ of those viruses. The nucleotide sequences of rhinovirus types 2 and 14 (Skern et al., 1985; Stanway et al., 1984) and of coxsackie virus B3 (Stalhandski et al., 1984) conform to this pattern of highly conserved polypeptides. This observation suggests that 2C, VPg, $3C^{pro}$ and $3D^{pol}$ have identical functions for all picornaviruses and possibly also for the comoviruses.

3. Cellular Components Implicated in Picornavirus RNA Replication
Apart from small molecules provided by the host cell, such as the nucleotide precursors, very few cellular components have been implicated in picornavirus RNA replication.

If the cytoplasmic extract of poliovirus-infected HeLa cells is subjected to centrifugation, all activity synthesizing poliovirus-specific RNA sediments with the membrane fraction. This fact, known for many years (Girard et al., 1967; Caliguiri and Tamm, 1970; Dorsch-Hasler et al., 1975; Takegami et al., 1983a; and references therein), indicates that RNA replication complexes are associated with membranous structures. Further fractionation of the membranous material on a discontinuous sucrose gradient has shown that the highest RNA synthesizing activity bands in the fraction containing smooth membranes (Caliguiri and Tamm, 1970; Dorsch-Hasler et al., 1975; Takegami et al., 1983a). It has been shown that these membranous complexes can synthesize authentic (VPg-linked and polyadenylated) plus-strand RNA, but this capacity is destroyed if as little as 0.5% non-ionic detergent, such as NP40, is added (Takeda et al., 1986). It is therefore believed that poliovirus RNA replication requires a membranous environment.

Two cellular enzymes have been purified and characterized during the last six years of which one or the other (but not both) is thought to be involved in the initiation of RNA synthesis. The first of these cellular proteins, termed "host factor" (HF), was originally isolated from ribosome washes of uninfected HeLa cells (Dasgupta et al., 1980). HF appeared to allow initiation of RNA synthesis on poliovirus RNA by purified $3D^{pol}$ in the absence of an oligo(U) primer. HF, a 67 kD mol.wt. protein, has been extensively purified and antisera to it have been prepared. These anti-HF sera inhibited transcription in a mixture of RNA, $3D^{pol}$ and HF. Most recently, HF was described as a protein kinase that, apart from apparent autophosphorylation, is capable of specifically phosphorylating an initiation factor of translation (eIF-2) (Morrow et al., 1985, and references therein).

The other cellular enzyme supposedly involved in RNA replication has been described recently by Andrews et al. (1985), and is a terminal uridylyl transferase (TUT). Such enzyme activity has been known for some time to exist in eukaryotic cells (see references in Andrews et al., 1985). In a transcription experiment in vitro, TUT can add a few uridylate residues to the 3'-terminal poly(A) of any polyadenylated RNA; the product can then form a hairpin, whose 3' end can serve then as primer for $3D^{pol}$. As we shall see, the resulting transcript, covalently linked at the homopolymeric end, resembles a fraction of RF molecules found in infected cells.

Both HF and TUT, it should be stressed, are supposedly factors that allow the $3D^{pol}$ to initiate RNA synthesis in the absence of an oligonucleotide primer. Takeda et al. (1986) have recently described a different cellular activity that appears to stimulate chain elongation in vitro in a poliovirus replication complex (see below). No further details of this activity are yet known. Finally, antibodies against a fraction of a homogenate of uninfected Krebs II cells have been found to inhibit the RNA synthesizing ability of an EMCV-induced replication complex (Dmitrieva et al., 1979). Although no specific cellular component has been identified, this report also suggests that host-cellular protein(s) may be involved in picornaviral RNA synthesis.

2. The Replication of Picornaviruses

B. Mechanism of Viral Replication

1. RNA Synthesis in Vivo

Much information has been gathered regarding RNA synthesis *in vivo*. Since some of the observations have been reviewed previously (Levintow et al., 1974; Rekosh, 1977; Koch and Koch, 1985), we will briefly summarize some facts, conclusions and speculations.

(i) Very early in infection the RNA genome, after a few rounds of translation, must be utilized as a template for minus-strand RNA synthesis. Since translation and transcription are mutually exclusive (they proceed in opposite directions) the mRNA must be cleared of ribosomes. The mechanism leading to the switch from translation to transcription in the initial phase of replication is unknown. Later in infection, of course, translation and transcription do not compete for template strands, since the system produces enough RNA to serve both functions.

(ii) Viral RNA synthesis has been described as occurring in three phases (Baltimore et al., 1964): an exponential phase that begins 30–40 min postinfection (p.i.) and lasts up to 3 h p.i. until about 20% of the final RNA yield has been synthesized. The system then switches over to a phase in which the kinetics show a linear, rather than an exponential, increase of synthesized RNA. During this linear phase of approximately $1-1\frac{1}{2}$ h, the bulk of RNA synthesis occurs. Four to five hours p.i. the rate of synthesis rapidly declines. The events leading to the change in the observed kinetics of RNA accumulation are obscure.

(iii) Plus strands and minus strands are synthesized throughout the replication cycle, but plus strands are synthesized in at least 10-fold excess over minus strands. This, of course, makes sense, since the plus strands function as templates for translation and transcription and they are also encapsidated to form progeny virions. The molecular basis of asymmetric RNA synthesis is obscure. A quantitative analysis of RNA synthesis and of the accumulation of different RNA structures in the poliovirus-infected cell has been reported by Hewlett et al. (1977).

(iv) All RNA replication appears to occur in membranous complexes that are considered to be free of ribosomes (Girard et al., 1967; Caliguiri and Tamm, 1970; Dorsch-Hasler et al., 1975; Takegami et al., 1983a). Yet inhibition of protein synthesis rapidly (within 5 min) inhibits RNA replication (see, for example, Ehrenfeld et al., 1970), as if there is a need for the continuous supply of a viral gene product. Moreover, the site of synthesis of minus strands has not been clearly established (membranes versus free cytoplasm).

(v) Whereas the RI is clearly the structure responsible for plus-strand RNA synthesis, this may not be the case for minus strands. It is possible that the synthesis of minus strands is inefficient and leads to predominantly RF molecules ("single run" model; Perez-Bercoff, 1979). If so, then the RF may be an intermediate, and not a by-product, of replication. On the other hand, RF molecules accumulate at a faster rate late in infection than they do early in infection. In this case RF molecules may be formed owing to an overall decrease in the rate of initiation of plus-strand synthesis. It remains to be seen whether the different structures of RF molecules discussed in Section III.A.1. are a reflection of different mechanisms of RF synthesis at various stages of the infection.

2. Studies of RNA Replication in Vitro

A genetic analysis of picornavirus RNA replication based on suitable mutants has not been possible, because picornavirus mutants, for unknown reasons, do not form complementation groups (Cooper, 1977). Cooper and his colleagues (reviewed in Cooper, 1977), however, have obtained a genetic recombination map of poliovirus and correctly suggested that a region coding for RNA replication function(s) was located at the opposite end of the genome from that coding for the capsid proteins.

In the absence of genetic information, numerous biochemical studies have been performed whose objective was to separate and purify components of the infected cell that are involved in viral RNA synthesis. The ultimate goal of all these experiments is to reconstitute *in vitro* a replication complex. We define replication *in vitro* of RNA as the synthesis of progeny RNA that has the same polarity as the input template RNA. Starting with virion RNA, that implies the pathway: plus strands→minus strands→plus strands. The synthesis of complementary RNA is, of course, not replication, although in many publications this distinction has not been made. Thus, an enzyme that can merely transcribe a template should not be called a replicase. Purified $3D^{pol}$, therefore, is a primer-dependent, template-dependent RNA polymerase (indeed the enzyme will transcribe any RNA under specific conditions), but is not a replicase.

In spite of enormous efforts over a period of 25 years, no replication system of picornavirus RNA has been developed *in vitro*. It should be noted that most investigators concerned with the replication of any RNA virus genome have tried to reproduce the classical experiments of S. Spiegelman and his collaborators: the incubation of a "purified viral protein" with genome RNA, in this case with bacteriophage Qβ RNA and nucleotide precursors that yielded authentic phage RNA synthesized *de novo* (Spiegelman and Hayashi, 1963). However, the "purified viral protein" used in the replication reaction turned out to be a subtle complex of the phage-encoded RNA polymerase with three bacterial proteins, the latter normally being involved in the translation of cellular mRNA. This important discovery by Kamen, Weismann, and Weber and their colleagues that cellular proteins unrelated to nucleic acid metabolism are recruited by viral RNA-synthesizing machinery has enormously influenced all subsequent research on RNA virus replication (Kamen, 1975). It suddenly appeared possible that any cytoplasmic polypeptide of the host cell has the potential to be involved in picornavirus RNA replication, either in a complex with $3D^{pol}$ or as a separate entity. This kind of reasoning has, in fact, led to the study of the above mentioned "host factor" HF.

It should be emphasized that we consider it impossible to develop a replication system *in vitro* for picornavirus RNAs similar to that of the RNA phages. The simple reason is that the involvement of proteins in picornavirus replication is not solely catalytic, since progeny RNA is VPg-linked. Moreover, it is likely that membranes are an intrinsic component of the picornavirus replication machinery.

Two different strategies have been followed in attempts to elucidate the mechanism of poliovirus RNA replication. The first involves the purification of polypeptides from virus-infected or uninfected cells. The various polypeptides and factors are then added to virion template RNA and the resulting products are analysed. The second is a study of a membranous replication complex isolated from infected cells that is capable of synthesizing authentic virion RNA. Both strategies, to be discussed separately, have yielded valuable results.

(a) Soluble systems of RNA replication. When the large, membranous replication complex that accumulates in picornavirus-infected cells (Girard *et al.*, 1967; Takegami *et al.*, 1983a,b) is extracted with a mixture of detergents and the solubilized material is sedimented through a sucrose gradient, smaller aggregates (70–130S) can be obtained that are free of membranes yet incorporate precursors into viral RNA (Arlinghaus and Polatnick, 1969; Ehrenfeld *et al.*, 1970). This "soluble" polymerase complex has been further purified by precipitation in 2 M LiCl, a treatment that did not inactivate the polymerase activity (Lundquist *et al.*, 1974). Subsequent analysis of the proteins contained in the partially purified material showed "NCVP4" to be its predominant virus-encoded polypeptide (Lundquist *et al.*, 1974). NCVP4 was later shown to be the virus-encoded RNA polymerase $3D^{pol}$ (Flanegan and Baltimore, 1979). An analysis of the products of polymerization revealed that both plus- and minus-stranded RNA is synthesized in the solubilized complexes (Caliguiri, 1974; Lundquist and Maizel, 1978). Although the separation of active $3D^{pol}$ from these replication complexes failed, these experiments provided very important information: (i) the poliovirus-induced RNA polymerase can catalyse RNA synthesis in the absence of membranes and is resistant to treatment with detergents such as deoxycholate/NP40; (ii) once bound, the polymerase has high affinity to its template and resists washes with strong salt; (iii) synthesis of both plus and minus strands is most likely catalysed by a single polymerase, the virus-encoded polypeptide $3D^{pol}$.

Using poly(A) as template and testing oligo(U) as a potential primer for RNA synthesis, Flanegan and Baltimore (1977) developed a new assay that they used to identify a poliovirus-induced "poly(U) polymerase". This enzyme proved to be the poliovirus-encoded enzyme $3D^{pol}$, a primer-dependent, template-dependent RNA polymerase (Flanegan and Baltimore, 1979; Van Dyke and Flanegan, 1980, and references therein). These reports considerably advanced the knowledge of picornavirus RNA replication, because not only was a simple assay at hand, but more importantly, a crucial and novel property of the viral RNA polymerase was discovered: the dependence of the enzyme upon a primer. Given this information it was established that $3D^{pol}$ can transcribe any polyadenylated RNA in the presence of oligo(U); thus the expected specificity for its natural template, the poliovirus genome, was missing. It is unlikely that oligo(U)-primed RNA synthesis occurs during poliovirus replication *in vivo*, since the cell cytoplasm contains numerous molecules of polyadenylated mRNAs far in excess of viral RNA, and free oligo(U) or poly(U) molecules have not been observed in the uninfected or infected cell (they would most likely be bound to the abundant polyadenylate tails anyway). Thus, although the enzyme responsible for chain elongation in picornavirus RNA replication was identified and characterized, the mechanism of initiation of RNA synthesis and the mechanism of the linking of VPg to progeny strands was still obscure. Dasgupta *et al.* (1980), meanwhile, described the discovery of the host factor (HF, mol.wt 67 kD) that allowed initiation of RNA synthesis on a poliovirus RNA template with $3D^{pol}$ in the absence of oligo(U). The mechanism by which HF leads to the priming of $3D^{pol}$ is obscure. The recent description of HF being a protein kinase (Morrow *et al.*, 1985) added little to the solution of this question, although phosphorylation of some protein component in the reconstituted reaction mixture appeared to be a requirement in the priming reaction.

Another unsolved question was how the terminal protein VPg is linked to the RNA. If HF can stimulate priming of RNA synthesis in the absence of oligo(U) as well as in the absence of VPg or its precursors, at what step of RNA synthesis are the 5' ends of the nascent RNA strands modified? An answer to this important question appeared at hand when it was reported that VPg-related proteins (determined by immunoprecipitation) were found covalently bound to the RNA synthesized *in vitro* (Baron and Baltimore, 1982; Morrow *et al.*, 1984). This result was very surprising, because the RNA synthesis was carried out with only virion RNA, purified $3D^{pol}$ and purified HF. It was speculated, however, that small amounts of VPg-containing polypeptides, undetectable by PAGE analysis, copurified with $3D^{pol}$.

Most recently, J. B. Flanegan (personal communication) and Andrews *et al.* (1985) have provided evidence that, in the reactions used by them, the appearance of newly synthesized VPg-linked RNA strands is the result of end-labelling of fragmented VPg-linked template RNA, and thus an artefact. This interpretation is not shared by Dasgupta and his colleagues, since in their transcription system *in vitro* the RNA products are linked to a 45 kD mol.wt., VPg-related protein that cannot originate from the virion RNA template.

A novel idea of the priming of $3D^{pol}$ was then introduced by Flanegan and his colleagues: a mechanism of elongation of the 3' end of the template RNA (Young *et al.*, 1985). Such a mechanism has a precedent in papovavirus DNA replication (Berns *et al.*, 1985). Flanegan's hypothesis is based upon the observation that in his transcription system up to one-half of the reaction product synthesized in the presence of virion RNA, $3D^{pol}$ and HF is covalently linked at one end, presumably by a hairpin-like structure to the template RNA. HF was required for the cross-link to occur. Thus, Young et al. (1985) speculated that HF might be able to force the 3'-terminal poly(A) to bend so that it could be used by $3D^{pol}$ for the elongation reaction (see Fig. 2.3). The model of hairpin priming received strong support from Andrews et al. (1985) who reported that their HF preparation could be further purified to yield a terminal uridylate transferase (TUT) that could substitute HF in the transcription reaction in vitro. It was proposed that virion template RNA is uridylylated by TUT at the 3'-terminal poly(A); the resulting oligo(U) tail then forms a snap-back structure whose 3' end serves as primer for $3D^{pol}$.

The addition of U residues to 3' termini of RNA was found to occur most efficiently on poly(A) tails. This fits well with the 3'-terminal poly(A) of poliovirus plus strands, but makes it difficult to explain initiation of plus strands at the 3' end of minus strands. In fact, the 3' ends of minus-stranded RNA in RF is...$UUUU(A)_n$, where $n=2$ in the majority of molecules that form a blunt end, although a small fraction of RF molecules may carry a small number of protruding adenylate residues. No extra U residues have ever been observed at these ends. Unfortunately, the 3' termini of the minus-stranded template RNA in RI have not been analysed.

Although the self-priming model is appealing, it does not provide any clue as to how VPg is eventually linked to the nascent RNA strand. Young et al. (1985) and Andrews et al. (1985) propose that VPg or a precursor polypeptide to VPg may nick the hairpin with the concomitant formation of a phosphodiester bond with the newly synthesized RNA strand (see Fig. 2.3). Such a mechanism of nucleic acid modification has been observed previously in interactions between supercoiled DNA strands with topoisomerases, and also in the replication of parvoviruses (Berns et al., 1985).

Whereas all transcription reactions discussed so far involved virion RNA as template and thus represented only minus-strand RNA synthesis, Kaplan et al. (1985) have recently reported that $3D^{pol}$ and HF can transcribe a minus-strand RNA template prepared in vitro with the SP6 polymerase. The resulting RNA was of plus-strand polarity and presumably contained non-viral sequences at either terminus. In spite of this, the RNA was infectious. This work showed that $3D^{pol}$ is capable of efficiently transcribing templates that are not polyadenylated, provided HF is added, an observation made previously also by Young et al. (1985). However, the data do not illuminate the mechanism of initiation of RNA synthesis. Interestingly, plus-stranded transcripts, prepared from the corresponding infectious cDNA clone with SP6 polymerase, were not infectious (Kaplan et al., 1985). Mizutani and Colonno (1985), on the other hand, obtained infectious rhinovirus RNA from cDNA clones by transcription with SP6 RNA transcriptase. A very efficient transcription system in vitro from infectious cDNA clones of poliovirus RNA has been constructed by van der Werf et al. (1986). In this case transcription is under the control of the phage T7 promoter and is catalysed by phage T7 RNA polymerase. The transcripts contain only two non-viral G residues at the 5' end and, at most, nine non-viral nucleotides following the 3'-terminal poly(A). This system can produce very large amounts of highly infectious ($>10^5$ PFU/μg of RNA) poliovirus RNA (van der Werf et al., 1986). For comparison, the specific infectivity of the most efficient cDNA clones of poliovirus is over 1000 PFU/μg of DNA (Semler et al., 1984) and that of virion RNA is 10^6 PFU/μg of RNA.

(b) Membranous replication system. None of the soluble systems described so far has yielded authentic poliovirus RNA. The major deficiency is the linking of VPg to the newly synthesized RNA, a process that requires not only an enzymatic activity catalysing the uridylylation of the phenolic hydroxy group of VPg, but possibly also the proteolytic processing enzyme $3C^{pro}$.

If membranes of infected HeLa cell cytoplasm are precipitated by centrifugation, a "crude replication complex" (CRC) cosediments that contains all virus-specific activity to incorporate UTP into viral RNA (Table 2.1). CRC has been shown to produce viral RNA whose "fingerprint" is indistinguishable from authentic virion RNA (Etchison and Ehrenfeld, 1981; Takeda et al., 1986). The two-dimensional fingerprints can be prepared such that the 5'-terminal RNase T1-resistant oligonucleotide (VPg-pUUAAAACAGp), which barely migrates in the first dimension, becomes visible in the autoradiogram (Takeda et al., 1986). The RNA produced *in vitro* contained newly synthesized 5' ends, an observation suggesting that RNA synthesis in CRC resembles the events occurring *in vivo* (Takeda et al., 1986). This encouraging result is the basis for the

Table 2.1. Incorporation of [^{3}H]UTP into RNA[a]

	Crude membrane fraction (pmol/mg protein)	Supernatant fraction (pmol/mg protein)
Uninfected HeLa cells	50	36
Infected HeLa cells	403	44
After sucrose density gradient centrifugation[b]		
Fraction I	1880	
Fraction II	853	
Fraction III	750	
Fraction IV	511	
Fraction V	386	

[a]Assayed by adsorption to DEAE paper.
[b]Discontinuous sucrose gradient.

strategy of dissecting and reassembling components of CRC, thereby learning the mechanism of picornavirus RNA replication.

Early studies with CRC revealed that the virus-induced polymerase activity resides mainly in the smooth membrane fraction (Caliguiri and Tamm, 1970). This is documented in Table 2.1; CRC was fractionated by isopycnic sucrose gradient centrifugation (fraction RC I; Caliguiri and Tamm, 1970; Takegami et al., 1983a). The membrane-associated replication complexes can synthesize all forms of poliovirus-specific RNA structures: RI, RF and ssRNA. Addition of detergent (for example, the non-ionic detergent NP40 at 0.5%) profoundly alters the labelling pattern, in that now only RF is synthesized (Etchison and Ehrenfeld, 1980; Takeda et al., 1986). CRC contains a good portion of all known virus-specific proteins, that is, all precursor polypeptides and their processing products seen in the infected cell; and that is true also for the very active, partially purified replication complex called RC I contained in the smooth membrane fraction (Semler et al., 1982; Takegami et al., 1983a; Tershak, 1984). The association of the viral proteins with the membrane(s) is very strong: high-ionic-strength buffer containing EDTA or 4 M urea does not dislodge them with the possible exception of $3C^{pro}$. Upon inspection of their amino acid sequences only 3A and 3AB, and precursors thereof, contain a 22-amino-acid-long hydrophobic region flanked by two basic amino acids that has been shown to serve as an anchor in the membrane (Semler et al., 1982; Takegami et al., 1983a). Tight interaction of 3AB with other replication polypeptides (e.g. 2C, $3D^{pol}$) is therefore thought to direct the viral replication complex to the membranes.

A key problem using CRC was solved by Takegami et al. (1983b), who developed conditions that allowed the synthesis in vitro of VPg-pU and VPg-pUpU. Indeed, each fraction of the sucrose-gradient-separated CRC was found capable of synthesizing VPg-pUpU, but RC I again had the highest activity. Similar results were obtained by Vartapetian et al. (1984), who analysed a CRC from EMCV-infected Ehrlich ascites cells. Crawford and Baltimore (1983) then reported that small amounts of VPg-pUpU can be found in poliovirus-infected HeLa cells if specific attention is given to the elusiveness of the nucleotidyl-oligopeptide owing to its size and charge.

Interestingly, the synthesis of VPg-pUpU was completely inhibited when small amounts of non-ionic detergent (0.5% NP40) were added. Moreover, the formation of VPg-pUpU was found to be most efficient under conditions that are less favourable to the RNA elongation reaction (Takegami et al., 1983b; Takeda et al., 1986). These data suggested that synthesis of the uridylyl-oligopeptide and the $3D^{pol}$-catalysed elongation of RNA strands are distinct processes.

Attempts to "chase" preformed VPg-pUpU into longer RNA strands have met with difficulties owing to the low yields of labelled nucleotidyl protein. The efficiency of the system was greatly stimulated when the membrane fractions (CRC or RC I) were treated with DEAE-cellulose to remove endogenous nucleotide triphosphates and an ATP regenerating system was included into the reaction mixture (Takeda et al., 1986). Under these conditions individual steps in the synthesis of VPg-pU→VPg-pUpU→VPg-pUUAAAACAGp can be analysed. An important result of this study was that preformed [^{32}P]VPg-pUpU, when incubated with nucleoside triphosphates (and UTP in large excess), can be "chased" into the elongation product VPg-pUUAAAACAGp. This reaction, but not the formation of VPg-pUpU, is stimulated by a cell extract of uninfected HeLa cells (Takeda et al., 1986). These data support a model in which preformed VPg-pUpU is the primer for 3D^{pol} (Takegami et al., 1983b). The data do not explain how VPg is being uridylylated. It is not even known whether the synthesis of VPg-pUpU is template-dependent because the RNAs contained in CRC or RC I are very resistant to digestion with micrococcal nuclease (or other RNases) (Takeda et al., 1986).

3. Models of Picornaviral RNA Replication

Any model for picornaviral RNA replication must take into account the following conditions. (i) The product RNAs, regardless of their polarity, must be VPg-linked. (ii) The process of initiation must accommodate two different termini: poly(A) at the 3' end of plus strands and...UUUUAA$_{OH}$ at the 3' end of minus strands. (iii) The bulk of minus and plus strand RNA synthesis must occur on membranes. (iv) The poliovirus-encoded RNA polymerase 3D^{pol} must be provided with a primer.

The models that in part accommodate these requirements are shown in figures 2.3, 2.4 and 2.5. Figure 2.3, kindly provided by Dr. J. B. Flanegan, describes the hairpin, snap-back model. Whether host factor or TUT provides the means of the formation of a structure acceptable to 3D^{pol} as primer is, for the purpose of this discussion, not important. This model, however, predicts that VPg, or one of its precursors, might act as a nicking enzyme with concomitant linkage of the Tyr residue of VPg to the 5' phosphate of the nascent RNA strand. So far, there is no evidence for such a reaction. The major problem with this model is that the nicking activity must recognize two very different hairpins. Moreover, if TUT is involved one would expect to find extra uridylate residues at the end of the minus strands in RF. Instead, the heteropolymeric end of RF is mostly blunt-ended and only a small fraction carry extra adenylate residues. The model is supported by the fact that RF molecules covalently linked at one end can be found in picornavirus-infected cells (Agol et al., 1970; Romanova and Agol, 1979; Senkevich et al., 1980; Robberson et al., 1982; Young et al., 1985) and that they can be synthesized in the polymerization reaction *in vitro*. The results of Andrews et al. (1985), who identified TUT in their preparations, provide some explanation as to how such end-linked RF molecules can arise. It should be stressed, however, that the biological significance of these structures is still obscure, since there exists no evidence linking them to RNA replication *in vivo*. Finally, the model predicts that both homo-linked and hetero-linked RF molecules (see Fig.

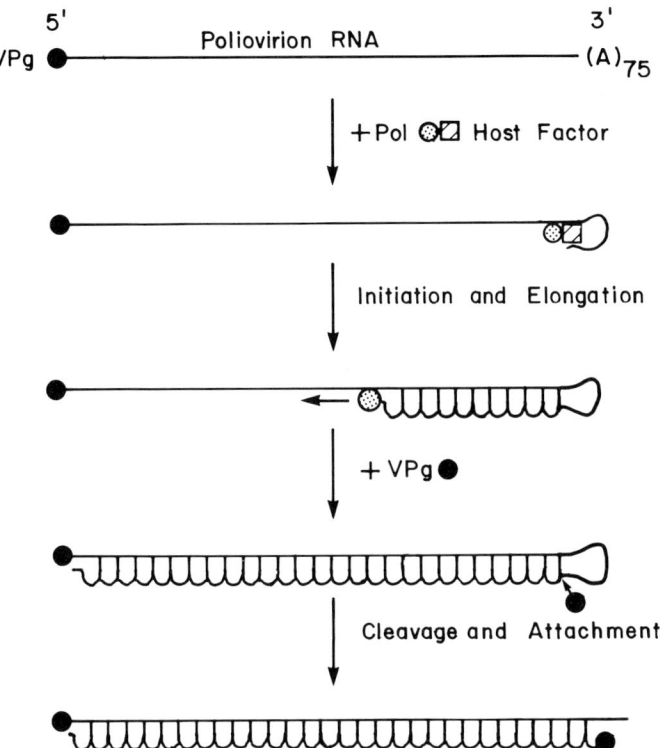

Figure 2.3. Self-priming by template RNA. A host factor (either a kinase or a terminal uridylate transferase) produces a hairpin at the 3' end of the template RNA that serves as a primer for 3D^{pol}. After some elongation has taken place, VPg (or its precursor) will cleave the hairpin thereby attaching itself to the 5' end of the nascent RNA strand. If nicking does not occur, an end-linked RF molecule is formed; as shown here, the homopolymeric segments of the RF would be covalently bound. We have named those structures "homo-linked" RF. Hairpin-mediated initiation and complete transcription of minus-stranded template would yield "hetero-linked" RF molecules. (Drawing kindly provided by Dr. J. B. Flanegan.)

2.3 for our definition of end-linked double-stranded poliovirus RNA) should occur in the infected cell. An analysis to test this prediction has not been done.

The other model, shown in Fig. 2.4, describes the formation of VPg-pU(pU) in a membranous replication complex. VPg-pUpU then primes for 3D^{pol} (Takegami et al., 1983b). This scenario, first proposed in 1977 (Lee et al., 1977; Nomoto et al., 1977a; Flanegan et al., 1977), resembles the events that lead to the initiation of DNA synthesis in adenoviruses and certain bacteriophages, that is, certain viruses with linear, double-stranded

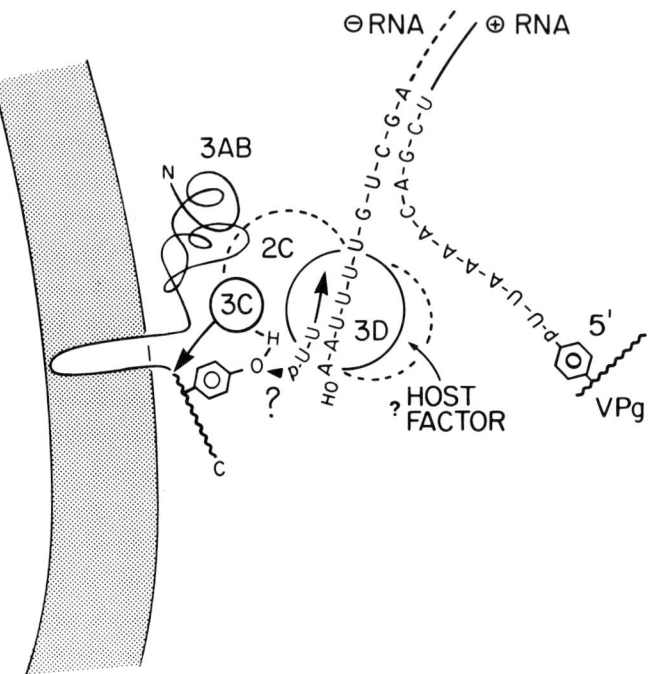

Figure 2.4. Model of VPg-pUpU primed initiation of RNA synthesis. 3AB is the membrane-bound poliovirus protein whose COOH terminus (wavy line) is VPg, 3C is the virus-encoded proteinase (responsible for the cleavage between 3A and 3B), 3D the primer-dependent RNA polymerase, and 2C an auxiliary viral protein, mutations of which (g^r) lead to an altered phenotype of RNA synthesis *in vivo*. The possibility of the involvement of a "host factor" is indicated, although there is no evidence to support such a factor in the membrane-bound replication complex. This model can account for the initiation of both plus and minus strand RNA.
(Modified from Takegami *et al.*, 1983b.)

genomes whose 5' termini are also protein-linked (Challberg and Kelly, 1982; Penalva and Salas, 1982; Lichy *et al.*, 1981; Hermoso and Salas, 1980; Pincus *et al.*, 1981; Desiderio and Kelly, 1981; and see a discussion by Wimmer, 1982). In the case of these viruses, a terminal protein is first linked to dAMP or dCMP, a reaction catalysed by the viral DNA polymerase. The resulting deoxynucleotidylyl protein then serves as a primer for DNA synthesis. Hence, the discovery of the occurrence of VPg-pUpU in infected HeLa cells (Crawford and Baltimore, 1983) and the efficient synthesis of VPg-pUpU *in vitro* (Takegami *et al.*, 1983b) that can be chased into longer strands (Takeda *et al.*, 1986), can clearly be taken as strong support for the priming mechanism shown in Fig. 2.4. This model, of course, does not require a cellular enzyme such as TUT. Andrews *et al.* (1985), however, cautioned that the occurrence of VPg-pUpU *in vitro* and *in vivo* may simply be the result of an aborted nicking/elongation reaction.

A solution to this problem will come if minus strands synthesized *in vitro* can be used in a reconstitution experiment in which the 3' termini of exogenously added RNA can be analysed after the reaction. Nevertheless, preformed VPg-pUpU *can* be utilized in the elongation reaction (Takeda et al., 1986).

A model that accommodates both modes of RNA syntheses is shown in Fig. 2.5. Formation of hairpin structures (possibly as a result of the action of TUT) may occur early in infection without the involvement of membranes and would lead to a homo-linked RF molecule. Indeed, this might be the mechanism by which the first minus RNA strands are synthesized (Fig. 2.5(*a*)). Proteins synthesized prior to the synthesis of the

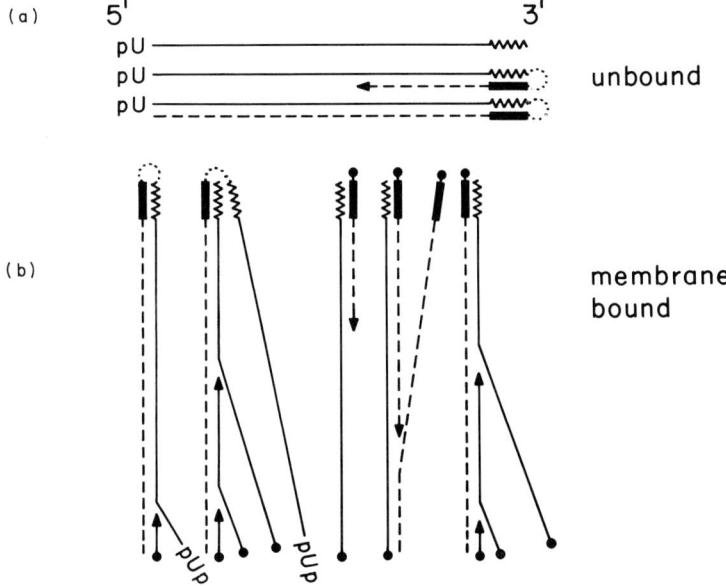

Figure 2.5. Model of the involvement of intracellular structures of picornavirus RNA in RNA replication. (*a*) Very early in infection, transcription of viral mRNA through TUT-mediated self-priming leads to homo-linked RF molecules (double-stranded RNA linked at the poly(A)-poly(U) homopolymeric duplex). Attachment of virus-encoded protein transports this molecule to the membrane. (*b*) Initiation of VPg-pUpU-primed synthesis of plus strands followed by strand replacement in the elongation reaction. Termination of transcription of the minus strands occurs at the end of the homopolymeric region; should read-through occur, the resulting structures would be twice the length of genome RNA and would consist of inverted repeats of two RF molecules. Such structures have been found in infected cells. Minus strand synthesis in the membrane-bound complex commences by a VPg-pUpU primed reaction. Note that this model does not require a nicking activity. It also predicts that hetero-linked RF molecules may not be formed. It should be stressed that the majority of RF molecules found in HeLa cells are unlinked at either end (see Fig. 2.2). (Modified from Takeda *et al.*, 1986.)

homo-linked RF might then attach to the RNA and position it in membranes. Initiation of plus strands will occur by VPg-pUpU priming (Fig. 2.5(b)). Elongation of these first plus strands proceeds by strand replacement (a model greatly favoured by experiments of Lundquist and Maizel, 1978); the plus strands, in turn, will then serve as templates for VPg-pUpU-primed minus strand synthesis (Fig. 2.5(b)). Should readthrough occur in the transcription reaction of the homo-linked RF molecules (e.g. if $3D^{pol}$ fails to terminate transcription at the hairpin), RF molecules twice the size of the genome would be formed. Such linear dimers of RF have indeed been observed in EMCV-infected cells (Senkevich et al., 1980). This model predicts that no hetero-linked RF molecules will be found in the infected cell. It does not require a nicking/linking activity of VPg.

IV. CONCLUSIONS

In the early 1970s very few molecular biologists anticipated that the replication of picornaviruses would continue to pose so many fascinating and unique problems. Some of these unsolved mysteries include the mechanism of uptake of the virion into the cell, the mechanisms of uncoating and of initiation of protein synthesis, the mechanism of polyprotein processing, the mechanism of RNA replication and the mechanism of morphogenesis. We know much but we can explain little, let alone reproduce these steps of replication *in vitro*. Basic research on picornaviruses has not ceased to be a challenge for the molecular biologist; moreover, it will remain important for those concerned with public health, for picornaviruses are the cause of frequent and serious diseases in man.

ACKNOWLEDGEMENTS

We thank our colleagues, particularly Naokazu Takeda, Chen-Fu Yang, Haruka Toyoda, Steven Pincus, Martin Nicklin, Aniko Paul, David Diamond and Michael Murray for many stimulating discussions. We are grateful to Sue Studier for the preparation of the manuscript. Work described here was supported in part by NIH grants AI 15122 and CA 28146. R.J.K. is a member of the Molecular Biology Graduate Training Program at Stony Brook.

REFERENCES

1. Adler, C. J., M. Elzinga and E. Wimmer (1983). *J. Gen. Virol.* **64**, 349–355.
2. Agol, V. I., Y. F. Drygin, L. I. Romanova and A. A. Bogdanov (1970). *FEBS Lett.* **8**, 13–16.
3. Alexander, H. E., G. Koch, I. Morgan Mountain and O. Van Damm (1958). *J. Exp. Med.* **108**, 493.
4. Alquist, P. and P. Kaesberg (1979). *Nucl. Acids Res.* **1**, 1195–1204.
5. Ambros, V. and D. Baltimore (1978). *J. Biol. Chem.* **253**, 5263–5266.

6. Ambros, V., R. F. Pettersson and D. Baltimore (1978). *Cell* **15**, 1439–1446.
7. Anderson-Sillman, K., S. Bartal and D. R. Tershak (1984). *J. Virol.* **50**, 922–928.
8. Andrews, N. C., D. Levin and D. Baltimore (1985). *J. Biol. Chem.* **260**, 7628–7635.
9. Argos, P., G. Kamer, M. J. H. Nicklin and E. Wimmer (1984). *Nucl. Acids Res.* **12**, 7251–7267.
10. Arlinghaus, R. B. and J. Polatnick (1969). *Proc. Natl Acad. Sci. USA* **62**, 821–828.
11. Armstrong, J. A., M. Edmonds, H. Nagazato, B. A. Phillips and M. H. Vaughan (1972). *Science* **176**, 526–528.
12. Baltimore, D., R. M. Franklin, H. J. Eggers and I. Tamm (1963). *Proc. Natl Acad. Sci. USA* **49**, 843–849.
13. Baltimore, D. (1964). *Proc. Natl Acad. Sci USA* **51**, 450–456.
14. Baltimore, D. (1966). *J. Mol. Biol.* **18**, 421–428.
15. Baltimore, D. (1968). *J. Mol. Biol.* **32**, 359–368.
16. Baltimore, D. (1969). *In* "The Biochemistry of Viruses" (Ed. H. Levi), pp. 101–176. Marcel Dekker, New York.
17. Baltimore, D., M. Girard and J. E. Darnell (1964). *Science* **143**, 1034.
18. Baltimore, D. and M. Girard (1966). *Proc. Natl Acad. Sci. USA* **56**, 741.
19. Baron, M. H. and D. Baltimore (1982). *Cell* **30**, 745–752.
20. Berns, K. I., N. Muzyczka and W. W. Hauswirth (1985). *In* "Virology" (Ed. B. N. Fields), pp. 415–432. Raven Press, New York.
21. Best, M., B. Evan and J. M. Bishop (1972). *Virology* **47**, 592–603.
22. Bishop, J. M. and G. Koch (1967). *J. Biol. Chem.* **242**, 1736.
23. Bishop, J. M., G. Koch, B. Evans and M. Merriman (1969). *J. Mol. Biol.* **46**, 235–249.
24. Brown, F., J. Newman, J. Stott, A. Porter, D. Frisby, C. Newton, N. Carey and P. Fellner (1974). *Nature* **251**, 342–344.
25. Burroughs, J. N. and F. Brown (1978). *J. Gen. Virol.* **41**, 443–446.
26. Butterworth, B. E., E. J. Shinshick and F. H. Yin (1976). *J. Virol.* **19**, 457–466.
27. Caliguiri, L. A. (1974). *Virology* **58**, 526–538.
28. Caliguiri, L. A. and I. Tamm (1968). *Virology* **35**, 408–417.
29. Caliguiri, L. A. and I. Tamm (1970). *Virology* **42**, 100–111.
30. Caliguiri, L. A. and A. G. Mosser (1971). *Virology* **46**, 375–386.
31. Campbell, B. A. and C. E. Cords (1983). *J. Virol.* **48**, 561–564.
32. Carroll, A. R., D. J. Rowlands and B. E. Clarke (1984). *Nucl. Acids Res.* **12**, 2461–2472.
33. Challberg, M. D. and T. J. Kelly (1982). *Annu. Rev. Biochem.* **51**, 901–934.
34. Chumakov, K. M. and V. I. Agol (1976). *Virology* **73**, 528–531.
35. Colonno, R. J., P. L. Callahan and W. J. Long (1986). *J. Virol.* **57**, 7–12.
36. Cooper, P. D. (1977). *In* "Comprehensive Virology" (Eds H. Fraenkel-Conrat, R. R. Wagner), Vol. 9, pp. 133–297. Plenum, New York.
37. Crawford, N. M. and D. Baltimore (1983). *Proc. Natl Acad. Sci. USA* **80**, 7452–7455.
38. Crowther, D. and J. L. Melnick (1961). *Virology* **15**, 65–74.
39. Dasgupta, A., M. H. Baron and D. Baltimore (1979). *Proc. Natl Acad. Sci. USA* **76**, 2679–2683.
40. Dasgupta, A., P. Zabel and D. Baltimore (1980). *Cell* **19**, 423–429.

41. Daubert, S. D., G. Bruening and R. C. Najavian (1978). *Eur. J. Biochem.* **92**, 45–51.
42. Dawson, W. O. (1975). *Intervirology* **6**, 83–89.
43. Desiderio, S. V. and T. J. Kelly (1981). *J. Mol. Biol.* **145**, 319–337.
44. Detjen, B. M., J. J. Lucas and E. Wimmer (1978). *J. Virol.* **27**, 582–586.
45. Dmitrieva, T. M., M. V. Shcheglova and V. I. Agol (1979). *Virology* **92**, 271–277.
46. Dorner, A. J., P. G. Rothberg and E. Wimmer (1981). *FEBS Lett.* **132**, 219–223.
47. Dorner, A. J., L. F. Dorner, G. R. Larsen, E. Wimmer and C. W. Anderson (1982). *J. Virol.* **42**, 1017–1028.
48. Dorner, A. J., B. L. Semler, R. J. Jackson, R. Hanecak, E. Duprey and E. Wimmer (1984). *J. Virol.* **50**, 507–514.
49. Dorsch-Hasler, K., Y. Yogo and E. Wimmer (1975). *J. Virol.* **16**, 1512–1527.
50. Edery, I., M. Hubelin, A. Darveau, K. Lee, S. Milburn, J. Hershey, H. Trachsel and N. Sonenberg (1983). *J. Biol. Chem.* **258**, 11398–11403.
51. Ehrenfeld, E., J. V. Maizel and D. F. Summers (1970). *Virology* **40**, 840–846.
52. Emini, E. A., M. Elzinga and E. Wimmer (1982). *J. Virol.* **42**, 194–199.
53. Emini, E. A., J. Leibowitz, D. Diamond, J. Bonin and E. Wimmer (1984). *Virology* **137**, 74–85.
54. Etchison, D. and E. Ehrenfeld (1980). *Virology* **107**, 135–142.
55. Etchison, D. and E. Ehrenfeld (1981). *Virology* **111**, 33–46.
56. Etchison, D., S. C. Milburn, I. Edery, N. Sonenberg and J. W. B. Hershey (1982). *J. Biol. Chem.* **257**, 14806–14810.
57. Etchison, D. and S. Fout (1985). *J. Virol.* **54**, 634–638.
58. Fellner, P. (1979). In "The Molecular Biology of Picornaviruses" (Ed. R. Perez-Bercoff), pp. 25–47. Plenum Press, New York.
59. Flanegan, J. B. and D. Baltimore (1977). *Proc. Natl Acad. Sci. USA* **74**, 3677–3680.
60. Flanegan, J. B., R. F. Pettersson, V. Ambros, M. J. Hewlett and D. Baltimore (1977). *Proc. Natl Acad. Sci. USA* **74**, 961–965.
61. Flanegan, J. B. and D. Baltimore (1979). *J. Virol.* **29**, 352–360.
62. Forss, S. and H. Schaller (1982). *Nucl. Acids Res.* **10**, 6441–6450.
63. Franssen, H., J. Leunissen, R. Goldbach, G. Lomonossoff and D. Zimmern (1984). *EMBO J.* **3**, 855–861.
64. Friedman, R. M. (1970). *J. Virol.* **6**, 628–636.
65. Ganem, D., L. Greenbaum and H. E. Varmus (1982). *J. Virol.* **44**, 374–383.
66. Gerlich, W. H. and W. S. Robinson (1980). *Cell* **21**, 801–809.
67. Ghosh, A., R. Dasgupta, T. Salerno-Rife, T. Rutgers and P. Kaesberg (1979). *Nucl. Acids Res.* **7**, 2137–2146.
68. Girard, M., D. Baltimore and J. E. Darnell (1967). *J. Mol. Biol.* **24**, 59–74.
69. Golini, F., A. Nomoto and E. Wimmer (1978). *Virology* **89**, 112–118.
70. Golini, F., B. L. Semler, A. J. Dorner and E. Wimmer (1980). *Nature* **287**, 600–603.
71. Grifo, J. A., S. M. Tahara, M. A. Morgan, A. J. Shatkin and W. C. Merrick (1983). *J. Biol. Chem.* **258**, 5804–5810.
72. Hanecak, R., B. L. Semler, C. W. Anderson and E. Wimmer (1982). *Proc. Natl Acad. Sci. USA* **79**, 3973–3977.
73. Hanecak, R., B. L. Semler, H. Ariga, C. W. Anderson and E. Wimmer (1984). *Cell* **37**, 1063–1073.

74. Hansen, J. and E. Ehrenfeld (1981). *J. Virol.* **38**, 438–445.
75. Harding, N., J. Ito and G. S. David (1978). *Virology* **84**, 279–292.
76. Harris, T. J. R. and F. Brown (1977). *J. Gen. Virol.* **34**, 87–105.
77. Hermoso, J. M. and M. Salas (1980). *Proc. Natl Acad. Sci. USA* **77**, 6425–6428.
78. Hewlett, M. J., S. Rozenblatt, V. Ambros and D. Baltimore (1977). *Biochemistry* **16**, 2763–2767.
79. Hogle, J. M., M. Chow and D. J. Filman (1985). *Science* **229**, 1358–1365.
80. Hruby, D. E. and W. K. Roberts (1978). *J. Virol.* **25**, 413–415.
81. Ito, J. (1978). *J. Virol.* **28**, 895–904.
82. Jackson, R. (1986). *Virology* **149**, 144–127.
83. Jacobson, M. F. and D. Baltimore (1968). *Proc. Natl Acad. Sci. USA* **61**, 77–84.
84. Jen, G. and R. E. Thach (1982). *J. Virol.* **43**, 250–261.
85. Kamen, R. I. (1975). In "RNA Phages" (Ed. N. D. Zinder), pp. 203–234. Cold Spring Harbor Monographs.
86. Kaplan, G., J. Lubinski, A. Dasgupta and V. R. Racaniello (1985). *Proc. Natl Acad. Sci. USA* **82**, 8424–8428.
87. Keegstra, W., P. S. Van Wielink and J. S. Sussenbach (1977). *Virology* **76**, 444–474.
88. King, A. M. Q., D. V. Sangar, T. J. R. Harris and F. Brown (1980). *J. Virol.* **34**, 627–634.
89. Kitamura, N., C. Adler and E. Wimmer (1980). *Ann. NY Acad. Sci.* **345**, 183–201.
90. Kitamura, N., B. L. Semler, P. G. Rothberg, G. R. Larsen, C. J. Adler, A. J. Dorner, E. A. Emini, R. Hanecak, J. J. Lee, S. van der Werf, C. W. Anderson and E. Wimmer (1981). *Nature* **291**, 547–553.
91. Koch, G. and F. Koch (1985). "The Molecular Biology of Poliovirus". Springer-Verlag, Wien and New York.
92. Koch, G., N. Quentrell and J. M. Bishop (1967). *Virology* **31**, 388.
93. Kozak, M. (1983). *Microbiol. Rev.* **47**, 1–45.
94. Larsen, G. R., B. L. Semler and E. Wimmer (1980). *Nucl. Acids Res.* **8**, 1217–1229.
95. Larsen, G. R., C. W. Anderson, A. J. Dorner, B. L. Semler and E. Wimmer (1982). *J. Virol.* **41**, 340–344.
96. Lee, K. A. W., I. Edery and N. Sonenberg (1985). *J. Virol.* **54**, 515–524.
97. Lee, Y. F., A. Nomoto, B. M. Detjen and E. Wimmer (1977). *Proc. Natl Acad. Sci. USA* **74**, 59–63.
98. Levintow, L. (1974). In "Comprehensive Virology" (Eds H. Fraenkel-Conrat and R. R. Wagner, Vol. 2, pp. 106–169. Plenum Press, New York.
99. Lichy, J. H., M. S. Horwitz and J. Hurwitz (1981). *Proc. Natl Acad. Sci. USA* **78**, 2678–2682.
100. Lloyd, R. E., D. Etchison and E. Ehrenfeld (1985). *Proc. Natl Acad. Sci. USA* **82**, 2723–2727.
101. Loddo, B., W. Ferrari, G. Brotzu and A. Spanedda (1962). *Nature (London)* **193**, 97–98.
102. Loesch, W. T. and R. B. Arlinghaus (1975). *Arch. Virol.* **47**, 201–215.
103. Lundquist, R. E., E. Ehrenfeld and J. V. Maizel (1974). *Proc. Natl Acad. Sci. USA* **71**, 4773–4777.
104. Lundquist, R. E. and J. V. Maizel (1978). *Virology* **85**, 434–444.

105. Madshus, I. H., S. Olsnes and K. Sandvig (1984). *J. Cell. Biol.* **98**, 1194–1200.
106. Mapoles, J. E., D. L. Krah and R. L. Crowell (1985). *J. Virol.* **55**, 560–566.
107. Mayo, M. A., H. Barker and B. O. Harrison (1982). *J. Gen. Virol.* **59**, 1149–1162.
108. Miller, R. L. and P. G. W. Plagemann (1972). *J. Gen. Virol.* **17**, 349–353.
109. Minor, P. D., P. A. Pipkin, D. Hockley, G. C. Schild and J. W. Almond (1984). *Virus Res.* **1**, 203–212.
110. Mizutani, S. and R. J. Colonno (1985). *J. Virol.* **56**, 628–632.
111. Molnar-Kimber, K. L., J. Summers, J. Taylor and W. S. Mason (1983). *J. Virol.* **45**, 165–172.
112. Montagnier, L. and F. K. Sanders (1963). *Nature* **199**, 664–669.
113. Morrow, C. D., M. Navab, E. Peterson, J. Hocko and A. Dasgupta (1984). *Virus Res.* **1**, 89–100.
114. Morrow, C. D., G. F. Gibbons, and A. Dasgupta (1985). *Cell* **40**, 913–921.
115. Mosenkis, J., S. Daniels-McQueen, S. Janovec, R. Duncan, J. B. W. Hershey, J. A. Grifo, W. C. Merrick and R. E. Thach (1985). *J. Virol.* **54**, 643–645.
116. Newton, S. E., A. R. Carroll, R. O. Campbell, B. E. Clarke and D. J. Rowlands (1985). *Gene* **40**, 331–336.
117. Nicklin, M. J. H., H. Toyoda, M. G. Murray and E. Wimmer (1986). *Biotechnology* **4**, 33–42.
118. Nilsen, T. W., D. L. Wood and C. Baglioni (1981). *Virology* **109**, 82–93.
119. Nomoto, A., Y. J. Lee and E. Wimmer (1976). *Proc. Natl Acad. Sci. USA* **73**, 375–380.
120. Nomoto, A., B. Detjen, R. Pozzatti and E. Wimmer (1977a). *Nature (London)* **268**, 208–213.
121. Nomoto, A., N. Kitamura, F. Golini and E. Wimmer (1977b). *Proc. Natl Acad. Sci. USA* **74**, 5345–5349.
122. Opperman, H. and G. Koch (1973). *Biochem. Biophys. Res. Commun.* **52**, 635–640.
123. Pallansch, M., O. M. Kew, B. L. Semler, D. R. Omilianowski, C. W. Anderson, E. Wimmer and R. R. Rueckert (1984). *J. Virol.* **49**, 873–880.
124. Palmenberg, A. C. and R. R. Rueckert (1982). *J. Virol.* **41**, 244–249.
125. Palmenberg, A. C., E. M. Kirby, M. R. Janda, G. M. Drake, K. F. Potratz and M. S. Collett (1984). *Nucl. Acids Res.* **12**, 2969–2985.
126. Penalva, M. A. and M. Salas (1982). *Proc. Natl Acad. Sci. USA* **79**, 5522–5526.
127. Penman, S., Y. Becker and J. E. Darnell (1964). *J. Mol. Biol.* **8**, 541–555.
128. Perez-Bercoff, R. (1979). *In* "The Molecular Biology of Picornaviruses" (Ed. R. Perez-Bercoff), pp. 293–318. Plenum, New York.
129. Pettersson, R. F., J. B. Flanegan, J. K. Rose and D. Baltimore (1977). *Nature* **268**, 270–272.
130. Pettersson, R. F., V. Ambros and D. Baltimore (1978). *J. Virol.* **27**, 357–365.
131. Pincus, S., W. Robertson and D. Rekosh (1981). *Nucl. Acids Res.* **9**, 4919–4938.
132. Pincus, S. E., D. C. Diamond, E. A. Emini and E. Wimmer (1986a). *J. Virol.* **57**, 638–646.
133. Pincus, S. E., H. Rohl and E. Wimmer (1987). *Virology* **157**, 83–88.
134. Pollack, R. and R. Goldman (1973). *Science* **179**, 915–916.

135. Porter, A., N. H. Carey and P. Fellner (1974). *Nature* **248**, 675–678.
136. Pringle, C. R. (1964). *Nature (London)* **204**, 1012–1013.
137. Putnak, J. R. and B. A. Phillips (1981). *Microbiol. Rev.* **45**, 287–315.
138. Racaniello, V. R. and D. Baltimore (1981). *Proc. Natl Acad. Sci. USA* **78**, 4887–4891.
139. Rekosh, D. M. K. (1977). *In* "The Molecular Biology of Animal Viruses" (Ed. D. P. Nayak), Vol. 1, pp. 63–110. Dekker, New York.
140. Rekosh, D., H. F. Lodish and D. Baltimore (1969). *Cold Spring Harbor Symp. Quant. Biol.* **34**, 747–753.
141. Rekosh, D. M. K., W. C. Russell, A. J. D. Bellet and A. J. Robinson (1977). *Cell* **11**, 283–295.
142. Revet, B. and E. Delain (1982). *Virology* **123**, 29–44.
143. Richards, O. C. and E. Ehrenfeld (1980). *J. Virol.* **36**, 387–394.
144. Richards, O. C., E. Ehrenfeld and J. Manning (1979). *Proc. Natl Acad. Sci. USA* **76**, 676–680.
145. Richards, O. C., S. C. Martin, H. G. Jense and E. Ehrenfeld (1984). *J. Mol. Biol.* **173**, 235–240.
146. Rightsel, W. A., J. R. Dice, R. J. McAlpine, E. A. Timm, I. W. McLean, S. J. Dixon and F. M. Schabel (1961). *Science* **134**, 558–559.
147. Robberson, D. L., G. B. Thornton, M. V. Marshall and R. B. Arlinghaus (1982). *Virology* **116**, 454–467.
148. Romanova, L. I. and V. I. Agol (1979). *Virology* **93**, 574–577.
149. Rossmann, M. G., E. Arnold, J. W. Erickson, E. A. Frankenberger, J. P. Griffith, H.-J. Hecht, J. E. Johnson, G. Kamer, M. Luo, A. G. Mosser, R. R. Rueckert, B. Sherry and G. Vriend (1985). *Nature* **317**, 145–153.
150. Rothberg, P. G., T. J. R. Harris, A. Nomoto and E. Wimmer (1978). *Proc. Natl Acad. Sci. USA* **75**, 4868–4872.
151. Roy, P. and D. H. L. Bishop (1970). *J. Virol.* **6**, 604–609.
152. Rueckert, R. R. (1976). *In* "Comprehensive Virology" (Ed. H. Fraenkel-Conrat and R. R. Wagner), Vol. 6, pp. 131–213. Plenum Press, New York.
153. Rueckert, R. R. and E. Wimmer (1984). *J. Virol.* **50**, 957–959.
154. Salas, M., R. P. Mellado, E. Vineula and J. M. Sogo (1978). *J. Mol. Biol.* **119**, 269–291.
155. Sangar, D. V., D. J. Rowlands, T. J. R. Harris and F. Brown (1977). *Nature* **268**, 648–650.
156. Sangar, D. V., D. N. Black, D. J. Rowlands, T. J. R. Harris and F. Brown (1980). *Virology* **33**, 59–68.
157. Sangar, D. V., J. Bryant, T. J. R. Harris, F. Brown and D. J. Rowlands (1981). *J. Virol.* **39**, 67–74.
158. Saunders, K. and A. M. Q. King (1982). *J. Virol.* **42**, 389–394.
159. Semler, B. L., R. Hanecak, C. W. Anderson and E. Wimmer (1981a). *Virology* **114**, 589–594.
160. Semler, B. L., C. W. Anderson, N. Kitamura, P. G. Rothberg, W. L. Wishart and E. Wimmer (1981b). *Proc. Natl Acad. Sci. USA* **78**, 3464–3468.
161. Semler, B. L., C. W. Anderson, R. Hanecak, L. F. Dorner and E. Wimmer (1982). *Cell* **28**, 405–412.
162. Semler, B. L., R. Hanecak, L. F. Dorner, C. W. Anderson and E. Wimmer (1983). *Virology* **126**, 624–635.
163. Semler, B. L., A. J. Dorner and E. Wimmer (1984). *Nucl. Acids Res.* **12**, 5123–5141.

164. Senkevich, T. G., I. M. Cumakov, G. Y. Lipskaya and V. I. Agol (1980). *Virology* **102**, 339–348.
165. Skern, T., W. Sommergruber, D. Blass, P. Greuendler, F. Fraundorfer, C. Pieler, I. Fogy and E. Kuechler (1985). *Nucl. Acids Res.* **13**, 2111–2127.
166. Sonenberg, N., D. Skup, H. Trachsel and S. Millward (1981). *J. Biol. Chem.* **256**, 4138–4141.
167. Spector, D. H. and D. Baltimore (1974). *Proc. Natl Acad. Sci. USA* **71**, 2983–2987.
168. Spector, D. H. and D. Baltimore (1975a). *J. Virol.* **15**, 1418–1431.
169. Spector, D. H. and D. Baltimore (1975b). *Virology* **67**, 498–505.
170. Spiegelman, S. and M. Hayashi (1963). *Cold Spring Harbor Symp. Quant. Biol.* **28**, 161.
171. Stalhandski, P. O. K., M. Lindberg and U. Pettersson (1984). *J. Virol.* **51**, 742–746.
172. Stanley, J., P. Rottier, J. W. Davies, P. Zabel and A. Van Kammen (1978). *Nucl. Acids Res.* **5**, 4505–4522.
173. Stanway, G., A. J. Cann, R. Hauptmann, P. Hughes, L. D. Clarke, R. C. Mountford, P. D. Minor, G. C. Schild and J. W. Almond (1983). *Nucl. Acids Res.* **11**, 5629–5643.
174. Stanway, G., P. J. Hughes, R. C. Mountford, P. D. Minor and J. W. Almond (1984). *Nucl. Acids Res.* **12**, 7859–7877.
175. Strebel, K. and E. Beck (1986). *J. Virol.* **58**, 893–899.
176. Summers, D. F. and L. Levintow (1965). *Virology* **27**, 44–53.
177. Summers, D. F. and J. V. Maizel Jr. (1968). *Proc. Natl Acad. Sci. USA* **59**, 966–971.
178. Summers, D. F., J. V. Maizel Jr. and J. E. Darnell (1967). *Virology* **31**, 427–435.
179. Svitkin, Y. V., S. V. Maslova and V. I. Agol (1985). *Virology* **147**, 243–252.
180. Tahara, S. M., M. A. Morgan and A. J. Shatkin (1981). *J. Biol. Chem.* **256**, 7691–7694.
181. Takeda, N., R. J. Kuhn, C.-F. Yang, T. Takegami and E. Wimmer. (1986) *J. Virol.* **60**, 43–53.
182. Takegami, T., B. L. Semler, C. W. Anderson and E. Wimmer (1983a). *Virology* **128**, 33–47.
183. Takegami, T., R. J. Kuhn, C. W. Anderson and E. Wimmer (1983b). *Proc. Natl Acad. Sci. USA* **80**, 7447–7451.
184. Tershak, D. R. (1982). *J. Virol.* **41**, 313–318.
185. Tershak, D. R. (1984). *J. Virol.* **52**, 777–783.
186. Toyoda, H., M. Kohara, Y. Kataoka, T. Suganuma, T. Omata, N. Imura and A. Nomoto (1984). *J. Mol. Biol.* **174**, 561–585.
187. Toyoda, H., M. J. H. Nicklin, M. G. Murray and E. Wimmer (1986a). In "Protein Engineering: Applications in Science, Medicine and Industry" (Eds M. Inouye and R. Sarma), pp. 314–337. Academic Press, London and Orlando.
188. Toyoda, H., M. J. H. Nicklin, M. G. Murray, C. W. Anderson, J. J. Dunn, F. W. Studier and E. Wimmer (1986b). *Cell* **45**, 761–770.
189. Traub, A., B. Diskin, H. Rosenberg and E. Kalmar (1976). *J. Virol.* **18**, 375–382.
190. van der Werf, S., J. Bradley, E. Wimmer, F. W. Studier and J. J. Dunn (1986). *Proc. Natl Acad. Sci. USA*, **83**, 2330–2334.

191. Van Dyke, T. A. and J. B. Flanegan (1980). *J. Virol.* **35**, 732–740.
192. Varma, J. P. (1968). *Virology* **36**, 305–308.
193. Vartapetian, A. B., Y. F. Drygin, K. M. Chumakov and A. A. Bogdanov (1980). *Nucl. Acids Res.* **8**, 3729–3741.
194. Vartapetian, A. B., E. V. Koonin, V. I. Agol and A. A. Bogdanov (1984) *EMBO J.* **3**, 2583–2588.
195. Wimmer, E. (1979). *In* "The Molecular Biology of Picornaviruses" (Ed. R. Perez-Bercoff), pp. 175–190. Plenum Press, New York/London.
196. Wimmer, E. (1982). *Cell* **28**, 199–201.
197. Wright, P. J. and P. D. Cooper (1976). *J. Gen. Virol.* **30**, 63–71.
198. Wu, M., N. Davidson and E. Wimmer (1978). *Nucl. Acids Res.* **5**, 4711–4723.
199. Yehle, C. D. (1978). *J. Virol.* **27**, 776–783.
200. Yogo, Y. and E. Wimmer (1972). *Proc. Natl Acad. Sci. USA* **69**, 1877–1882.
201. Yogo, Y. and E. Wimmer (1973). *Nature New Biology* **242**, 171–174.
202. Yogo, Y. and E. Wimmer (1975). *J. Mol. Biol.* **92**, 467–477.
203. Yogo, Y., M.-H. Teng and E. Wimmer (1974). *Biochem. Biophys. Res. Commun.* **61**, 1101–1109.
204. Young, D. C., D. M. Tuschall and J. B. Flanegan (1985). *J. Virol.* **54**, 256–264.
205. Zabel, P., M. Moerman, G. Lomonossoff, M. Shanks and K. Beyreuther (1984). *EMBO J.* **3**, 1629–1634.
206. Zimmerman, E. F., M. Heeter and J. E. Darnell (1963). *Virology* **19**, 400–408.

3. Investigation of the Molecular Basis of Attenuation in the Sabin Type 3 Vaccine Using Novel Recombinant Polioviruses Constructed from Infectious cDNA

Gareth D. Westrop*†[a], David M. A. Evans†, Philip D. Minor†, David Magrath†, Geoffrey C. Schild†, Jeffrey W. Almond[b]

*The University of Leicester, University Road, Leicester LE1 7RH, U.K., and †National Institute of Biological Standards and Control, Holly Hill, Hampstead, London NW3 6RB, U.K.

I. Introduction	53
II. Recovery of Parental viruses from cDNA	54
III. Construction of Progenitor/vaccine recombinants via cDNA	55
IV. Antigenic properties of recovered viruses	57
V. Neurovirulence properties	57
VI. Temperature sensitivity	58
VII. Conclusions	59
References	59

I. INTRODUCTION

Polioviruses occur as three distinct serotypes, each of which must be represented in a comprehensive poliomyelitis vaccine. An attenuated strain

---

[a]*Present address:* The University of Glasgow, Institute of Virology, Church Street, Glasgow G11 5JR, U.K.
[b]*Present address:* The University of Reading, London Road, Reading RG1 5AQ, U.K.

of each serotype was developed by Sabin in the 1950s and these form the basis of the trivalent live-attenuated oral vaccine used in much of the world. Although these vaccine strains have proved highly successful, there is good evidence that in rare instances the type 2 and type 3 strains can cause paralysis in recipients (Cossart, 1977; Nottay et al., 1981; Minor, 1980). Moreover, in spite of the fact that they have been in widespread use since the early 1960s, the molecular basis of their attenuation is still poorly understood. Attenuation was achieved empirically by multiple passage of wild-type virus in monkey tissue *in vivo* and *in vitro* (Sabin and Boulger, 1973).

Polioviruses possess single-stranded positive-sense RNA genome of approximately 7400 nucleotides. A relatively long 5′ non-coding region of 743 nucleotides precedes a single major open reading frame from which all virus proteins are translated via proteolytic cleavage of a polyprotein (see Kuhn and Wimmer, this volume, Chapter 2). The virus particle is composed of 60 copies each of four virus proteins VP1-VP4 surrounding the RNA genome (Hogle et al., 1985).

Our previous work has concentrated on poliovirus type 3, the most problematical of the three strains. Determination of the complete nucleotide sequence of the vaccine strain P3/Leon/12a$_1$b and that of its neurovirulent progenitor P3/Leon/37 indicated that these strains differ by just 10 point mutations out of the 7432 nucleotides making up their genomes (Stanway et al., 1983a, 1984). Comparison of these data with the complete sequence of a neurovirulent revertant of the vaccine, strain P3/119, suggested that mutations important to the attenuated phenotype are likely to include a change in the 5′ non-coding region at position 472, one or more of the mutations affecting coding in the structural protein region, and possibly a mutation just prior to the poly A tract (Cann et al., 1984). Further evidence in support of these ideas has come from the partial sequence analysis of other revertant strains (Almond et al., 1985; Evans et al., unpublished).

The possibility of recovering infectious virus from complete cloned cDNAs, as shown by Racaniello and Baltimore (1981), provides an opportunity to construct a series of progenitor/vaccine recombinants and to assess the level of attenuation of such viruses in a neurovirulence test. This chapter describes results obtained on a series of recombinants designed to assess the individual and possible synergistic effects on neurovirulence of mutations present in the Sabin type 3 vaccine genome.

II. RECOVERY OF PARENTAL VIRUSES FROM cDNA

The construction of an *E. coli* plasmid carrying a full-length DNA copy of the genome of the neurovirulent progenitor strain, P3/Leon/37, has been described elsewhere (Stanway et al., 1984). A similar procedure was used to construct a full-length cDNA with a sequence identical to that previously published for the vaccine strain (Westrop et al., unpublished). From each of these full-length cDNAs, virus was recovered by transfection of Hep 2C cells in culture (Racaniello and Baltimore, 1981). Recovered

Table 3.1. Comparison of properties of viruses recovered from cDNA (prefix r) with strains from which they were derived

	Neurovirulence		Temperature sensitivity	Antigenicity
	No. of animals paralysed	Mean histological lesion score	Reduction in infectivity titre 35/40°C (log 10)	[a]Neutralization by mcAb 138
P3/Leon/37	6/6	2.91	0.00	—
rpOLIO Leon	4/4	2.71	0.00	—
P3/Leon 12a$_1$b	0/18	0.72	6.40	+
rpOLIO Sabin	0/4	0.11	6.25	+

[a]Minor (1985).

viruses were shown to be indistinguishable from the parent stocks of the two strains in terms of neurovirulence, temperature sensitivity (Table 3.1), and, using monoclonal antibody 138 (Minor et al., 1985; Minor, this volume, Chapter 15), by antigenic analysis. The genetic composition of recovered viruses was also verified by determining the sequence of certain regions of their genomes. These experiments showed that the viruses recovered from cDNA clones were of the correct phenotype, and that the 10 mutations in the vaccine previously reported (Stanway et al., 1983a, 1984) must account for the attenuated properties of this strain.

III. CONSTRUCTION OF PROGENITOR/VACCINE RECOMBINANTS VIA cDNA

A panel of recombinants between the neurovirulent progenitor strain, P3/Leon/37, and the attenuated vaccine, P3/Leon/12a$_1$b, was constructed in an attempt to identify the mutations responsible for the attenuated phenotype (Fig. 3.1). Construction of the recombinants made use of restriction endonuclease cleavage sites in the cDNA falling between the mutations of interest, e.g. *Aat* II site at position 1809, *Sma* I at 2769 and *Bgl* II at 3421. The first recombinant, SLR1, which involves a *Sma* 1/*Sal*I double digestion of the cDNA plus vector, has been described elsewhere (Almond et al., 1985). Further recombinants, derived from pSLR1 and plasmid pOLIO/Leon are designated ST/L and SV3/L (Fig. 3.1). These were designed to investigate the individual effects of the mutations carried by SLR1, and considered likely to be important to the attenuated phenotype, i.e. the C-U mutation at position 472 in the 5' non-coding region of the genome and the C-U mutation at position 2034 giving rise to a Ser-Phe amino acid substitution in VP3. Recombinants to assess the effects of the three Sabin mutations affecting coding together (SCC/L) and individually (SP2/L, SV1/L, together with SV3/L) were also constructed.

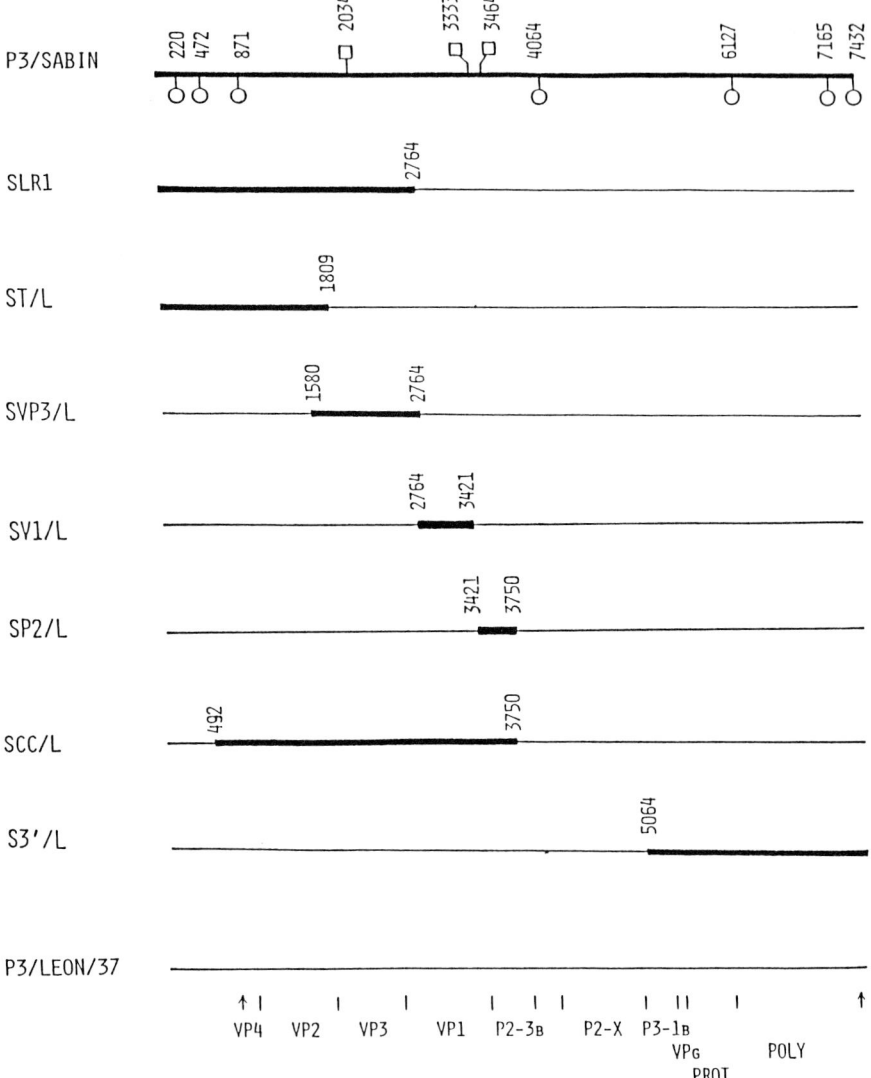

Figure 3.1. Progenitor/vaccine poliovirus type 2 recombinants constructed via cDNA. Sites of mutations are indicated on the Sabin vaccine virus genome. □ = mutations affecting coding changes. ○ = silent mutations. Positions of cross-over points are indicated and correspond to restriction endonuclease cleavage sites. All viruses recovered by transfection of recombinant cDNAs had their structures verified by primer extension RNA sequencing at each side of cross-over points.

3. Novel Recombinant Polioviruses from cDNA

The final recombinant designated S3′L was designed to assess the significance of the non-coding change just prior to the poly(A) tract, plus the two silent changes at 6127 and 7165. DNAs from each recombinant plasmid were prepared in *E. coli* JA221 (Cann *et al.*, 1983) and purified by centrifugation on CsCl gradients.

From each recombinant cDNA virus was recovered by transfection of Hep 2C cells in culture and its structure was confirmed by primer extension sequence analysis of virus RNA on each side of the recombination sites. The viruses were also characterized by neutralization with a type 3 specific polyclonal antiserum and monoclonal antibody 138 (Minor *et al.*, 1985), and tested for temperature sensitivity using the *rct* 40°C marker test. Neurovirulence in cynomolgus monkeys was assayed for all viruses as described previously (Stanway *et al.*, 1983b).

IV. ANTIGENIC PROPERTIES OF RECOVERED VIRUSES

All recovered viruses were neutralized by the polyclonal antiserum and viruses SV1/L and SCC/L were neutralised by monoclonal antibody 138, consistent with previous findings that this monoclonal reacts with an antigenic site which includes amino acids 286–288 of VP1 (Minor *et al.*, 1985) and that these two recombinants derive this region of their genome from the Sabin strain.

V. NEUROVIRULENCE PROPERTIES

The results of the neurovirulence tests carried out to date using four monkeys per recombinant virus are illustrated in Table 3.2, together with the results of the *rct* 40°C marker test (see below). The data indicate that the mutation in the 5′ non-coding region of the genome, most likely that at

Table 3.2. Results of neurovirulence and *rct* 40°C marker tests on recombinant viruses

	Number of animals paralysed	Mean histological lesion score	Reduction in infectivity titre 35°C/40°C (log 10)	
rpOLIO SABIN	0/4	0.11	6.25	ts
rSLR1	0/7	1.23	4.38	ts
rST/L	1/4	1.11	1.75	ts$^+$
rSV3/L	0/4	1.37	5.00	ts
rSV1/L	3/4	2.68	2.25	ts$^+$
rSP2/L	4/4	2.51	2.00	ts$^+$
rSCC/L	2/4	2.06	5.39	ts
rS3′/L	4/4	2.40	2.00	ts$^+$
rpOLIO LEON	4/4	2.71	1.75	ts$^+$

position 472 (Evans *et al.*, 1985), leads to a partial attenuation of the virus but does not account for the full attenuated phenotype of the vaccine. A similar result was obtained with virus SV3/L, suggesting that the mutation at position 2034 causing a Ser-Phe change in virus structural protein VP3, also attenuates the virus. Neither of these recombinants was as highly attenuated as the vaccine, suggesting that their effects are probably additive in the vaccine. Viruses SV1/L and SP2/L, containing the mutation at 3333 (Lys-Arg in VP1) and 3464 (Thr-Ala in P2-A) respectively, showed a similar level of neurovirulence as the recovered P3/Leon/37, suggesting that neither of these mutations contributed to the attenuated phenotype. This conclusion was supported by the observation that the recombinant SCC/L, containing the 3333 and 3464 mutations plus the mutation at 2034 affecting VP3 discussed above, was no more attenuated than SV3/L which possesses only the 2034 change. Virus recovered from recombinant S3'/L initially gave an anomalous response. A plaque-purified isolate from the pool of virus recovered from a single transfection experiment proved to be attenuated, suggesting that this mutation may reduce neurovirulence. However, this result was not confirmed by analysis of the pool of virus from the same transfection experiment and from material recovered in other transfection experiments. We therefore conclude that the plaque-purified isolate had accumulated additional, fortuitously attenuating, mutations and was atypical. The pool result (presented in Table 3.2) suggests that the 3' end change is probably not important to the attenuated phenotype.

VI. TEMPERATURE SENSITIVITY

The viruses illustrated in Fig. 3.1 were subjected to the *rct* 40°C marker test. The results are presented in Table 3.2. The only virus carrying a single Sabin-specific mutation that was found to be temperature-sensitive was SV3/L, indicating that this mutation (C-U at 2034, causing a Ser-Phe in VP3) as well as being attenuating, confers a temperature-sensitive phenotype on the virus. Two other recombinants, SLR1 and SCC/L, were shown to be temperature-sensitive. Both of these contain the VP3 mutation present in SV3/L. The recombinants ST/L, SV1/L and SP2/L all behave as wild-type in this test, indicating that the mutations at positions 472, 3233 and 3464 do not confer a *ts* phenotype. The recombinant S3'/L again produced an anomalous result. The plaque-purified isolate discussed above which possessed an attenuated phenotype was also temperature-sensitive, whereas the pool of virus from this and other transfection experiments was wild-type, providing further evidence that the plaque-purified material was untypical of the majority population.

Our conclusions are that the Sabin vaccine virus is temperature-sensitive because of a single mutation at position 2034. We cannot firmly rule out the possibility that the mutation at 4064 is also temperature-sensitive, but the conservation of this mutation in revertant P3/119 and the fact that it is silent with respect to amino acid sequence make this very unlikely.

VII. CONCLUSIONS

Our conclusions from the data presented here are that the Sabin vaccine is attenuated by at least two point mutations. The first of these is a C-U change at position 472 in the 5′ non-coding region of the genome. This mutation has been shown to revert rapidly upon passage of the virus in the human gut — an event that is associated with an increase in neurovirulence (Evans et al., 1985). Our results with recombinants do not formally rule out a role for the mutations at positions 220 and 871, but the fact that these changes are conserved in the revertant of the vaccine, P3/119, suggests that they are unimportant. The second mutation affecting neurovirulence is the C-U at position 2034, which results in a serine-phenylalanine substitution in virus structural protein VP3. The rct 40°C marker test indicates that this mutation also confers a temperature-sensitive phenotype on the virus.

The physiological basis of attenuation remains unclear, particularly in the case of the 472 mutation. It is possible that a mutation that gives rise to temperature sensitivity may *ipso facto* be attenuating, as has been suggested previously (Lwoff, 1959). The anomalous result with the plaque-purified derivate of S3′/L, which was seen to be temperature-sensitive and attenuated, is consistent with this idea.

Further recombinants are under construction to verify the results presented here. We anticipate that a complete understanding of the genetic basis of attenuation in this vaccine will suggest new ways in which more stable strains can be constructed. The recombinants outlined here, particularly those carrying single attenuating mutations, will be useful in experiments designed to look at the physiological effects of the mutations.

REFERENCES

1. Almond, J. W., G. D. Westrop, A. J. Cann, G. Stanway, D. M. A. Evans, P. D. Minor and G. C. Schild (1985). *In* "Vaccines 85" (Eds R. A. Lerner, R. M. Channock and F. Brown), pp. 271–278. Cold Spring Harbor Laboratory.
2. Cann, A. J., G. Stanway, R. Hauptmann, P. D. Minor, G. C. Schild, L. D. Clarke, R. C. Mountford and J. W. Almond (1983). *Nucl. Acids Res.* 11(5), 1267–1281.
3. Cann, A. J., G. Stanway, P. J. Hughes, P. D. Minor, D. M. A. Evans, G. C. Schild, and J. W. Almond (1984). *Nucl. Acids Res.* 12(20), 7787–7792.
4. Cossart Y. E. (1977). *Br. Med. J.* 1, 1621–23.
5. Evans, D. M. A., G. Dunn, P. D. Minor, G. C. Schild, A. J. Cann, G. Stanway, J. W. Almond, K. Currey and J. V. Maizel Jr. (1985). *Nature, (London)* 314(6011), 548–550.
6. Hogle J., M. Chow and D. J. Filman (1985). *Science* 229, 1359–1363.
7. Lwoff, A. (1959). *Bacteriol. Rev.* 23, 109–24.
8. Minor, P. D. (1980). *J. Virol.* 34, 73–84.

9. Minor, P. D., D. M. A. Evans, M. Ferguson, G. C. Schild, G. Westrop and J. W. Almond, (1985). *J. Gen. Virol.* **65**, 1159–1165.
10. Nottay, B. K., O. M. Kew, M. H. Hatch, J. Heyward and J. Obijeski (1981). *Virology* **108**, 405–23.
11. Racaniello, V. R. and D. Baltimore (1981). *Science* **214**, 916–919.
12. Sabin, A. B. and L. R. Boulger (1973). *J. Biol. Stand.* **1**, 115–18.
13. Stanway, G., A. J. Cann, R. Hauptmann, P. Hughes, L. D. Clarke, R. C. Mountford, P. D. Minor, G. C. Schild and J. W. Almond (1983a). *Nucl. Acids Res.* **11**, 5629–43.
14. Stanway, G., A. J. Cann, R. Hauptmann, R. C. Mountford, L. D. Clarke, P. Reeve, P. D. Minor, G. C. Schild and J. W. Almond (1983b). *Eur. J. Biochem.* **135**, 529–33.
15. Stanway, G., R. C. Mountford, S. D. J. Cox, G. C. Schild, P. D. Minor and J. W. Almond (1984). *Arch. Virol.* **81**, 67–78.

4. Molecular Biological Aspects of Human Rhinovirus

Tim Skern[a,b], Wolfgang Sommergruber, Peter Gruendler, Dieter Blaas and Ernst Kuechler

Institut für Biochemie der Universität Wien, Währinger Strasse 17, A-1090 WIEN, Austria

---

I. Introduction 61
II. The 5' untranslated region 62
III. Homology at the amino acid level between HRV2, HRV14 and poliovirus type 1 63
IV. Relationship between rhinoviruses and polioviruses 64
References 66

I. INTRODUCTION

Human rhinoviruses, members of the picornavirus family, are the main causative agents of the common cold (Gwaltney, 1975). Although the disease is not in itself serious, it is of considerable economic importance and can lead to secondary infection, especially in the elderly (Stott and Killington, 1972). The existence of more than a hundred antigenically different serotypes is an important factor in precluding the acquisition of immunity to rhinoviral infection (Fox, 1976). Elucidation of the nature of this serotypic diversity is at present not understood and represents one of the most important challenges in rhinovirus research.

The prospect of a deeper understanding of this phenomenon has been brought closer, however, by the work carried out in several laboratories during the last two years. The solving of the crystal structure of human rhinovirus 14 (HRV14) by Rossmann and co-workers (1985) represents a major breakthrough in the rhinovirus and picornavirus fields, showing the

---

[a]*Present address:* Transgene S.A., 11 rue de Molsheim, 67082 Strasbourg, France.
[b]Author to whom all correspondence should be addressed.

contribution of each amino acid to the three-dimensional structure of the virus (see Chapter 17, this volume). At the nucleotide level, the availability of complete sequences from two serotypes, human rhinovirus 2 (HRV2) and HRV14, has enabled detailed comparisons to be carried out (Skern et al., 1985; Stanway et al., 1984; Callahan et al., 1985). In contrast to the serotypic diversity, work on receptor utilization has shown that all rhinoviruses use one of only two receptors (Abraham and Colonno, 1984). Of these two receptors one is used by all except four rhinovirus serotypes so far tested.

We have previously reported the complete nucleotide sequence of HRV2, derived from the cloned cDNA, and presented some comparisons with HRV14 and poliovirus type 1 (Skern et al., 1985). In this report, further, more detailed comparisons are presented that shed light on the relationships between these viruses.

II. THE 5' UNTRANSLATED REGION

It is accepted that the information encoded in picornaviral RNAs is contained in one large open reading frame, with translation commencing between 600 and 1200 bases from the 5' end and ending some 40–100 bases from the poly(A) tract. These regions at the 5' and 3' ends have been termed noncoding, though formal proof for this assumption does not exist. It seems probable that if the coding capacity of these regions were utilized, the encoded protein would be conserved in all picornaviruses. To this end, the open reading frames present in the 5' regions of HRV2, HRV14 and poliovirus type 1 Mahoney (Skern et al., 1985; Stanway et al., 1984; Kitamura et al., 1981) were examined; a comparison is shown diagrammatically in Fig. 4.1.

It can be seen that several open reading frames are present in the 5'

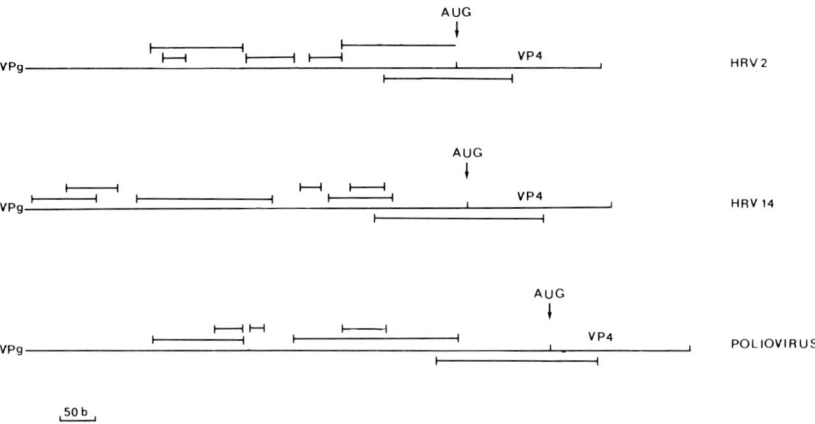

Figure 4.1. Open reading frames in the 5' regions of HRV2, HRV14 and poliovirus type 1. The AUG codon at the start of the large open reading frame is marked in each case.

region of each of the three viruses examined; however, they all appear to be located randomly and no evidence of conservation is seen. The open reading frame spanning the ATG of the major open reading frame does seem at first sight to be conserved throughout; however, the reading frame used is not the same in HRV2 and HRV14 and the amino acid sequences thus generated are not related. Furthermore, the poliovirus sequence is quite different in this region. In HRV2, translation from the ATG codon at position 449 would continue into the large open reading frame. Two lines of evidence imply that it is not used *in vivo*. Firstly, the size of VP4 would be approximately twice that actually detected on SDS-polyacrylamide gels, and secondly, the only AUG codon possessing an exact fit with Kozak's rules (Kozak, 1983) is that at HRV2 map position 611.

Despite the fact that this region is not itself translated, it is quite possible that it may play a role in translation. Upon entry of a virus particle into a host cell, the viral RNA must be translated as soon as possible before being degraded or inactivated; it must also compete with cellular mRNAs for ribosomes. Modification of initiation factors enabling preferential translation of viral RNA only takes place after some viral protein synthesis has occurred. The 5' untranslated region may therefore enhance ribosome binding and in this way increase the chances of the viral RNA being translated in a hostile environment. As only one AUG codon has an optimal sequence for initiation, correct initiation is ensured.

That the 5' untranslated region is involved in carrying out essential functions in the virus life cycle is shown by comparisons of this region from different viruses. Extensive homology between the 5' untranslated regions of poliovirus and HRV14 was first observed by Stanway *et al.* (1984) and was later extended to HRV2 by our work (Skern *et al.*, 1985). Interestingly, homology is not found throughout the region, but occurs in blocks, as shown diagrammatically in Fig. 4.2. Five blocks of 16 or more consecutively homologous nucleotides are conserved between HRV2 and HRV14; two of these blocks are also found in all three serotypes of poliovirus (Toyoda *et al.*, 1984; Stanway *et al.*, 1983). At present, one can only speculate as to the function of this region, although recent work has implicated the 5' untranslated region of poliovirus type 3 in neurovirulence (Evans *et al.*, 1985).

III. HOMOLOGY AT THE AMINO ACID LEVEL BETWEEN HRV2, HRV14 AND POLIOVIRUS TYPE 1

HRV2 shows about 50% homology at the amino acid level to HRV14 and to poliovirus serotype 1. Is this homology also arranged in blocks as in the 5' untranslated region? Are those regions conserved between HRV2 and HRV14 also conserved between HRV2 and poliovirus? To answer these questions, a computer was employed to search the polyproteins of HRV2 and HRV14 for stretches of eight or more consecutively homologous amino acids. This was then repeated with HRV2 and poliovirus type 1, with the results being shown diagrammatically in Fig. 4.3.

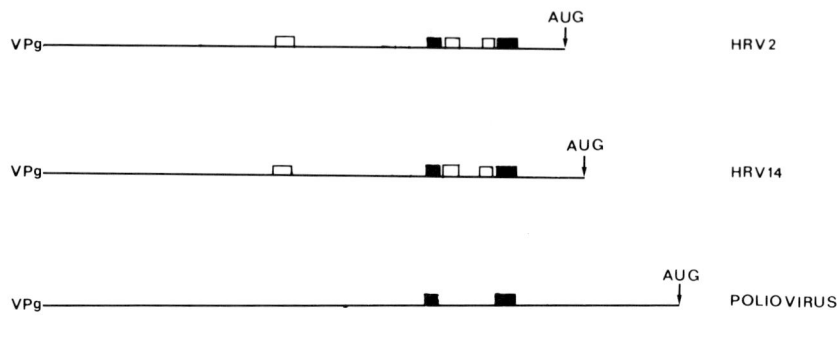

Figure 4.2. Homology within the 5' regions of HRV2, HRV14 and poliovirus. A computer was employed to search for stretches of 16 or more consecutively homologous nucleotides. Solid rectangles: regions of 16 or more consecutively homologous nucleotides found between HRV2, HRV14 and all three serotypes of poliovirus. Open rectangles: regions of 16 or more consecutively homologous nucleotides found only between HRV2 and HRV14.

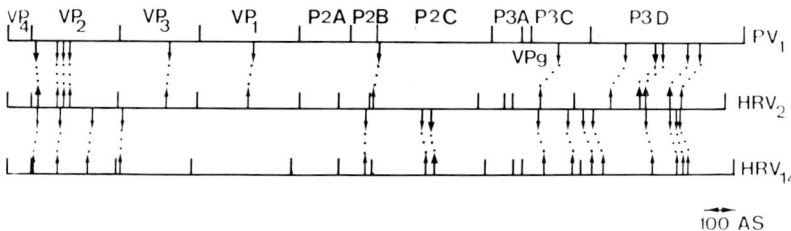

Figure 4.3. Comparison of the polyproteins of HRV2, HRV14 and poliovirus type 1. A computer search for regions of eight or more consecutively homologous amino acids was performed; their positions are arrowed.

Fifteen such blocks can be found between HRV2 and HRV14 and 13 between HRV2 and poliovirus 1. However, only five blocks of eight or more consecutively homologous amino acids are conserved between HRV2, HRV14 and poliovirus type 1. Thus, there appears to be little selective pressure to conserve amino acids residues within the polyprotein, an observation that makes the conservation of the two blocks of nucleotides in the 5' untranslated region between HRV2, HRV14 and all three poliovirus serotypes all the more intriguing.

IV. RELATIONSHIP BETWEEN RHINOVIRUSES AND POLIOVIRUSES

It has been suggested that given the degree of homology between members of the rhinovirus and poliovirus genera at the amino acid and nucleotide levels, the possibility of amalgamating the two genera should be considered

```
GLY PRO TYR SER GLY GLU PRO LYS PRO --- LYS THR LYS ILE PRO --- GLU ARG ARG VAL VAL THR GLN    HRV2 VPG
 *   *   *   *   *   *       *   *       *   *   *   *   *           *   *   *   *
GLY PRO TYR SER GLY ASN PRO PRO HIS ASN LYS LEU LYS ALA PRO THR LEU ARG PRO VAL VAL VAL GLN    HRV14 VPG

GLY ALA TYR THR GLY LEU PRO ASN LYS LYS PRO ASN VAL PRO THR ILE ARG THR ALA LYS VAL GLN        POLIO VPG
```

Figure 4.4. The sequences of the protein VPg from HRV2, HRV14 and poliovirus type 1. Two insertions have been made within the HRV2 sequence to optimize homology; amino acids conserved between HRV2 and HRV14 are indicated with an asterisk.

(Stanway et al., 1984). However, as only two sequences of rhinoviruses are at present available, it is felt that such a decision may be premature; it may be that the two strains whose sequences have been determined represent the extreme ends of the rhinovirus spectrum. It is hoped the availability of further sequence data will help to clarify the picture.

Furthermore, careful comparison of sequences for functional domains, instead of using the percentage homology as a criterion may reveal combinations that are unique to rhinoviruses. As an example, the sequences of VPg from HRV2, HRV14 and poliovirus type 1 are outlined in Fig. 4.4. The first five amino acids are conserved between HRV2 and HRV14, in contrast to only three with poliovirus type 1. There is also more overall homology between the VPgs of HRV2 and HRV14 than that of poliovirus, even after optimizing the homology. It is therefore clear that the basis of rhinovirus-specific sequence motifs is established; work on further strains to deepen our knowledge of the relationships between rhinoviruses is at present underway.

ACKNOWLEDGEMENTS

We thank A. Gupta and L. Frasel for technical assistance, and A. Barta and Z. Rattler for helpful discussions. Financial support for this project was provided by the "Osterreichischer Fonds zur Förderung der wissenschaftlichen Forschung" and by Boehringer Ingelheim.

REFERENCES

1. Abraham, G. and R. J. Colonno (1984). *J. Virol.* **51**, 340–345.
2. Callahan, P.L., S. Mizutani and R. J. Colonno (1985). *Proc. Natl Acad. Sci. USA* **82**, 732–736.
3. Evans, D., G. Dunn, P. Minor, G. Schild, A. Cann, G. Stanway, J. Almond, K. Currey, J. Maizel (1985). *Nature (London)* **315**, 548–550.
4. Fox, J. P. (1976). *Amer. J. Epidemiol.* **103**, 345–354.
5. Gwaltney, J. M., Jr. (1975). *Yale J. Biol. Med.* **48**, 17–45.
6. Kitamura, N., B. L. Semler, P. G. Rothberg, D. R. Larsen, C. J. Adler, A. J. Dorner, E. A. Emini, R. Hanecak, J. L. Lee, S. van der Werf, C. W. Anderson and E. Wimmer (1981). *Nature (London)* **291**, 547–553.
7. Kozak, M. (1983). *Microbiol. Rev.* **47**, 1–45.
8. Rossman, M. G., E. Arnold, J. W. Erickson, E. A. Frankenberger, J. P. Griffith, H.-J. Hecht, J. E. Johnson, G. Kramer, M. Luo, A. G. Mosser, R. R. Rueckert, B. Sherry and G. Vriend (1985). *Nature (London)* **317**, 145–153.
9. Skern, T., W. Sommergruber, D. Blaas, P. Gruendler, F. Fraundorfer, C. Pieler, I. Fogy and E. Kuechler (1985). *Nucl. Acids Res.* **13**, 2111–2126.
10. Stanway, G., A. J. Cann, R. Hauptmann, P. Hughes, L. D. Clarke, R. C. Mountford, P. D. Minor, G. C. Schild and J. W. Almond (1983). *Nucl. Acids Res.* **11**, 5629–5643.
11. Stanway, G., P. J. Hughes, R. C. Mountford, P. D. Minor and J. W. Almond (1984). *Nucl. Acids Res.* **12**, 7859–7875.
12. Stott, E. J. and R. A. Killington, (1972). *Ann. Rev. Microbiol.* **26**, 503–525.
13. Toyoda, H., M. Kohara, Y. Katoaka, T. Suganuma, T. Omata, A. Imur and Nomoto, A. (1984). *J. Mol. Biol.* **174**, 561–586.

5. Insect Picornaviruses

N. F. Moore[a], L. A. King[b] and J. S. K. Pullin

NERC Institute of Virology, Mansfield Road, Oxford OX1 3SR, U.K. and Department of Biochemistry, University of Oxford, South Parks Road, Oxford OX1 3QU, U.K.

I.	Introduction	67
II.	Replication of DCV and CrPV in insect hosts	68
III.	CrPV: Replication in mammals?	68
IV.	Replication of DCV and CrPV in D. Melanogaster cells	69
V.	Organization of the CrPV genome	72
VI.	Replication studies on other insect picornaviruses	72
VII.	Discussion	72
	References	73

I. INTRODUCTION

More than 20 viruses from insects have been described as picornavirus-like or enterovirus-like. The majority of these viruses have been isolated from laboratory and field populations of *Drosophila melanogaster* or from *Apis mellifera* and *A. cerana*. Insufficient physicochemical information is available to include many of the insect "picornaviruses" in the *Picornaviridae* family. On the basis of the molecular weight of RNA (2.4–3.0 MD), number and size of capsid proteins (three of 20–40 kD), buoyant density in CsCl (1.32–1.37), size of virus as measured by electron microscopy (27–32 nm) and sedimentation coefficient (150–170S), relatively few insect viruses can be classified as members of the *Picornaviridae*. These are *Drosophila* C virus (DCV), cricket paralysis virus (CrPV), infectious flacherie virus from *Bombyx mori*, *Rhopalosiphum padi* virus and Sacbrood

[a]*Present address:* Chemical Defence Establishment, Porton Down, Salisbury, Wiltshire SP4 0JQ, U.K.
[b]*Present address:* Department of Biology, Oxford Polytechnic, Gypsy Lane, Headington, Oxford, U.K.

virus from honey bees (for reviews see Moore and Tinsley, 1982; Reavy and Moore, 1982; Moore *et al.*, 1985).

II. REPLICATION OF DCV AND CrPV IN INSECT HOSTS

CrPV was originally identified in laboratory colonies of *Teleogryllus oceanicus* and *T. commodus*. Symptoms were initially identified in early instar nymphs, which became uncoordinated, this being partly attributable to paralysis of hind limbs. The transmissible disease led to mortality (Reinganum *et al.*, 1970). *Drosophila* C virus was first identified in populations of *D. melanogaster* by Jousset *et al.* (1977), and was also transmissible and highly pathogenic when injected.

In the insect, the site of the virus multiplication varies with the host. Both viruses are found in the cytoplasm of infected cells, often in the form of large paracrystalline arrays. In infected crickets, CrPV is detected in a variety of cells including those of the gut and ganglia. Examination of DCV-infected *D. melanogaster* shows the presence of virions in the tracheal cells and cells surrounding the cerebral ganglia (Scotti *et al.*, 1981).

DCV and CrPV have been found in widespread populations of insects mainly by serological identification. In the majority of instances the viruses identified in serosurveys have not been characterized in detail and it is not possible to say how similar they are to the viruses against which antisera were raised. However, Reinganum (1975) used CrPV in infectivity studies with numerous orthoptera and lepidoptera and found that the virus replicated in virtually all of the insects studied. DCV has been detected in many laboratory stocks of *D. melanogaster* and hence is present in tissue culture systems derived from fly tissue explants (Plus, 1978; 1980).

III. CrPV: REPLICATION IN MAMMALS?

A serological relationship has been demonstrated between DCV and CrPV (Scotti, 1976; Plus *et al.*, 1978) and, interestingly, CrPV has been found to react with the sera of cattle and several other mammalian species (Scotti and Longworth, 1980; F.O. MacCallum and J. S. Robertson, personal communication; Moore *et al.*, 1981a). The levels of reaction were low; cattle serum from one site in New Zealand had titres of 10^{-1} to 10^{-2}, whereas rabbit hyperimmune serum was shown to have a neutralization titre of 10^{-5} (Scotti and Longworth, 1980; Scotti *et al.*, 1981). It has not been elucidated why these animals should have antibodies against CrPV. The antibodies were of the IgM type, which suggests that the antigenic protein was present at only a low level. Tinsley *et al.* (1984) found that CrPV shared common antigen(s) with encephalomyocarditis (EMC) virus. They demonstrated that EMC virus and CrPV reacted with CrPV antiserum, with the formation of precipitant lines. Spur formation indicated that the viruses were similar but not identical. With EMC virus antiserum and both viral antigens, confluence of the bands was obtained (Tinsley et al., 1984). These workers further demonstrated the serological relationship between the two viruses by studying immunological enhancement. By exposing guinea pigs to first one virus and then the other they demonstrated slight enhancement of the antibody response to EMC virus by previous or subsequent infection of CrPV. This further indicates an antigenic relationship between the viruses.

Only two of these viruses, DCV and CrPV have been demonstrated to replicate in tissue culture (*D. melanogaster* cells) permitting analysis of their replication mechanism and hence a more detailed comparison with the mammalian picornaviruses.

IV. REPLICATION OF DCV AND CrPV IN *D. MELANOGASTER* CELLS

Like mammalian picornaviruses CrPV RNA has a 5' genome-linked protein (VPg) and is polyadenylated at the 3' end (Eaton and Steacie, 1980; Pullin et al., 1982; King, 1985). Both CrPV and DCV grow to high titres in *D. melanogaster* cells, in excess of 10^9 pfu/ml or 10^8 $TCID_{50}$/ml (Scotti, 1976, 1977; Moore et al., 1981b; Moore and Pullin, 1982). In infected cells virus-induced RNAs were detected by radiolabelling between 3 h and 5 h post-infection. No subgenomic mRNAs were identified; virus-induced RNAs were of genome size or were high molecular weight dsRNA (Eaton and Steacie, 1980; Reavy et al., 1983a). Using radiolabelled amino acid precursors, similar virus-induced proteins were detected in both CrPV-infected and DCV-infected cells. The shut-off of host cell protein synthesis is very marked with CrPV and considerably more virus-induced proteins are apparent than can be seen in DCV-infected cells. As with mammalian picornaviruses, more proteins are apparent than can be accounted for by the coding potential of the virus RNA. More than 20 induced polypeptides are seen in CrPV-infected cells with molecular weights between 144 kD and 12 kD (see Fig. 5.1) (Moore et al., 1980). The identity of precursor proteins was established by pulse-chase experiments and immunoprecipitation of the induced protein by antiserum raised against mature virions (Moore et al., 1981b; Reavy, 1982). Processing events were inhibited by introducing protease inhibitors and amino acid analogues into the virus-infected cells (Moore et al., 1981d,e). From these data it appears that the genome of CrPV functions as a polycistronic mRNA in the same manner as that of mammalian picornaviruses. Reavy and Moore (1983b) further supported this by using pactamycin to deplete the RNA of ribosomes at the 5' end and thus demonstrated that the structural proteins were coded at the 5' end in a similar order to those of the mammalian viruses. However, with CrPV it was not possible to identify a protein analogous to the VP4 of the mammalian picornaviruses.

In DCV-infected cells fewer virus-induced proteins are apparent because the shut-off of host cell proteins is poor and possibly there is more efficient processing of precursor proteins. To suppress the cellular protein

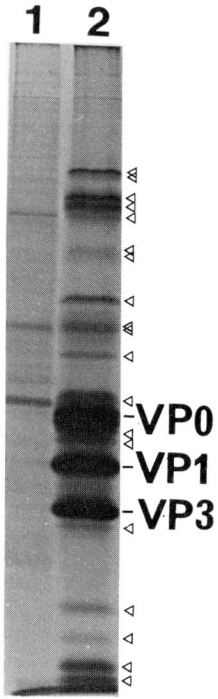

Figure 5.1. A 12.5% polyacrylamide gel of the S^{35}-methionine labelled proteins of D. melanogaster cells infected with cricket paralysis virus (track 2) or mock infected (track 1). Major viral structural proteins are labelled (VP0,1,3) and other induced polypeptides are indicated by open arrow heads.

synthesis, DCV-infected D. melanogaster cells were heat-shocked from 28°C to 37°C prior to pulsing with radiolabelled amino acid precursors. Heat-shocking of D. melanogaster cells results in the appearance of a set of proteins called heat-shock proteins. Synthesis of these proteins can be suppressed by adding Actinomycin D. Heat-shocking resulted in the suppression of the normal pattern of host cell protein synthesis and the appearance of several more high molecular weight DCV-induced proteins (see Fig. 5.2) (Moore et al., 1981c; Moore and Pullin, 1983; Reavy et al., 1983b). These proteins could be chased into virus structural proteins by reducing the temperature from 37°C to 28°C. The proteins were similar to those found by inhibiting proteolytic processing by iodoacetamide treatment of infected cells before pulse-labelling. After chasing, a protein of molecular weight approximately 10 kD in DCV-infected cells was readily identifiable. This protein was also detected in stained or radiolabelled purified preparations of DCV (Jousset et al., 1977; Moore et al., 1981f) and probably corresponds to the VP4 of mammalian picornaviruses.

The intracellular proteins induced by several isolates of DCV from

5. Insect Picornaviruses

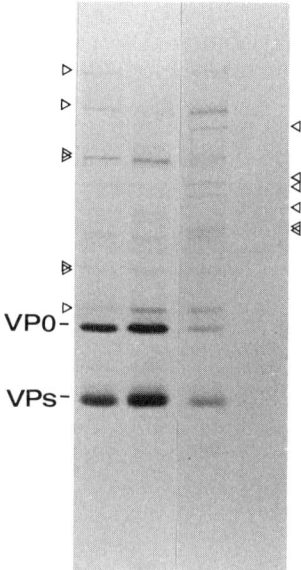

Figure 5.2. A 12.5% polyacrylamide gel of the S^{35}-methionine labelled proteins of *D. melanogaster* cells infected with *Drosophila* C virus (B) (tracks 1–3) or mock-infected (track 4). The cells were pulsed at 4–5 h post-infection at 28°C (track 2), at 28°C following iodoacetamide treatment (track 1) or after heat-shock at 37°C (track 3). All samples contained actinomycin D at 5 µg/ml to inhibit cellular protein synthesis at both 28°C and 37°C. Major viral structural proteins are labelled and other induced polypeptides are indicated by open arrow heads.

Morocco, France and the French Antilles were examined, and minor differences in the molecular weights of viral structural and virus-induced proteins were detected (Moore *et al.*, 1982). However, oligonucleotide mapping showed several major differences between their genomes (Clewley *et al.*, 1983). Oligonucleotide mapping and sequence homology determined by complementary DNA liquid hybridization have also demonstrated that CrPV is only distantly related to the DCV isolates (Pullin *et al.*, 1982; King *et al.*, 1984).

The proteins encoded by DCV and CrPV RNA were also examined by translation *in vitro* using rabbit reticulocyte lysates and the products were found to be very different from those in infected cells. Using pulse-chase analyses it was possible to chase high molecular weight proteins into viral structural proteins (Reavy and Moore, 1981a,b, 1983c). A series of "dilution" experiments demonstrated that in addition to a virus-coded protease, a cellular protease was involved in the processing of the virus-induced proteins and that CrPV protease could act on DCV-induced protein and *vice versa* (Reavy and Moore, 1983a,c).

V. ORGANIZATION OF THE CrPV GENOME

Much of the CrPV genome has been cloned into bacteria using standard cDNA techniques and overlapping clones spanning about 7.5 kb from the 3' terminus have been obtained (King, 1985). Preliminary sequence data, covering about 1500 bases at the 3' terminus, demonstrate that CrPV has a 3' non-coding region of 242 bases, which is significantly longer than that of other picornaviruses so far examined. The 3' coding region has limited homology with other picornaviral polymerase sequences (Franssen et al., 1984) and it is therefore likely that the CrPV polymerase is coded at the 3' end of the polyprotein as in the mammalian viruses.

VI. REPLICATION STUDIES ON OTHER INSECT PICORNAVIRUSES

In the absence of suitable tissue culture systems, replication studies on other insect picornaviruses have been restricted to an examination of the sites of infection in insects. Only one other virus, infectious flacherie virus of the silkworm (Hashimoto and Kawase, 1983), has been examined to investigate which virus-induced proteins are produced and whether the viral RNA acts as a polycistronic messenger. Infectious flacherie virus particles contain a single piece of RNA of molecular weight 2.4 MD and there are five major capsid proteins in the 27 nm mature virions. These apparently correspond to VPO, and VP1-VP4 of the mammalian picornaviruses. Hashimoto et al. (1984) found that the sizes of the translation products considerably exceeded the coding capacity of the RNA and suggested that a 200 kD protein was the polyprotein.

VII. DISCUSSION

The evidence presented here indicates that the two most studied insect "picornaviruses", DCV and CrPV, have many of the characteristics normally associated with mammalian picornaviruses, the physicochemical properties, protein content and RNA content and replicative events being similar to those of mammalian viruses. There is no evidence that the insect picornaviruses can replicate in mammalian systems, although the experiments to introduce them into tissue culture cells are poorly documented. CrPV will not replicate in BHK or Vero cells or in suckling mice (Reinganum, 1973; J. S. K. Pullin and N. F. Moore, unpublished information). The potential for replication of the insect viruses in mammalian systems is of great importance when considering the possibility of the use of picornaviruses as field control agents against pest insect species. There has been only one instance of a candidate picornavirus being used in this way: Gonometa virus, which killed larval stages of *Gonometa podocarpi* feeding on exotic pines in Uganda (Harrap et al., 1966; Longworth et al., 1973). However, the question of possible mammalian infection does not only occur with the use of purified preparations of the insect picornaviruses as control agents. Currently, nuclear polyhedrosis viruses (NPVs) and a few cytoplasmic polyhedrosis viruses (CPVs) are used against insect pests, and as many of the laboratory strains of insects in which the viruses are produced harbour picornaviruses and other small RNA viruses, it is possible that contamination of the NPVs and CPVs will result. It is therefore necessary to develop sensitive screening procedures, most likely involving cDNA probes, against the commonly occurring small RNA viruses to test both the insects used for producing the insecticidal virus and the final product to be used in the field. Even if the insect picornaviruses are unable to replicate in man or his domestic animals, the contaminating virus may kill non-target, useful insects. Since much of the potential of biological control agents, such as insect viruses, is based on their specificity, this must be ensured. The identification of serological reaction between EMC virus and CrPV is of interest and may partly explain the serological reactions between CrPV and several mammalian sera.

ACKNOWLEDGEMENTS

We would like to thank Jackie Murdock for assistance in the preparation of this manuscript.

REFERENCES

1. Clewley, J. P., J. S. K. Pullin, R. J. Avery and N. F. Moore (1983). *J. Gen. Virol.* **64**, 503–506.
2. Eaton B. T. and A. D. Steacie (1980). *J. Gen. Virol.* **50**, 167–171.
3. Franssen, H., J. Leunissen, R. Goldbach, G. Lomonssoff and D. Zimmern (1984). *EMBO J.* **3**, 855–861.
4. Harrap, K. A., J. F. Longworth, T. W. Tinsley and K. W. Brown (1966). *J. Invertebr. Pathol.* **8**, 270–272.
5. Hashimoto, Y. and S. Kawase (1983). *J. Invertebr. Pathol.* **41**, 68–76.
6. Hashimoto, Y., A. Watanabe and S. Kawase (1984). *Biochim. Biophys. Acta.* **781**, 76–80.
7. Jousset, F.-X., M. Bergoin and B. Revet (1977). *J. Gen. Virol.* **34**, 269–285.
8. King, L. A. (1985). D.Phil Thesis, University of Oxford.
9. King, L. A., P. R. Massalski, J. I. Cooper and N. F. Moore (1984). *J. Gen. Virol.* **65**, 1193–1196.
10. Longworth, J. F., C. C. Payne and R. Macleod (1973). *J. Gen. Virol.* **18**, 119–125.
11. Moore, N. F. and J. S. K. Pullin (1982). *J. Invertebr, Pathol.* **39**, 10–14.
12. Moore, N. F. and J. S. K. Pullin (1983). *Ann. Virol.* **134E**, 285–292.
13. Moore, N. F. and T. W. Tinsley (1982). *Arch. Virol.* **72**, 229–245.
14. Moore, N. F., A. Kearns and J. S. K. Pullin (1980). *J. Virol.* **33**, 1–9.
15. Moore, N. F., L. McKnight and T. W. Tinsley (1981a). *Infect. Immun.* **31**, 825–827.
16. Moore, N. F., J. S. K. Pullin, W. A. L. Crump, B. Reavy and L. K. Greenwood (1981b). *Microbiologica* **4**, 359–370.

17. Moore, N. F., J. S. K. Pullin and B. Reavy (1981c). *FEBS Lett.* **128**, 93–96.
18. Moore, N. F., J. S. K. Pullin and B. Reavy (1981d). *Arch. Virol.* **70**, 1–9.
19. Moore, N. F., B. Reavy and J. S. K. Pullin (1981e). *Arch. Virol.* **68**, 1–8.
20. Moore, N. F., B. Reavy, J. S. K. Pullin and N. Plus (1981f). *Virology* **112**, 411–416.
21. Moore, N. F., J. S. K. Pullin, W. A. L. Crump and N. Plus (1982). *Arch. Virol.* **74**, 21–30.
22. Moore, N. F., B. Reavy and L. A. King (1985). *J. Gen. Virol.* **66**, 647–659.
23. Plus, N. (1978). *In Vitro* **14**, 1015–1021.
24. Plus, N. (1980). *In* "Invertebrate Systems in Vitro" (Eds E. Kurstak, K. Maramorosch, A. Dübendorfer), pp. 435–439. Elsevier/North Holland Biomedical Press, Amsterdam and New York.
25. Plus, N., G. Croizier, C. Reinganum and P. D. Scotti (1978). *J. Invertebr. Pathol.* **31**, 296–302.
26. Pullin, J. S. K., N. F. Moore, J. P. Clewley and R. J. Avery (1982). *FEMS Microbiol. Lett.* **15**, 215–218.
27. Reavy, B. (1982). D.Phil Thesis, University of Oxford.
28. Reavy, B. and N. F. Moore (1981a). *J. Gen. Virol.* **55**, 429–438.
29. Reavy, B. and N. F. Moore (1981b). *Arch. Virol.* **67**, 175–180.
30. Reavy, B. and N. F. Moore (1982). *Microbiologica* **5**, 63–84.
31. Reavy, B. and N. F. Moore (1983a). *J. Gen. Virol.* **64**, 1831–1833.
32. Reavy, B. and N. F. Moore (1983b). *Virology* **131**, 551–554.
33. Reavy, B. and N. F. Moore (1983c). *Arch. Virol.* **76**, 101–115.
34. Reavy, B., W. A. L. Crump and N. F. Moore (1983a). *J. Invertebr. Pathol.* **41**, 397–400.
35. Reavy, B., J. S. K. Pullin and N. F. Moore (1983b). *Microbios* **38**, 91–98.
36. Reinganum, C. (1973). M.Sc. Thesis, Monash University.
37. Reinganum, C. (1975). *Intervirology* **5**, 97–102.
38. Reinganum, C., G. T. O'Loughlin and T. W. Hogan (1970). *J. Invertebr. Pathol.* **16**, 216–220.
39. Scotti, P. D. (1976). *Intervirology* **6**, 333–342.
40. Scotti, P. D. (1977). *J. Gen. Virol.* **35**, 393–356.
41. Scotti, P. D. and J. F. Longworth (1980). *Intervirology* **13**, 186–191.
42. Scotti, P. D., J. F. Longworth, N. Plus, G. Croizier and C. Reinganum (1981). *Adv. Virus Res.* **26**, 117–143.
43. Tinsley, T. W., F. O. MacCallum, J. S. Robertson and F. Brown (1984). *Intervirology* **21**, 181–186.

6. The Genomes of Alphaviruses and Flaviviruses: Organization and Translation

James H. Strauss, Ellen G. Strauss, Chang S. Hahn and Charles M. Rice

Division of Biology 156-29, California Institute of Technology, Pasadena, California 91125, U.S.A.

I.	Introduction	75
II.	The genome organization of alphaviruses	76
III.	Cleavage of the polyprotein precursors	78
IV.	Expression of cloned copies of Sindbis structural proteins	83
V.	Conserved sequences in alphavirus RNAs	86
VI.	Codon usage in alphaviruses	87
VII.	Relationship of alphaviruses to certain plant viruses	91
VIII.	The genome organization of flaviviruses	96
IX.	Comparison of flavivirus sequences	99
X.	Structures in the yellow fever RNA	101
XI.	Concluding remarks	102
	References	102

I. INTRODUCTION

In the last few years our laboratory has succeeded in obtaining the complete sequence of the genome of Sindbis virus, the type alphavirus (Strauss *et al.*, 1984), and of yellow fever virus, the type flavivirus (Rice *et al.*, 1985a). N-terminal amino acid sequence analysis of a number of the virus-specific proteins has allowed us to position the nucleotide sequences encoding many or all of them along the genome, and thus to deduce the complete amino acid sequence of the viral proteins. In this process, many of the details of the translation and processing of the viral proteins, as well as of their structure and transport, have also been deduced (Rice and Strauss, 1981; Strauss et al., 1984; Rice et al., 1985a). Comparison of the amino acid sequences of the proteins of these viruses with those of other viruses has led to some interesting observations on possible sequences involved in RNA replication and packaging (Ou et al., 1982a,b, 1983; Strauss et al., 1983) and the evolution of RNA viruses (Kamer and Argos, 1984; Haseloff et al., 1984; Ahlquist et al., 1985).

II. THE GENOME ORGANIZATION OF ALPHAVIRUSES

The replication strategy of Sindbis virus is illustrated in Fig. 6.1. The virus genome is 11,703 nucleotides in length and has in addition a 5' cap of type 0 (Hefti et al., 1976; Hsu Chen and Dubin, 1976) and a 3' poly(A) tract (Eaton and Faulkner, 1972; Johnston and Bose, 1972) of about 70 nucleotides. The molecular weight of this RNA is 4.06 MD (expressed as the Na^+ form).

The deproteinized viral genome is infectious (Wecker, 1959; Frey and Strauss, 1978), as is the case for plus-stranded RNA viruses in general, and thus the incoming RNA must be translated to produce any proteins required to replicate or transcribe the RNA. Examination of the complete sequence of Sindbis RNA, together with comparative nucleotide sequence analysis of several alphaviruses, has led to the conclusion that Sindbis genomic RNA is translated beginning with an AUG codon at nucleotides 60–62 from the 5' end (Ou et al., 1983). There follows a long open reading frame of 7539 nucleotides, encoding a polyprotein of 2513 amino acids, that has within it a single opal codon (UGA) at position 1897 of the polyprotein (Strauss et al., 1984). Translation studies in vivo (reviewed in Schlesinger and Kääriäinen, 1980; Lopez et al., 1985) and in vitro (Collins et al., 1982) have led to the conclusion that the Sindbis genome is translated as two polyproteins. The first is 1896 amino acids in length and terminates at the opal termination codon. This polyprotein is processed by proteolytic cleavage to produce three proteins, which we refer to as nsP1, nsP2, and nsP3 in order of their location in the genome. A second polyprotein is translated from the RNA by readthrough of the opal codon; this polyprotein is 2513 amino acids long and is processed by proteolytic cleavage to produce a fourth non-structural protein, nsP4 (presumably nsP1, nsP2, and nsP3 are also produced but do not contribute significantly to the pool). Because readthrough is inefficient, relatively small amounts of this second polyprotein and its unique end product, nsP4, are produced.

The mechanism by which readthrough occurs is unknown at present. Naturally occurring opal suppressor tRNAs have been described in birds (Hatfield et al., 1983) and mammals (Geller and Rich, 1980; Diamond et al., 1981; Pratt et al., 1985; O'Neill et al., 1985), and it is possible that these suppressor tRNAs mediate suppression. Alternatively, it is possible that suppression results from occasional misreading of the opal codon by $tRNA^{Trp}$, as occurs in bacteria (Hirsh and Gold, 1971; Weiner and Weber,

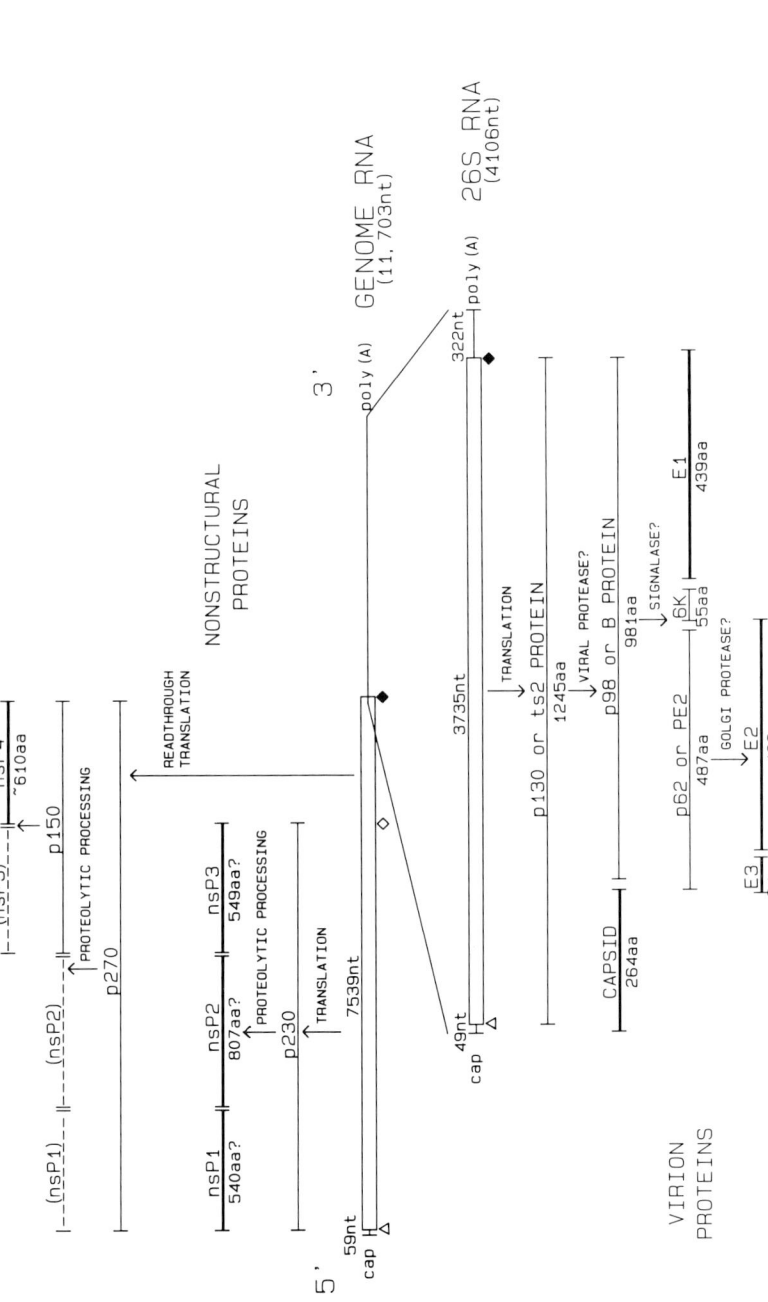

Figure 6.1. Translation strategy of Sindbis virus. Untranslated regions of the genomic RNA are shown as single lines, and the translated region as an open rectangle. The subgenomic RNA region is expanded below using the same convention. Translation products are indicated and the final protein products, both virion and non-structural, are indicated by heavy lines. Open triangles are initiation codons, solid diamonds are termination codons. The open diamond is the UGA codon read through to produce nsP4. (Reproduced from Strauss et al. (1984) with permission of Academic Press.)

1971). Readthrough of an opal termination codon to produce nsP4 also occurs in Middelburg virus (Strauss et al., 1983). Surprisingly, however, Semliki Forest virus does not contain a corresponding opal codon (Takkinen, 1986) and nsP4 is produced in larger relative amounts (Keränen and Ruohonen, 1983). It is not clear why alphaviruses differ in what would be expected to be a fundamental mechanism of translation control for regulating amounts of replicase proteins; all alphaviruses are closely related and show 30–90% amino acid sequence homology in their proteins (Bell et al., 1984).

The structural proteins are translated from a subgenomic 26S mRNA. This RNA is 4106 nucleotides long in the case of Sindbis virus, is 3' coterminal with the genomic RNA (Ou et al., 1981), beginning at nucleotide 7598 of this RNA (Strauss et al., 1984), and is capped (Dubin et al., 1977) and polyadenylated (Clegg and Kennedy, 1974). It has been shown to be transcribed from a full length minus-strand copy of the genomic RNA, using an internal transcription initiation site (Simmons and Strauss, 1972). Translation begins with an AUG codon at nucleotides 50–52 of 26S RNA and proceeds down the RNA to an opal stop codon at nucleotides 3785–3787, such that a polyprotein of 1245 amino acids is produced, as is the case for the non-structural proteins. Cleavage occurs while the polyprotein is nascent, to produce the basic nucleocapsid protein and two integral membrane glycoproteins that form, ultimately, the outer spikes of the virus. Cleavage of the capsid protein from the nascent polyprotein is believed to be an autocatalytic event, as discussed in more detail below. Cleavage of the glycoprotein precursors is thought to be effected by two organelle-bound enzymes of the host cell: signalase acting in the lumen of the endoplasmic reticulum and an enzyme possessing a tryptic-like activity found in the Golgi complex (Garoff et al., 1980b; Rice and Strauss, 1981).

III. CLEAVAGE OF THE POLYPROTEIN PRECURSORS

Many of the plus stranded RNA animal viruses produce polyprotein precursors that are cleaved proteolytically to produce the final protein products (see Chapter 1, this volume). At least some of these cleavages are known to be performed by virus-encoded enzymes (Palmenberg et al., 1979; von der Helm, 1977; Bhatti and Weber, 1979), and we reasoned some time ago that in all probability all such cleavages that occur in the cytosol (as opposed to within subcellular organelles such as the lumen of the endoplasmic reticulum or the Golgi apparatus) would be carried out by virus-encoded enzymes (Rice and Strauss, 1981). The reasoning for this was twofold. Firstly, it seemed intuitively unlikely that so many different virus-encoded enzymes would have evolved if cellular enzymes were readily available to process viral polyproteins. Secondly, no cellular proteases active in the cytosol that could perform such cleavages have been described to date. Thus one of the key elements in a virus replication strategy is the nature, activity and origin of virus-encoded proteases.

We believe that at least two virus proteases are active in processing alphavirus precursors, one to release the capsid protein from the structural polyprotein precursor and a second to process the non-structural polyprotein precursor. The capsid protein has long been thought to be an autoprotease. We first postulated this more than 10 years ago, based upon the observation that cleavage of the capsid from the polyprotein precursor was quantitative when small amounts of 26S messenger RNA were translated in extracts of rabbit reticulocytes (Simmons and Strauss, 1974). The results of one of these experiments are shown in Fig. 6.2. In this experiment 26S mRNA from a temperature-sensitive mutant of Sindbis virus, *ts*2, was translated in extracts of rabbit reticulocytes at 27°C (the permissive temperature) or at 37°C (a non-permissive temperature) as shown by the solid lines. The phenotype of *ts*2 is such that in cells infected at the non-permissive temperature the polyprotein precursor is not cleaved, and the broken line shows the protein pattern from *ts*2-infected cells at 40°C. The proteins labelled 1, 2, and Hb are made in response to endogenous rabbit mRNAs, whereas those labelled *ts*2, B and C are virus-specific proteins. The *ts*2 protein is the product of complete translation of 26S RNA, the B protein is the polyprotein after removal of the capsid protein (because no membranes are present in this reaction, it is not further processed), and C is the capsid protein. At 37°C no C protein is found and only the *ts*2 protein is produced, whereas at 27°C the major products are the B protein and the C protein (in the case of *ts*2, cleavage is not quantitative even at a permissive temperature).

Additional evidence for the autoproteolytic nature of the capsid protein comes from data suggesting that when cells are co-infected with a *ts*2-like mutant and a second mutant of different phenotype, the *ts*2 polyprotein can be cleaved by a diffusible factor (Scupham et al., 1977), and from translation data *in vitro* that show that cleavage of the capsid protein from the polyprotein can be inhibited with amino acid analogues (Aliperti and Schlesinger, 1978).

In an attempt to probe further the nature of the capsid autoprotease, we have sequenced the capsid proteins and areas of proteins E3 and E2 from three *ts*2-like mutants and compared these sequences with those of the parental strain and of temperature-insensitive revertants from each mutant. We found that all three mutants had a single amino acid substitution in the C-terminal half of the capsid protein that led to temperature-sensitive cleavage of the capsid protein (Hahn et al., 1985). *ts*2 and *ts*5 were independent isolates of the same mutation in which proline-218 of the wild-type capsid is substituted by serine; the mutation in *ts*13 was substitution of lysine-138 of the capsid protein by isoleucine. The ts^+ revertants all showed same site reversion to restore the parental amino acid. Thus these results are consistent with the hypothesis that the C-terminal half of the capsid protein possesses proteolytic activity.

In Fig. 6.3 is shown an homology plot in which the amino acid sequences of the Sindbis capsid protein and that of Ross River virus, two different

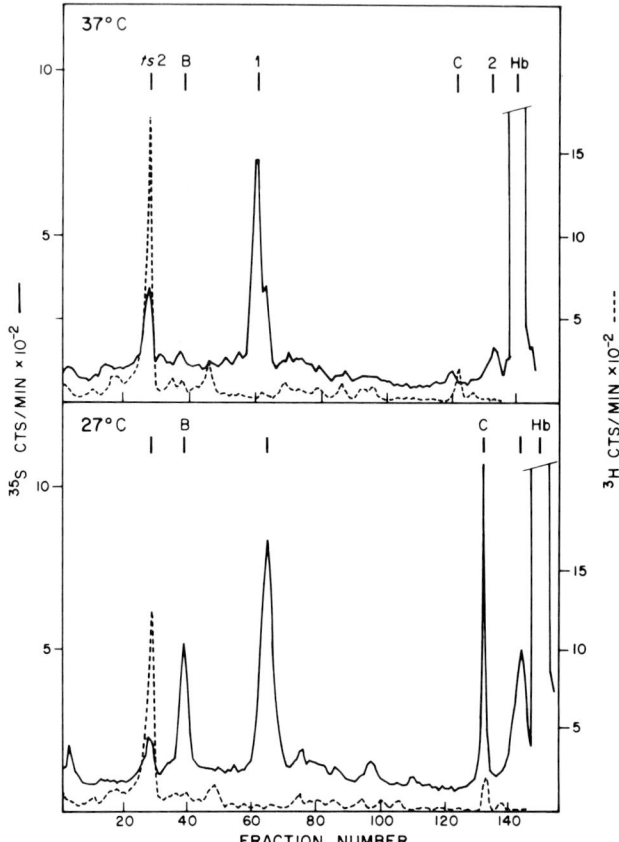

Figure 6.2. Acrylamide gel electropherogram of polypeptides synthesized *in vitro* in response to 26S RNA of *ts*2. The solid line in the upper panel shows the polypeptides synthesized at 37°C (a non-permissive temperature) in a rabbit reticulocyte cell-free system primed with *ts*2 26S RNA. The "*ts*2 protein" is the only prominent Sindbis-specific peak, since the peaks labelled 1, 2, and Hb are rabbit proteins. The dotted line is a marker from chick cells infected with *ts*2 at 40°C and labelled with ^{3}H-amino acids. The lower panel shows the products *in vitro* when translation is at 27°C (a permissive temperature), using the same RNA preparation (solid line). A small amount of the "*ts*2 protein" is made but the major products are the capsid protein (C) and the B protein. The marker (dotted line) is the same as in the upper panel. (Modified from Simmons and Strauss (1974) and reproduced with permission of Academic Press.)

alphaviruses, are compared. The percent homology is plotted as a moving average with a string length of 20. The location of the *ts*2, *ts*5 and *ts*13 mutations are indicated, and the shaded areas will be discussed in more detail shortly. The *C*-terminal halves of the capsid protein demonstrate

6. Organization and Translation in Alpha- and Flaviviruses

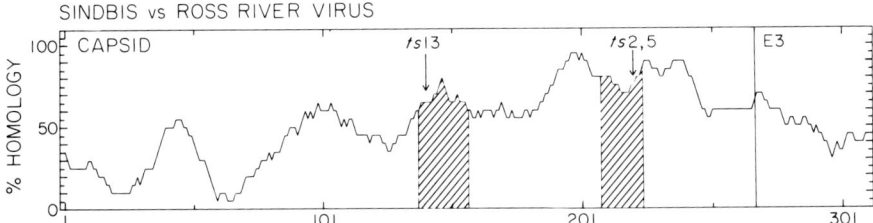

Figure 6.3. Homology between the capsid proteins and E3 proteins of Sindbis virus and Ross River virus plotted as a moving average with a string length of 20. The first shaded region is the possible serine protease active site between Lys-135 and Lys-155, which contains the His-141 and Asp-147 residues (see Fig. 6.4). The second shaded region is the second part of the possible serine protease active site between Pro-206 and Asn-222, which contains the Gly(213)-Asp-Ser-Gly(216) conserved sequence. The arrows indicate the location of the mutations in *ts*2, *ts*5 and *ts*13. (Reproduced from Hahn et al. (1985).)

extensive similarity, with up to 95% homology in a string of 20 amino acids. This, and the fact that two of the mutants sequenced represent independent isolations of the same mutation, suggest that the target window for mutation to temperature-sensitivity in the region is narrow. Furthermore, the location of the two mutations implicate the shaded regions in the proteolytic activity.

Because the cleavage of the capsid protein occurs after tryptophan-264 of the capsid protein, we reasoned that the protease might be a chymotryptic-like enzyme, consistent with some early inhibitor studies of Pfefferkorn and Boyle (1972). In addition, Boege et al. (1981) had pointed out that the alphavirus capsid protein contained the sequence Gly-Asp-Ser-Gly characteristic of serine proteases. We therefore searched for possible amino acid homology between the capsid proteins of alphaviruses and serine proteases, with emphasis on the regions probably involved in proteolysis; the results are shown in Fig. 6.4. Serine proteases are a class of enzymes that appear to be derived from a common ancestral protein whose active site contains a histidine, an aspartate, and a serine residue, shown by the solid triangles in the figure. Partial amino acid sequences are shown for five animal serine proteases: mammalian trypsin, chymotrypsin, and elastase, and two insect serine proteases, one from hornet and one from silkworm. Three regions of suggestive homology with the three alphavirus capsid proteins are boxed. These three regions encompass the three amino acids forming the catalytic triad of the serine proteases and include the amino acids changed in the Sindbis *ts* mutants sequenced to date. The lysine changed in the *ts*13 mutation is three residues removed from the putative histidine active in proteolysis, and the proline changed in *ts*2 and *ts*5 is three residues removed from the hypothetically active serine. The sequence homology overall is limited, but fairly convincing around the serine residue, and the case for the activities of these amino acids is bolstered by

```
              ts13
         135  ↓       145            155           165
         |    |       |     |        |     |       |     |
SIN   ..KVMKPLHV KGTI DHPVLSKLK FTKSSAYDMEFAQLP...
SF    ..KVMKPAHV KGVI DNADLAKLA FKKSSKYDLECAQIP...
RR    ..KVMKPAHV KGTI DNPDLAKLT YKKSSKYDLECAQIP...

HRN   ..YVLTAAHC ....DIG  LIRVSKDISFTQLVQPVKLP...
TRP   ..WVVSAAHC ....DIM  LIKLKSAASLNSRVASISLP...
CHY   ..WVVTAAHC ....DIT  LLKLSTAASFSQTVSAVCLP...
ELA   ..WVMTAAHC ....DIA  LLRLAQSVTLNSYVQLGVLP...
           *     *▲       ▲   * **      *            *
           |     |        |         |           |
          51    56       102       110         120
```

```
                    ts2, ts5
         210       ↓ 220                230              240
         |    |    |    |            |     |          |     |
SIN   ..PRGVGGR GDSGRPIMD   NSG R  VVAIVLGGADEGTR   TALSVVT...
SF    ..PTGAGKP GDSGRPIFD   NKG R  VVAIVLGGANEGSR   TALSVVT...
RR    ..PTGAGKP GDSGRPIFD   NKG R  VVAIVLGGANEGAR   TALSVVT...

HRN   ..GEGACH  GDSGGPLVA   N G V  QIGIVSYGHP   CAIGS PNVFT...
SLK        DACQ GDSGGPVQ    NAG R

TRP   ..GKDSCQ  GDSGGPVVC   S G K  LQGIVSWGS   GCAQKNKPGVYT...
CHY   ..V SSCM  GDSGGPLVCKKN GAWTLVGIVSWGS STCSTST PGVYA...
ELA   ..VRSGCQ  GDSGGPLHCLVN GGYAVHGVTSFVSRLGCNVTRKPTVFT...
                **▲* *      *  *      **  *            * *
           |         |             |           |
          190       200           210         220
```

Figure 6.4. Homology between the highly conserved COOH-terminal half of alphavirus capsid protein and mammalian and insect serine protease active sites. Top three lines: partial amino acid sequences of Sindbis virus (SIN), Semliki Forest virus (SF) and Ross River virus (RR) capsid proteins. Sequences are from Rice and Strauss (1981), Garoff et al. (1980a) and Dalgarno et al. (1983), respectively; numbering of amino acids is for SIN capsid protein. Middle lines: partial sequences of hornet chymotrypsin (HRN) (Jany et al., 1983) and silkworm cocoonase (SLK) (Kramer et al., 1973). Bottom three lines: active site domains of three mammalian serine proteases: bovine trypsin (TRP), bovine chymotrypsin (CHY), and porcine elastase (ELA), respectively; the lower numbers are the standard numbering of amino acids for chymotrypsinogen (Greer, 1981; Dayhoff, 1978). Solid triangles are three active amino acids which form the catalytic triad (His-57, Asp-102 and Ser-195) and asterisks are amino acids highly conserved between the capsid proteins and the animal serine proteases; regions of highest homology are boxed. Solid arrows indicate the locations of mutations in $ts2$, $ts5$, and $ts13$. The single-letter amino acid code is used. A = Ala, C = Cys, D = Asp, E = Glu, F = Phe, G = Gly, H = His, I = Ile, K = Lys, L = Leu, M = Met, N = Asn, P = Pro, Q = Gln, R = Arg, S = Ser, T = Thr, V = Val, W = Trp, Y = Tyr. (Reproduced from Hahn et al. (1985).)

the locations of the ts mutations. We hypothesize that the alphavirus capsid protein is a serine protease that functions autocatalytically, and plan to test the importance of key residues in this domain of the protein more directly using site-directed mutagenesis. Although convergent evolution cannot be ruled out as an explanation for the amino acid similarities found between alphavirus capsid proteins and animal serine proteases, we feel it more likely that these sequence similarities reflect divergence from a common ancestral protein. This suggests that a protoalphavirus may have captured the gene for the enzyme from its host at some time in the past.

The nature of the protease involved in processing of the alphavirus non-structural proteins is unclear at present. It prefers to cleave after the sequence glycine-alanine and therefore has a distinctive activity. During *in vitro* translation of viral genomic RNA these non-structural cleavages appear to occur (Simmons and Strauss, 1974; Collins *et al.*, 1982) and we hypothesize that one of the non-structural proteins is also a protease. We plan to test this hypothesis by mapping *ts* mutations in the non-structural proteins which process their respective precursors inefficiently at the non-permissive temperature and by translation *in vitro* of selected portions of the non-structural region.

IV. EXPRESSION OF CLONED COPIES OF SINDBIS STRUCTURAL PROTEINS

In an effort to understand in more detail the processing pathways of the Sindbis structural proteins, and to facilitate future manipulations of the structural genes (including site-directed mutagenesis), we have obtained cDNA clones containing these genes and expressed them in a vaccinia vector (Rice *et al.*, 1985b). The recombinant vaccinia virus contained a cDNA copy of the 26S region of Sindbis virus inserted into the vaccinia *tk* gene under the control of the vaccinia promoter for the so-called 7.5 kD protein gene. This vaccinia recombinant expressed Sindbis-specific RNA constitutively in cells both permissive or non permissive for vaccinia growth, and all three Sindbis structural proteins were present at about 10% of the level of these proteins in Sindbis virus-infected cells. In the absence of the non-structural genes, the structural proteins were apparently processed, transported, and incorporated into virions normally. Figure 6.5 shows gel patterns of proteins produced after infection of cells by wild type vaccinia virus, by vaccinia recombinant VV3S-7, and by Sindbis virus, after immunoprecipitation with preimmune serum (P1), anti-E1, anti-E2, anti-C, or anti-Sindbis virus antisera. Structural proteins E1, E2, and C, as well as the precursor PE2, are present in cells infected with the vaccinia recombinant and these comigrate with the comparable proteins produced in Sindbis-infected cells. The glycoproteins E1 and E2 expressed in vaccinia are processed and transported to the cell surface; Fig. 6.6 shows the results of an immunofluorescence study in which anti-E1 or anti-E2 antibodies were used to demonstrate the presence of both E1 and E2 at the surface of cells infected with the vaccinia recombinant.

Finally, the expressed glycoproteins can be incorporated into infectious virions, as shown by a phenotypic mixing experiment (Table 6.1). A strain of Sindbis AR339, called SIN V33/50/23, which is resistant to monoclonal antibodies 50 and 23 (both reactive with E2) as well as 33 (reactive with E1) was isolated and supplied by Dr. Alan Schmaljohn. Our Sindbis HR strain

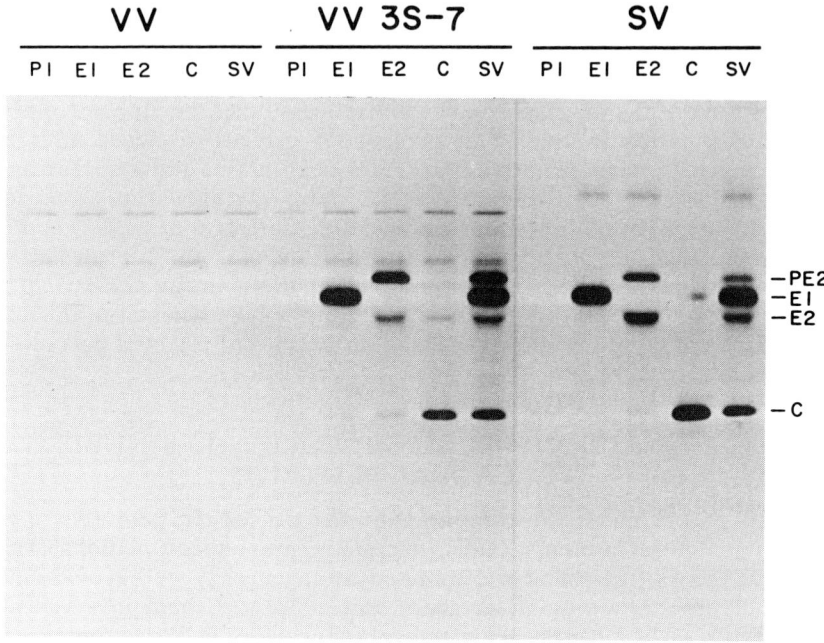

Figure 6.5. Expression of Sindbis proteins. Radioactively labelled proteins from vaccinia virus (VV), vaccinia recombinant VV3S-7, or Sindbis (SV) virus-infected cells were immunoprecipitated with pre-immune serum (PI), heterospecific anti-Sindbis serum (SV), or monospecific antisera directed towards Sindbis envelope (E1 or E2) or capsid (C) proteins. The immune precipitates were then subjected to SDS-polyacrylamide gel electrophoresis and autoradiography. Sindbis-specific proteins are identified in the right margin. (Reproduced from Rice et al. (1985b) with permission of the American Society for Microbiology.)

(SIN HRsp) is reactive with monoclonals 33 and 50, and thus the glycoproteins synthesized by the vaccinia recombinant VV3S-7 can be distinguished from those synthesized by Sindbis strain V33/50/23. In a phenotypic mixing experiment cells were infected with vaccinia recombinant VV3S-7 and superinfected with Sindbis V33/50/23. Virions produced in the infection were purified and precipitated with monoclonal antibodies 50 or 33 and the results are shown in Table 6.1. Although virions of the Sindbis variant V33/50/23 are not precipitable by monoclonals 50 or 33, virions produced after mixed infection with the vaccinia recombinant are precipitable, showing that the vaccinia-derived Sindbis proteins can be incorporated into virus particles.

Thus all of the processing signals and virus-encoded protease activities necessary for production and transport of the structural proteins are found within the structural proteins themselves; translation of virus non-structural proteins is not required for these events, nor is assembly of

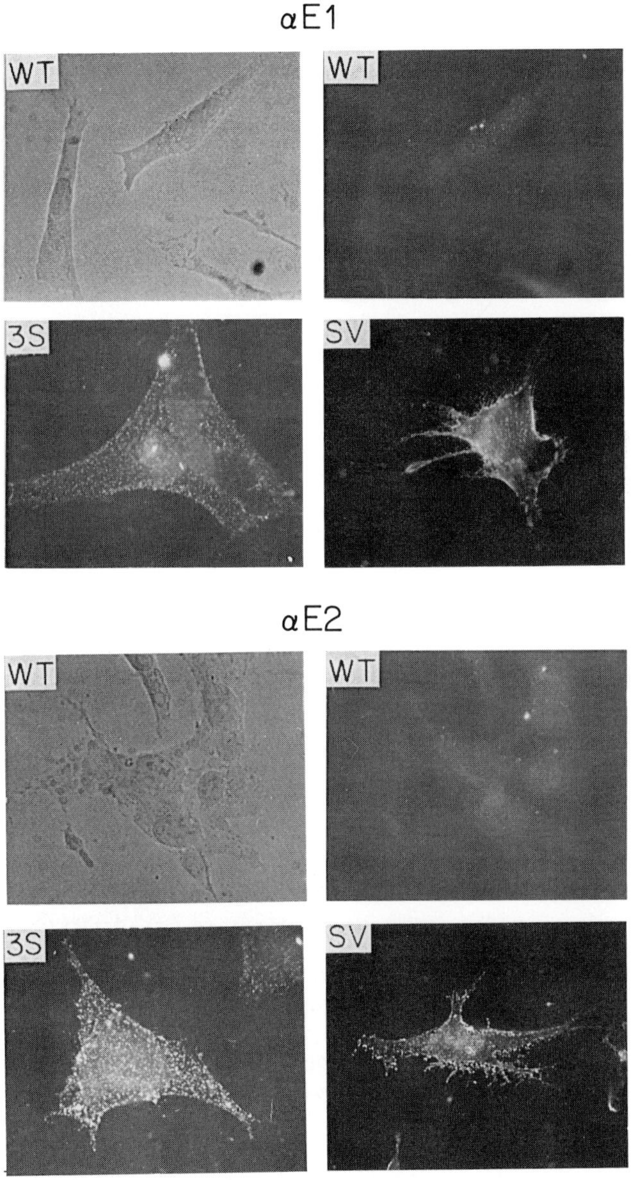

Figure 6.6. Immunofluorescence microscopy of VV3S-7-infected cells. BHK cells infected for 8 h with either wild-type vaccinia virus (WT), VV3S-7 (3S), or Sindbis virus (SV) were fixed, incubated with rabbit E1-specific (αE1) or E2-specific (αE2) antibodies and stained with fluorescein conjugated anti-rabbit immunoglobulin goat antisera. Immunofluorescence photographs are shown for all of these samples with phase contrast photographs included for wild-type vaccinia-infected cells only. (Reproduced from Rice et al. (1985b) with permission of the American Society for Microbiology.)

Table 6.1. Incorporation of vaccinia-encoded Sindbis glycoproteins into infectious Sindbis virus[a]

Mixed infection	% cpm precipitated by antiserum[b]		
	Non-immune	No. 33	No. 50
SIN V33/50/23 + VV wt	1	1	2
SIN V33/50/23 + VV3S-7	2	32	56
SIN HRsp + VV wt	0	85	78

[a]Cells were infected with either vaccinia wild-type (VV wt) or vaccinia recombinant VV3S-7 (which expresses Sindbis glycoproteins from Sindbis strain SIN HRsp) and superinfected with either Sindbis variant SIN V33/50/23 (resistant to monoclonal antibodies 33, 50, and 23) or Sindbis SIN HRsp. Sindbis virions produced in the mixed infection were purified and precipitated with either monoclonal antibody 33 (E1 specific) or 50 (E2 specific).
[b]Expressed as percent of input c.p.m. (Reproduced from Rice et al. (1985b) with permission of the American Society for Microbiology.)

nucleocapsids required. Similar conclusions were reached by Huth et al. (1984) who microinjected 26S RNA of Semliki Forest virus into *Xenopus* oocytes. These experiments with the vaccinia system demonstrate that biologically active viral proteins are produced in significant quantities. Thus, this system is ideally suited for studying the effects on protein function of mutations engineered into cloned DNA copies of alphavirus genomes.

V. CONSERVED SEQUENCES IN ALPHAVIRUS RNAs

Our laboratory, and those of Sondra Schlesinger and Henry Huang at Washington University of Saint Louis, are trying to develop expression systems in which the significance of RNA sequences in replication, packaging, and RNA function can be tested, and Schlesinger and Huang (Levis et al., 1986) have succeeded in expressing a cloned copy of a Sindbis defective interfering (DI) RNA. For this a cDNA copy of a Sindbis DI RNA was placed next to a promoter for the RNA polymerase of phage SP6 in a bacterial plasmid, and RNA was transcribed *in vitro* that contains the complete sequence of the DI RNA flanked at both the 5' and 3' ends by short exogenous sequences. The transcribed RNA was used to transfect cells simultaneously infected with helper Sindbis virus, and after two or three passages the transfected DI RNA becomes the predominant viral RNA species, behaving as a DI RNA and being amplified quickly while inhibiting growth of wild-type virus. This system thus allows testing of the importance of various regions for DI function, in that regions can be deleted or rearranged at the DNA level and the effect upon rescue of a functional DI RNA assayed. A number of such constructs have been tested, but only two of their results will be discussed here.

We have found that nucleotide sequences or structures are conserved in four regions of the alphavirus genome and have postulated that these sequences or structures play important roles in alphavirus RNA replication and packaging (Ou et al., 1982a,b, 1983). One such conserved sequence is shown in Fig. 6.7. In this figure the nucleotide sequences at the 3' end, immediately adjacent to the poly(A) tract, of eight different alphaviruses are presented, reading 3' to 5'. The first 19 nucleotides are highly conserved. In five of the viruses only a single nucleotide in the first 19, nucleotide 6, shows any variability at all. In three of the viruses, which represent a distinct branch of the alphavirus evolutionary tree, the sequence is more variable but nonetheless highly conserved overall.

The importance of these 19 nucleotides for RNA function has been tested in the DI expression system. The cloned DI DNA was subject to deletion analysis so that differing numbers of 3' terminal nucleotides derived from the parental sequence remained adjacent to the poly(A) tract. Constructs that retained 19 or more parental nucleotides adjacent to the poly(A) tract were active. However, a construct that retained only 17 parental nucleotides was inactive. Thus the conserved 19-nucleotide stretch must be present for DI RNA function, and probably acts as a recognition sequence for the viral replicase that attaches to the plus strand and makes a minus strand copy of it.

A second conserved nucleotide sequence is illustrated in Fig. 6.8. This

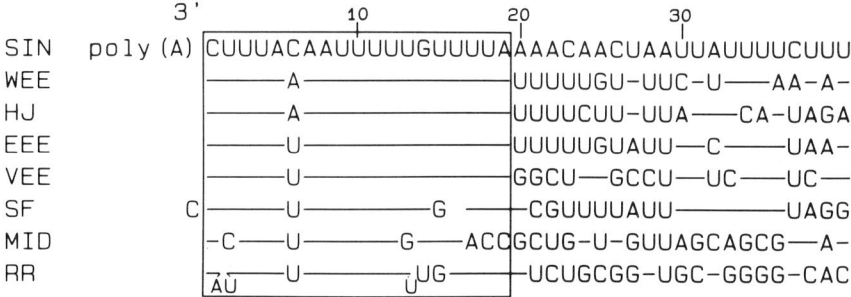

Figure 6.7. Conserved sequences in alphavirus RNAs. The 3'-terminal sequences of eight alphavirus genomic RNAs are shown from 3'(poly(A)) to 5'. The box encloses a region of 19 nucleotides (from Sindbis virus) that are highly conserved. A horizontal line indicates that the nucleotide at that position is identical to the one shown in the complete sequence at the head of the group. Gaps have been introduced for alignment. Data for the construction of this figure are from Ou et al. (1982b) and Dalgarno et al. (1983). SIN, Sindbis virus; WEE, Western equine encephalitis virus; HJ, Highlands J virus; EEE, Eastern equine encephalitis virus; VEE, Venezuelan equine encephalitis virus; SF, Semliki Forest virus; MID, Middelburg virus; RR, Ross River virus.

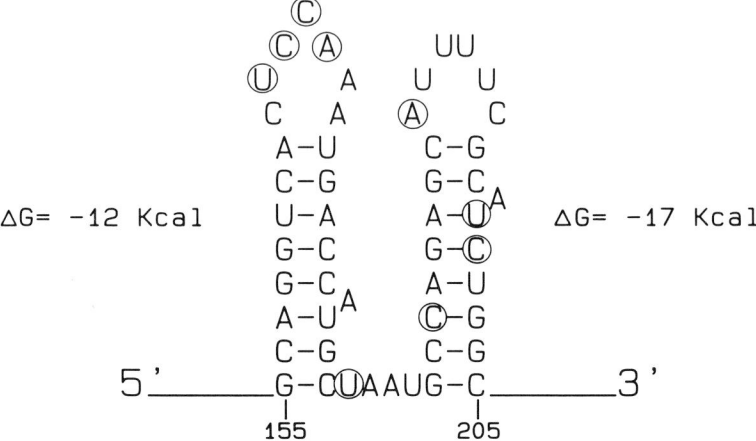

Figure 6.8. Conserved structure in alphavirus RNAs. A possible structure is shown for a 51-nucleotide conserved sequence found near the 5' end of alphavirus RNAs. The sequence shown is that of Sindbis virus, HR strain, and the nucleotides are numbered from the 5' end of this RNA. Nucleotides which vary in one or more of the sequences are circled; others are invariant for four viruses. (Reproduced from Ou et al. (1983) with the permission of Academic Press.) Free energy was calculated according to the method of Tinoco et al. (1973).

stretch of 51 nucleotides is found near the 5' end of Sindbis RNA, beginning about 150 nucleotides from the 5' end. It is conserved in five viruses examined, with only the circled nucleotides showing any variability whatsoever in these viruses. As can be seen, this sequence is also capable of forming a double-hairpin structure. We had originally postulated that this sequence was involved in RNA replication and packaging. However, the DI expression assay has shown that this sequence can be deleted and the construct still be active, indicating that this sequence is not required for DI RNA replication or packaging. This finding is somewhat puzzling because all naturally occurring DIs sequenced to date retain this sequence and some have even amplified it (i.e. they contain several copies of it). In any event, these exciting results show how much we have yet to learn and point out the power of recombinant DNA techniques in unravelling the puzzle of viral RNA replication.

VI. CODON USAGE IN ALPHAVIRUSES

We and others have sequenced regions of several different alphaviruses, so that comparative sequence data are available in both the structural and non-structural regions for a number of alphaviruses. The genome organization of all alphaviruses is the same, and each individual protein exhibits considerable sequence similarity with the corresponding protein of other

6. *Organization and Translation in Alpha- and Flaviviruses* 89

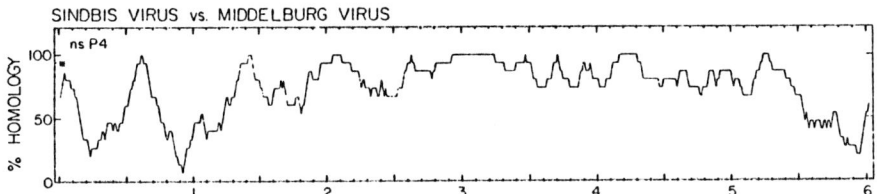

Figure 6.9. Graphic representation of amino acid homology between Sindbis virus and Middelburg virus for non-structural proteins nsP4. A string length of 15 was used. Amino acids are numbered in hundreds from the beginning of nsP4. The asterisk marks the position of the opal codon. Data from Strauss *et al.* (1983) were used to construct the figure.

alphaviruses. The amino acid sequence homologies found vary from 30% to 90% depending upon the protein and the viruses being compared. Overall, the non-structural proteins of any two alphaviruses exhibit considerably greater sequence homology than the structural proteins, although as discussed earlier the sequence homology is not uniform, with some domains being highly conserved and others less conserved. Fig. 6.3 presents a homology plot of the capsid proteins of two alphaviruses. Fig. 6.9 shows a similar homology plot for one of the non-structural proteins, nsP4, of alphaviruses Sindbis and Middelburg. The string length used is 15. The sequence homology in this protein for these distantly related alphaviruses is striking, averaging 73% overall. The central core of the protein is very highly conserved; there is a string of 98 amino acids near the middle in which only five amino acid differences are present.

Although amino acid sequence homology can be quite extensive, the nucleotide sequence encoding this amino acid sequence has been randomized between any two alphaviruses to the extent permitted by degeneracy in the genetic code. Figure 6.10 presents a short stretch of translated

```
          L    F    K    L    G    K    P    L
SIN      CUG  UUU  AAG  UUG  GGU  AAA  CCG  CUC
          L    F    K    L    G    K    P    L
MID      CUC  UUU  AAG  CUC  GGA  AAA  CCG  CUG
          L    F    K    L    G    K    P    L
RR       UUA  UUU  AAA  CUA  GGU  AAA  CCU  UUA
          L    F    K    L    G    K    P    L
SF       CUG  UUC  AAG  UUG  GGU  AAG  CCG  CUA
CODONS    3    2    2    3    2    2    2    4
```

Figure 6.10. Randomization of codons used for conserved amino acids in alphaviruses. A short stretch of conserved protein sequence found in nsP4 about 80 amino acids from the C-terminus, along with the nucleotide sequence encoding it, is shown for four alphaviruses (the first leucine is amino acid 2431 in the Sindbis non-structural polyprotein). The bottom line gives the number of different codons used at each position. Data from Ou *et al.* (1982a) and Riedel *et al.* (1982).

nucleotide sequence in nsP4 for four alphaviruses. The eight amino acids in this sequence are perfectly conserved in all four. However, the codons used for these conserved amino acids differ significantly; note in particular that for encoding the last leucine residue, a different codon is used in each of the four viruses.

A second way of visualizing this codon randomization is shown in Table 6.2. In the first line, the amino acid sequence of Sindbis virus nsP4 is compared with that of Middelburg nsP4. As discussed above, these two proteins exhibit 73% amino acid sequence similarity, with 448 of 626 amino acids being identical in the two nsP4s. The two nsP4s also exhibit perfect alignment with no gaps or deletions. Of the 448 amino acids that are identical, 181 are encoded by the same codon in the two viruses; for the rest different codons are used in the two viruses. On the basis of random chance, making allowance for the amino acid compositions and the codon preferences exhibited by alphaviruses, 171 matches would be expected. A similar comparison is shown for glycoproteins E1 of Semliki Forest and Ross River viruses in the second line. Semliki Forest and Ross River show much greater overall amino acid sequence homology than do Sindbis and Middelburg, and these two alphaviruses presumably diverged much more recently from one another, but the randomization of codons encoding conserved amino acids is again essentially complete. This analysis tells us something about evolution in RNA viruses; mutation in RNA genomes is so rapid that conservation is at the level of protein function (and thus amino acid sequence) rather than at the level of RNA sequence. The studies of Domingo *et al.* (1978) have led to the conclusion that in bacteriophage Qβ, silent changes are selected against weakly but detectably, so that the overall structure of the RNA must be optimal for replication and packaging. Such negative selection against silent changes is probably also present in alphaviruses and other plus-stranded viruses, but

Table 6.2. Codon randomization[a]

		Amino Acids		Codon Used[b]	
		Total	Conserved	Same[c]	Expected[d]
SIN/MID	nsP4	626	448	181	171
RR/SF	E1	438	345	143	134

[a] Comparison of Sindbis (SIN) and Middelburg (MID) protein nsP4 and of Ross River (RR) and Semliki Forest (SF) glycoprotein E1.
[b] Codons used for conserved amino acids.
[c] Number of conserved amino acids encoded by the same codon.
[d] Expected number of conserved amino acids encoded by the same codon based on random chance, calculated using the actual composition of conserved amino acids and the average codon usage distribution exhibited by the alphaviruses Sindbis, Semliki Forest and Ross River.

this relative pressure is so weak that the codon usage has nevertheless been randomized during evolution. Secondly, codon randomization tells us that alternative reading frames are not used in alphaviruses, at least not in the regions for which we have comparative sequence data to date. For if two overlapping reading frames are being used, any change in nucleotide sequence would result in a coding change, and in viruses in which overlapping reading frames are used the nucleotide sequence as well as the amino acid sequence is conserved. Use of a single reading frame appears to be a general characteristic of plus-stranded RNA viruses.

VII. RELATIONSHIP OF ALPHAVIRUSES TO CERTAIN PLANT VIRUSES

With the advent of rapid sequencing techniques leading to the accumulation of a large body of sequence data for RNA viruses, and the existence of better computer techniques to analyse these data, it has become possible to compare the amino acid sequences of viral proteins, not only from closely related viruses but from distantly related viruses as well. Haseloff *et al.* (1984) reported the exciting finding that one of the alphavirus replicase proteins, Sindbis virus nsP4, shared amino acid sequence homology with one of the replicase proteins of three distinct RNA plant viruses, representing three different morphological groups—tobacco mosaic virus (TMV), bromegrass mosaic virus (BMV), and alfalfa mosaic virus (AlMV). In a subsequent paper, Ahlquist *et al.* (1985) showed that these homologies extended to nsP1 and nsP2 of Sindbis virus as well. The genome organizations and translation strategies of these viruses are illustrated schematically in Figure 6.11 in which the homologous regions in the replicases are indicated by shading (see also Chapter 13 of this volume for discussion of these similarities).

These four groups of viruses all share similarities in their replication strategies, and the similarities between alphaviruses and TMV are particularly striking. Both produce two large proteins by translation of the infectious genomic RNA. The first protein terminates at an amber codon in TMV or at an opal codon in (at least) two alphaviruses and contains two of the conserved domains (Ahlquist *et al.*, 1985). In alphaviruses, but not in TMV, post-translational cleavage occurs to separate the two conserved domains into nsP1 and nsP2 and to produce a third non-structural protein nsP3 which shares no detectable sequence homology with the plant virus proteins. In both viruses a second non-structural protein is translated by readthrough of the termination codon (Strauss *et al.*, 1983; Pelham, 1978), and the readthrough portions share amino acid sequence homology (Haseloff *et al.*, 1984). Proteolytic cleavage separates the readthrough regions (as nsP4) in the alphaviruses but not in TMV. In both viruses subgenomic RNAs are transcribed which are translated into the virion structural proteins. In TMV there is also a second subgenomic RNA produced which is translated into a protein of unknown function. BMV

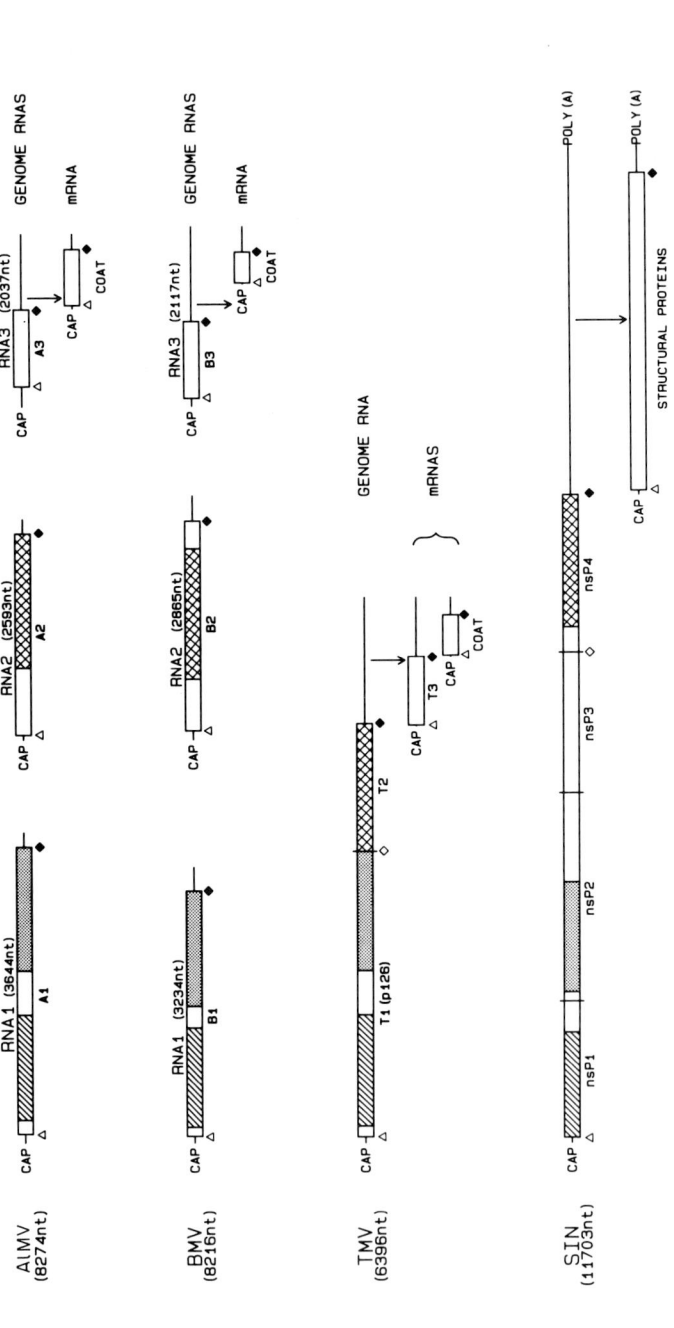

Figure 6.11. Genome organization and amino acid homologies of Sindbis virus and three plant viruses. Conventions are the same as those used in Fig. 6.1. Within the translated regions (open rectangles) there are three regions of homology indicated by different types of shading (diagonal hatching, dotted overlay and cross-hatching, respectively). All genomes are shown to scale. Virus abbreviations: SIN, Sindbis virus; AlMV, alfalfa mosaic virus; BMV, bromegrass mosaic virus; TMV, tobacco mosaic virus. (Reproduced from Ahlquist et al. (1985) with the permission of the American Society for Microbiology.)

and AlMV differ from this scheme only in having the genome divided into three segments. Thus, to produce the domain corresponding to nsP4 (A2 or B2) requires translation of a separate genomic segment rather than readthrough; regulation of the amount produced could be regulated by a differential efficiency of translation of RNA2. A subgenomic RNA is also transcribed that is translated into the structural protein.

The similarities in replication and translation strategy and the amino acid sequence homologies found suggest that these viruses have descended from a common ancestor and that certain domains of the viral replicases have been conserved because they serve equivalent functions. Even though evolution of RNA viruses is rapid, the internal milieu of eukaryotic cells changes only slowly with time, and once evolved to replicate virus RNA efficiently and rapidly in the cell cytoplasm the replicases might evolve much more slowly than structural proteins that interact with the external environment, including vertebrate immune systems.

These four groups of viruses differ markedly in their morphology, which has been the major characteristic used for virus classification. Alphaviruses possess an icosahedral nucleocapsid containing the RNA that is surrounded by a lipid bilayer containing two virus-specific glycoproteins; TMV is a rod-shaped virus built on principles of helical symmetry and only a single species of coat protein is used to construct the virion; BMV is a simple icosahedron that again is constructed with a single species of coat protein; and AlMV is bacilliform. Perhaps for this reason, at least in part, no sequence homologies can be detected among the virion structural proteins. The structural proteins of the alphaviruses are more divergent than the non-structural proteins and if these four groups of viruses are descended from a common ancestor it would appear that the structural proteins have diverged so extensively that new forms of virus structure have evolved and no residual sequence homology exists. In this process new genes could have been captured from the host cell, such as the virus encoded proteases. For this reason similarities in replication strategy and genome organization may prove to be a more accurate indicator of evolutionary relationships than virus structure (Strauss and Strauss, 1983).

The amino acid homologies among the conserved domains of the replicases for BMV, AlMV, TMV and Sindbis are illustrated in Figures 6.12 to 6.14, in which the homologies are plotted as a moving average in pairwise combinations. In each figure the top panel compares AlMV and BMV, the two tripartite viruses; the second panel compares BMV and TMV; and the bottom panel compares TMV and Sindbis. In each case the homology is by no means uniform along the region but rather consists of a number of conserved domains interspersed with regions of little or no homology. As expected, the homology is highest overall between the two tripartite plant viruses, followed by TMV and BMV; Sindbis has fewer regions in common with plant viruses. However, some interesting observations can be made. Firstly, the homology is not uniform between any two viruses but consists of conserved domains interspersed with nonconserved

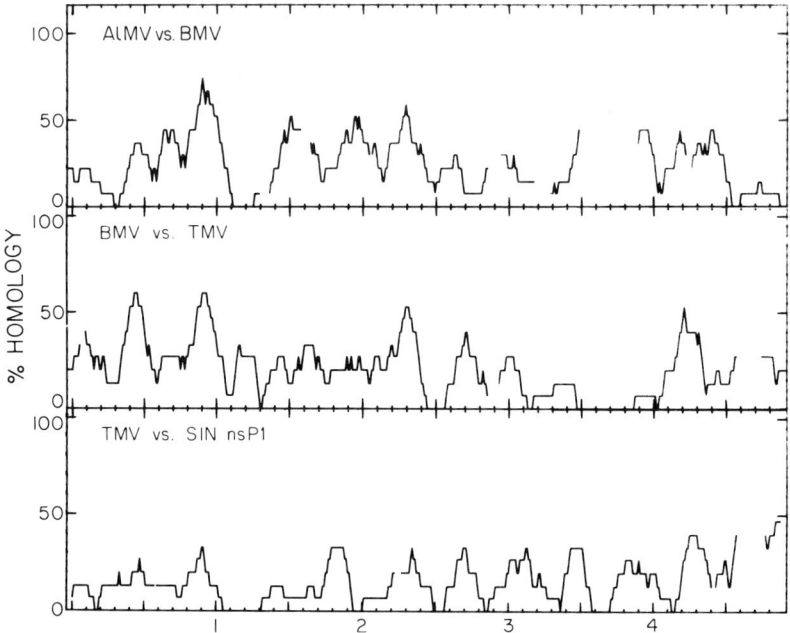

Figure 6.12. Graphic representations of amino acid homologies between nsP1 of Sindbis virus and comparable polypeptides encoded by three plant viruses. The regions shown are those that are diagonally hatched in Fig. 6.11. Virus abbreviations are the same as used in Fig. 6.11. The sequences reported for BMV (Ahlquist et al., 1984), AlMV (Cornelissen et al., 1983a,b), TMV (Goelet et al., 1982), and SIN (Strauss et al., 1984), were aligned as reported in Ahlquist et al. (1985). The AlMV and BMV sequences begin at amino acids 55 and 36 respectively in the open reading frame of RNA 1, the TMV sequence begins at amino acid 43 of the 126 kD protein and the SIN sequence begins with the initiating Met of nsP1. The ordinate is per cent homology as a moving average with a search string of 15, the abscissa is amino acid residues in hundreds. Gaps have been introduced for alignment and treated as non-matches; coincident gaps cause a break in the homology line.

domains. Secondly, the various proteins and the domains within those proteins exhibit different degrees of homology. Thus nsP1 demonstrates the least homology of the three proteins considered and the homologies with Sindbis are weak, whereas there are domains within nsP2 and nsP4 that are highly conserved (up to 80% sequence homology between the two tripartite plant viruses, up to 70% between TMV and BMV, and more than 50% sequence homology between TMV and Sindbis virus, using a window of 20 amino acids).

Although we believe it likely that the alphaviruses and these three groups of plant viruses diverged from a common ancestral protovirus, it is possible that these RNA viruses emerged independently and in so doing

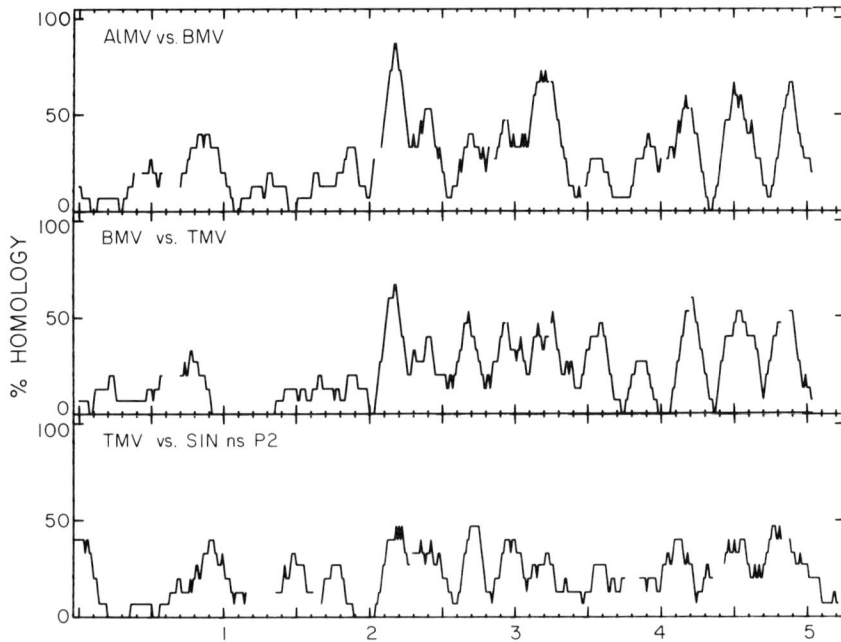

Figure 6.13. Graphic representation of amino acid homologies between nsP2 of Sindbis and portions of the plant virus replicases, in the region shown by dotted overlay in Fig. 6.11. Abbreviations, conventions, and sequence sources are the same as used in Figs. 6.11 and 6.12. The AlMV and BMV sequences begin at amino acids 641 and 510, respectively, of the open reading frame of RNA 1, the TMV sequence begins at amino acid 735 of the 126 kD protein and the SIN sequence begins with amino acid 30 of nsP2.

captured the same host polymerase or related host polymerases that had in turn evolved from a common ancestral protein. If we take the viewpoint that RNA viruses arose but once (or arose only a very limited number of times) the known rapidity of RNA evolution (Holland *et al.*, 1982) suggests that the alphaviruses and the plant viruses TMV, BMV, and AlMV diverged in the not too distant past and indeed that perhaps all the current eukaryotic RNA viruses are a relatively recent innovation. From the standpoint of radiation it is attractive to speculate that these viruses may have arisen in insects. Many viruses other than the alphaviruses are known that are able to replicate in insects and in vertebrates and, similarly, many plant viruses are known that can replicate in both plants and insects. Thus, an insect protovirus could radiate in principle to both plants and animals, and once radiation occurred, the viruses might in some cases retain the ability to replicate in insects as well as in their new hosts, whereas in other cases the ability to replicate in insects could have been lost.

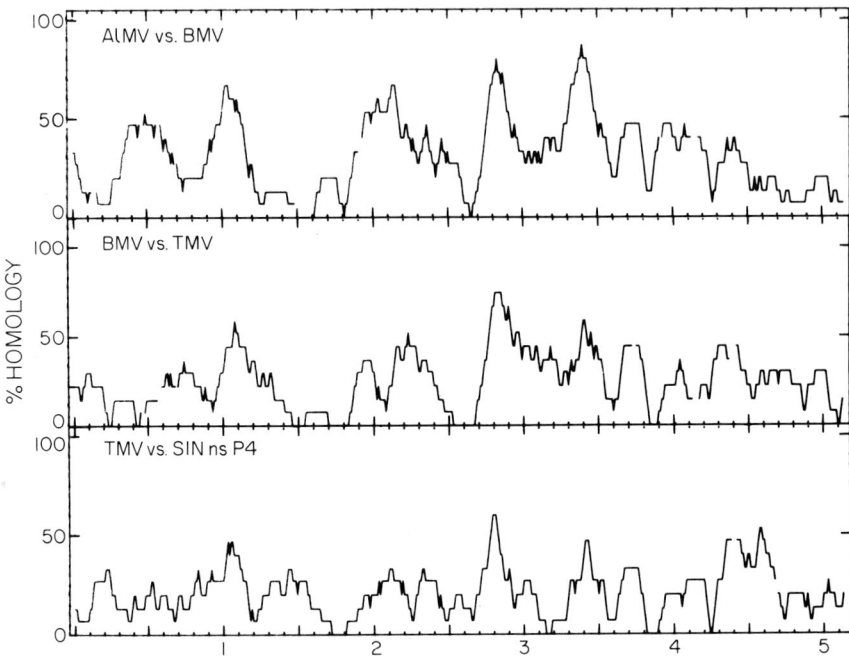

Figure 6.14. Graphic representation of amino acid homologies between nsP4 of Sindbis and those portions of the plant virus replicases indicated by cross-hatching in Fig. 6.11. Conventions and abbreviations are as in Figs 6.11 and 6.12. The AlMV and BMV sequences begin with amino acids 265 and 202, respectively, of the open reading frame of RNA 2. The TMV sequence begins at the amber codon punctuating the open reading frame, and the SIN sequence starts with amino acid 101 of nsP4. This protein contains the Gly-Asp-Asp domain common to many RNA virus polymerases (Kamer and Argos, 1984).

VIII. THE GENOME ORGANIZATION OF FLAVIVIRUSES

We have now determined the complete nucleotide sequence of the genome of yellow fever virus (17D strain) (Rice et al., 1985a) and the genome organization and translation strategy of yellow fever virus is illustrated in Figure 6.15. The RNA is 10,862 nucleotides in length (molecular weight 3.75 MD) and is capped at the 5′ end with a type 1 cap (Wengler et al., 1978; Cleaves and Dubin, 1979). The 3′ end lacks the poly(A) found in most plus-stranded RNA viruses, terminating instead with U (Wengler and Wengler, 1981). The most striking feature of the sequence is the presence of a single long open reading frame of 10,233 nucleotides that could encode a polyprotein of 380,763 daltons, leaving 5′ and 3′ non-coding regions of 118 and 511 nucleotides, respectively. In this long open reading frame, the structural proteins occupy the 5′ terminal quarter and the non-

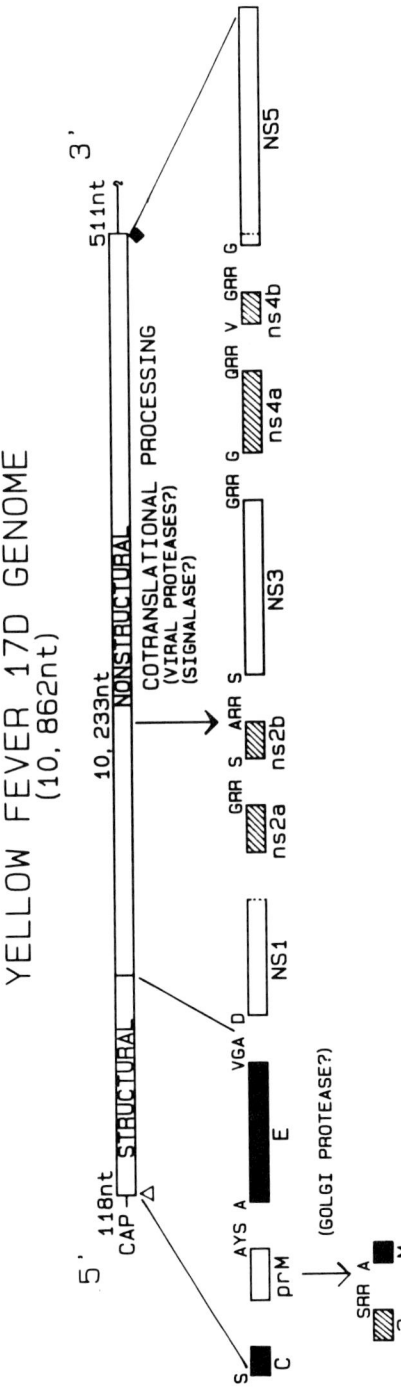

Figure 6.15. Organization and processing of proteins encoded by the yellow fever genome. Untranslated regions are shown as single lines and the translated region as an open rectangle. The open triangle is the initiation codon (AUG); the solid diamond the termination codon (UGA). The protein nomenclature is described in Rice et al. (1985a). In this nomenclature the non-structural proteins are named in order of their appearance in the genome. Capital NS is used for proteins identified in YF-infected cells (NS1 is also called NV3, NS3 is also called NV4, and NS5 is also called NV5), and lower case ns for non-structural proteins assigned only on the basis of characteristic cleavage sites. The structural proteins are C (capsid), E (envelope) and M (membrane); prM (also called NV2) is a precursor to M. The single-letter amino acid code is used for sequences flanking assigned cleavage sites (solid lines). Two other potential cleavage sites are shown as dotted lines. Structural proteins, identified non-structural proteins, and hypothesized non-structural proteins are indicated by solid, open, and hatched boxes, respectively. Other potential cleavage sites have been found and are described in Rice et al. (1985a). (Reproduced from Rice et al. (1985a) with permission of the American Association for the Advancement of Science. Copyright 1985 by the AAAS.)

structural proteins occupy the remainder. The presence of such a long open reading frame, the existence of a consistent set of potential cleavage sites that could explain the generation of the yellow fever polypeptides by post-translational cleavages, the gene order deduced from pactamycin runoff experiments of Westaway (1977), and recent evidence for polyprotein precursors in flavivirus-infected cells (G. Cleaves, personal communication; C. Blair, personal communication) support the view that translation of the flavivirus genome *in vivo* initiates with the capsid protein near the 5' end and progresses sequentially through the genome to produce one large precursor polyprotein. Cleavage of the polyprotein is very rapid, however, and for the most part occurs during translation so that the complete precursor polyprotein is not seen in its entirety and even partially processed precursors are seen only with difficulty. This hypothesis is at variance with the previous hypothesis of Westaway (1977; Westaway *et al.*, 1984) that during translation of flavivirus proteins a number of independent internal initiation events occur.

Proteolytic processing of the yellow fever structural proteins, including the non-structural glycoprotein NS1, appears to parallel in many ways the processing of the alphavirus structural proteins. The basic capsid protein is *N*-terminal; it is cleaved from the precursor after a series of uncharged amino acids, possibly by cellular signalase. There follow three glycoproteins that are apparently conducted into the membrane by internal (or *N*-terminal) signal sequences in the polyprotein and anchored in the membrane by *C*-terminal hydrophobic regions in the processed polypeptides; cleavages to separate the three glycoproteins may also be effected by signalase. The glycoprotein prM (also called NV2) is further processed during virus maturation in an event similar to that of alphavirus PE2 or of glycoproteins of influenza virus, paramyxoviruses, or retroviruses. These cleavages occur after the canonical sequence Arg-X-Arg/Lys-Arg (Dalgarno *et al.*, 1983), where X can vary, and may be effected by an enzyme associated with the Golgi complex that possesses a tryptic-like activity (reviewed in Strauss and Strauss, 1985). Cleavage of prM produces M, the "membrane-like" protein, that forms part of the flavivirion and is a small, non-glycosylated peptide of about 8 kD that probably traverses the lipid bilayer. The E protein forms the major antigenic spike of the flavivirion and is not always glycosylated. The function of NS1 (also called NV3) is unknown; because it is a glycoprotein and is found at least in part at the cell surface (J. Schlesinger, personal communication) it may be involved in virus maturation rather than RNA replication.

The remainder of the non-structural proteins follow. These proteins are presumably responsible for RNA replication. The two largest non-structural proteins, NS3 and NS5, are both fairly hydrophilic, carry net positive charges, and are presumably involved in RNA replication. NS5 contains at least one domain that is significantly homologous with the poliovirus RNA polymerase as well as non-structural proteins of other plant and animal RNA viruses, including the alphavirus nsP4 (Rice *et al.*, 1985a). Cleavages to separate the large non-structural proteins as well as several putative small proteins are all believed to occur following two basic residues in succession that are usually flanked by short-side-chain amino acids, often glycine or serine. Thus, it is necessary to postulate only one enzyme to carry out these cleavages, and we postulate that this protease is virus-encoded and resides in one of these seven non-structural polypeptides.

IX. COMPARISON OF FLAVIVIRUS SEQUENCES

We have obtained the sequence of approximately half of the genomic RNA of Murray Valley encephalitis virus (MVE) and compared this sequence with that of yellow fever RNA (Dalgarno et al., 1986). These two viruses represent two serological subgroups of the mosquito-borne flaviviruses, and the amino acid sequence homologies found between them are similar in many ways to those found between different alphaviruses. These sequence homologies are illustrated in Figure 6.16 in a moving-average plot with a window of 20 amino acids. The sequence homology is non-uniform. There are highly conserved domains and domains that are poorly conserved; and even at the level of complete proteins there are differences. E and NS1 exhibit 45% amino acid sequence similarity, whereas C and ns2a exhibit 27% similarity. In the E protein, which carries the major antigenic epitopes of flaviviruses, it is tempting to speculate that group cross-reactive epitopes are present in highly conserved domains and type-specific epitopes in poorly conserved domains. It is also of interest that the potential carbohydrate addition sites (of the type Asn-X-Ser/Thr and denoted by asterisks) in E are not conserved between yellow fever and MVE, and E is not always glycosylated, suggesting that glycosylation plays only a minor role in the function of E. In contrast, two glycosylation sites in NS1 and one glycosylation site in prM are conserved, suggesting that glycosylation is important for the function of these proteins.

Also shown in Figure 6.16 are hydrophobicity profiles of the yellow fever and MVE proteins, calculated by the methods of Kyte and Doolittle (1982). The hydrophobicity profiles for the two viruses are virtually superimposable, even in regions where the sequence homology is low, and the hydrophobicity or hydrophilicity of various domains as calculated in this fashion must therefore be important to the function of those domains. It is also of interest that the three glycoproteins, prM, E, and NS1, all possess hydrophobic domains at or near their C-termini that could function as membrane-spanning anchors and/or as signal sequences to insert the next protein in line. (It should be made clear, however, that we do not have C-terminal amino acid sequence of the various proteins and there may be interspersed sequences, analogous to the 6K protein of alphaviruses, positioned between any two proteins that are removed during processing.) Finally, the very hydrophobic nature of ns2a and ns2b is of interest. The function of the small polypeptides is unknown and, indeed,

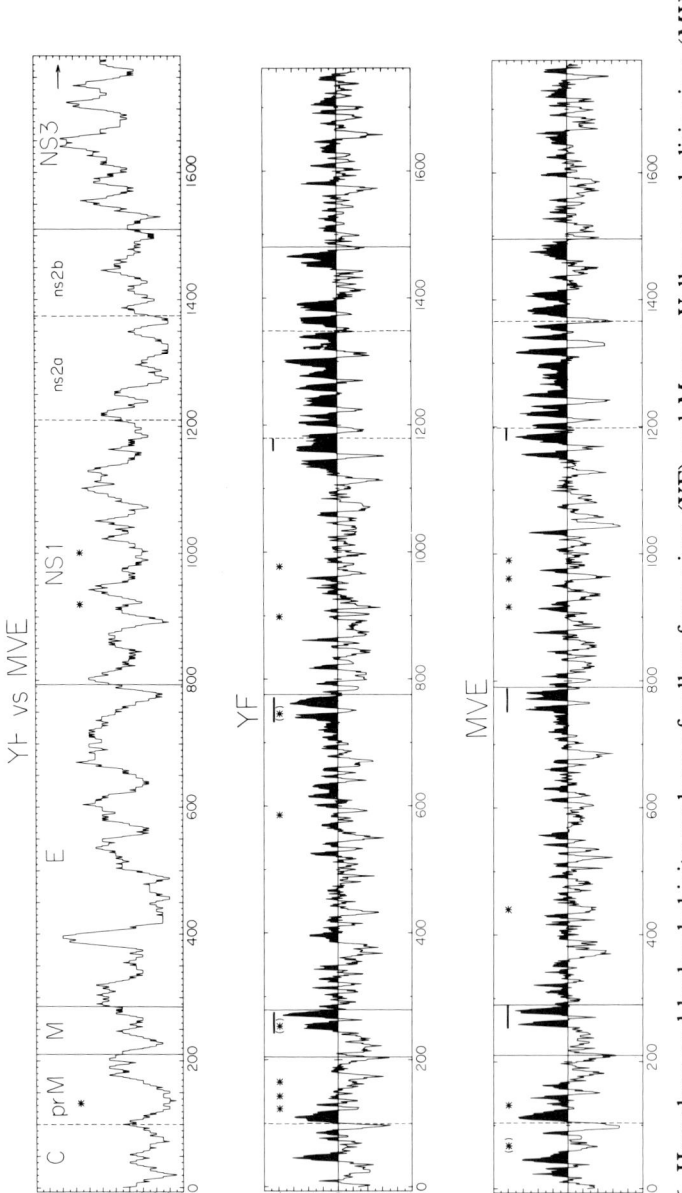

Figure 6.16. Homology and hydrophobicity analyses of yellow fever virus (YF) and Murray Valley encephalitis virus (MVE) proteins. The top panel shows a moving-average per cent homology plot of the aligned protein sequences using a window of 20 amino acids. The middle and bottom panels are hydrophobicity analyses of yellow fever and Murray Valley encephalitis proteins, respectively, using the program of Kyte and Doolittle (1982) and a search length of seven residues. Hydrophobic regions above the midline have been shaded. Putative hydrophobic, membrane-associated domains of the structural proteins and of NS1 are indicated by solid bars; potential N-linked glycosylation sites are denoted by asterisks; glycosylation sites conserved between YF and MVE are indicated on the homology profile. The nomenclature for the proteins is described in Rice et al. (1985a) and in Fig. 6.15. The arrow after NS3 indicates that not all of this protein sequence is available for comparison. Cleavage sites are indicated by solid vertical lines (confirmed sites) or broken vertical lines (tentative assignments). Data are from L. Dalgarno et al. (1986).

X. STRUCTURES IN THE YELLOW FEVER RNA

A number of possible secondary structures in the yellow fever genome have been identified, of which the one illustrated in Figure 6.17 is of particular interest. This secondary structure has a very large calculated free energy, -40 kcal for form 1 or -45.8 kcal for form 2. In form 2 the RNA structure is quite complex. One or both of these forms may exist in solution, given the potentially large free energy decrease associated with

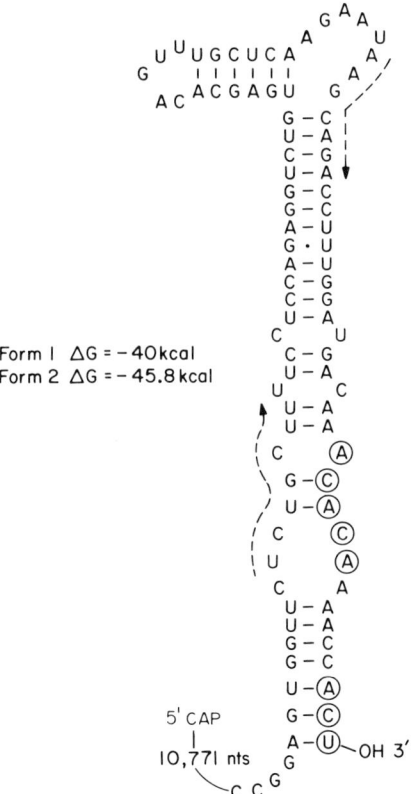

Figure 6.17. Possible secondary structures at the 3' terminus of yellow fever virus genomic RNA. Circled nucleotides are shared with the 3' terminus of the yellow fever ($-$) strand. ΔG values were calculated according to Tinoco et al. (1973). A more stable conformation than the one shown (form 1) can be formed if the two overlined sequences are base paired (form 2). (From Rice et al. (1985a) with permission of the American Association for the Advancement of Science.)

them. In both of these forms the 3'-terminal uridine is base-paired into the structure, which might account for the known difficulties in labelling of the 3' ends of flavivirus RNAs. This structure is probably involved in some way in RNA replication or packaging.

XI. CONCLUDING REMARKS

In this brief review, we have tried to summarize much of the new information about both alphaviruses and flaviviruses that arises from the sequencing being undertaken in our laboratory and in other laboratories. This information tells us not only about the details of the genome structure and translation strategy of these viruses but also much about the evolution of RNA viruses, and it opens up new avenues for further study and for vaccine development.

ACKNOWLEDGEMENTS

We are grateful to our colleagues Drs K. Takkinen, A. Schmaljohn, S. Schlesinger, H. Huang, G. Cleaves, C. Blair, and J. Schlesinger for sharing information with us prior to its publication. The work of the authors is supported by Grants AI 10793 and AI 20612 from the National Institutes of Health, and Grant DMB83-16856 from the National Science Foundation. C. S. H. is supported by a fellowship from the California Foundation for Biomedical Research.

REFERENCES

1. Ahlquist, P., R. Dasgupta and P. Kaesberg (1984). *J. Mol. Biol.* **172**, 369–383.
2. Ahlquist, P., E. G. Strauss, C. M. Rice, J. H. Strauss, J. Haseloff and D. Zimmern (1985). *J. Virol.* **53**, 536–542.
3. Aliperti, G. and M. Schlesinger (1978). *Virology* **90**, 366–369.
4. Bell, J. R., R. M. Kinney, D. W. Trent, E. G. Strauss and J. H. Strauss (1984). *Proc. Natl Acad. Sci. USA* **81**, 4702–4706.
5. Bhatti, A. R. and J. Weber (1979). *Virology* **96**, 478–485.
6. Boege, U., G. Wengler, G. Wengler and B. Wittman-Liebold (1981). *Virology* **113**, 293–303.
7. Cleaves, G. R. and D. T. Dubin (1979). *Virology* **96**, 159–165.
8. Clegg, J. C. S. and S. I. T. Kennedy (1974). *J. Gen. Virol.* **22**, 331–345.
9. Collins, P. L., F. J. Fuller, P. I. Marcus, L. E. Hightower and L. A. Ball (1982). *Virology* **118**, 363–379.
10. Cornelissen, B. J. C., F. T. Brederode, R. J. M. Moormann and J. F. Bol (1983a). *Nucl. Acids Res.* **11**, 1253–1265.
11. Cornelissen, B. J. C., F. T. Brederode, G. H. Veeneman, J. H. van Boom and J. F. Bol (1983b). *Nucl. Acids Res.* **11**, 3019–3025.
12. Dalgarno, L., C. M. Rice and J. H. Strauss (1983). *Virology* **129**, 170–187.
13. Dalgarno. L., D. W. Trent, J. H. Strauss, and C. M. Rice (1986). *J. Mol. Biol.* **187**, 309–323.
14. Dayhoff, M. O. (Ed.) (1978). "Atlas of Protein Sequence and Structure" Vol.

5, Suppl. 3, pp. 79–83. National Biomedical Research Foundation, Washington, D.C.
15. Diamond A., B. Dudock and D. Hatfield (1981). *Cell* **25**, 479–506.
16. Domingo, E., D. Sabo, T. Taniguchi and C. Weissmann (1978). *Cell* **13**, 735–744.
17. Dubin, D. T., V. Stollar, C.-C. HsuChen, K. Timko and G. M. Guild (1977). *Virology* **77**, 457–470.
18. Eaton, B. T. and P. Faulkner (1972). *Virology* **50**, 865–873.
19. Frey, T. K. and J. H. Strauss (1978). *Virology* **86**, 494–506.
20. Garoff, H., A.-M. Frischauf, K. Simons, H. Lehrach and H. Delius (1980a). *Proc. Natl Acad. Sci. USA* **77**, 6376–6380.
21. Garoff, H., A.-M. Frischauf, K. Simons, H. Lehrach and H. Delius (1980b). *Nature (London)* **288**, 236–241.
22. Geller, A. I. and A. Rich (1980). *Nature (London)* **283**, 41–46.
23. Goelet, P., G. P. Lomonossoff, P. J. G. Butler, M. E. Akam, M. J. Gait and J. Karn (1982). *Proc. Natl Acad. Sci. USA* **79**, 5818–5822.
24. Greer, J. (1981). *J. Mol. Biol.* **153**, 1027–1042.
25. Hahn, C. S., E. G. Strauss and J. H. Strauss (1985). *Proc. Natl Acad. Sci. USA* **82**, 4648–4652.
26. Haseloff, J., P. Goelet, D. Zimmern, P. Ahlquist, R. Dasgupta and P. Kaesberg (1984). *Proc. Natl Acad. Sci. USA* **81**, 4358–4362.
27. Hatfield, D. L., B. S. Dudock and F. C. Eden (1983). *Proc. Natl Acad. Sci. USA* **80**, 4940–4944.
28. Hefti, E., D. H. L. Bishop, D. T. Dubin and V. Stollar (1976). *J. Virol.* **17**, 149–159.
29. Hirsch, D. and L. Gold (1971). *J. Mol. Biol.* **58**, 459–468.
30. Holland, J., K. Spindler, F. Horodyski, E. Grabau, S. Nichol and S. Van de Pol (1982). *Science* **215**, 1577–1585.
31. Hsu Chen, C.-C. and D. T. Dubin (1976). *Nature (London)* **264**, 190–191.
32. Huth, A., T. A. Rapoport and L. Kääriäinen (1984). *EMBO J.* **3**, 767–771.
33. Jany, K. D., K. Bekelar, G. Pfleiderer and J. Ishay (1983). *Biochem. Biophys. Res. Commun.* **110**, 1–7.
34. Johnston, R. E. and H. R. Bose (1972). *Biochem. Biophys. Res. Commun.* **46**, 712–718.
35. Kamer, G. and P. Argos (1984). *Nucl. Acids Res.* **12**, 7269–7282.
36. Keranen, S. and L. Ruohonen (1983). *J. Virol.* **47**, 505–515.
37. Kramer, K. J., R. L. Felsted and J. H. Law (1973). *J. Biol. Chem.* **248**, 3021–3028.
38. Kyte, J. and R. F. Doolittle (1982). *J. Mol. Biol.* **157**, 105–132.
39. Levis, R., B. G. Weiss, M. Tsiang, H. Huang, and S. Schlesinger (1986). *Cell* **44**, 137–145.
40. Lopez, S., J. R. Bell, E. G. Strauss and J. H. Strauss (1985). *Virology* **141**, 235–247.
41. O'Neill, V. A., F. C. Eden, K. Pratt and D. L. Hatfield (1985). *J. Biol. Chem.* **260**, 2501–2508.
42. Ou, J.-H., E. G. Strauss and J. H. Strauss (1981). *Virology* **281**, 281–289.
43. Ou, J.-H., C. M. Rice, L. Dalgarno, E. G. Strauss and J. H. Strauss (1982a). *Proc. Natl Acad. Sci. USA* **79**, 5235–5239.
44. Ou, J.-H., D. W. Trent and J. H. Strauss (1982b). *J. Mol. Biol.* **156**, 719–730.
45. Ou, J.-H., E. G. Strauss and J. H. Strauss (1983). *J. Mol. Biol.* **168**, 1–15.

46. Palmenberg, A. A., M. A. Pallansch and R. R. Rueckert (1979). *J. Virol.* **32**, 770–778.
47. Pelham, H. R. B. (1978). *Nature (London)* **272**, 469–471.
48. Pfefferkorn, E. R. and M. K. Boyle (1972). *J. Virol.* **9**, 187–188.
49. Pratt, K., F. C. Eden, H. Y. Kwang, V. A. O'Neill and D. Hatfield (1985). *Nucl. Acids Res.* **13**, 4765–4775.
50. Rice, C. M. and J. H. Strauss (1981). *Proc. Natl Acad. Sci. USA* **78**, 2062–2066.
51. Rice, C. M., E. M. Lenches, S. R. Eddy, S. J. Shin, R. L. Sheets and J. H. Strauss (1985a). *Science*, **229**, 726–733.
52. Rice, C. M., C. A. Franke, J. H. Strauss and D. E. Hruby (1985b). *J. Virol.* **56**, 227–239.
53. Riedel, H., H. Lehrach and H. Garoff (1982). *J. Virol.* **42**, 725–729.
54. Schlesinger, M. J. and L. Kaariainen (1980). *In* "The Togaviruses" (Ed. R. W. Schlesinger), pp. 371–392. Academic Press, London and Orlando.
55. Scupham, R. K., K. J. Jones, B. P. Sagik and H. R. Bose Jr. (1977). *J. Virol.* **22**, 568–571.
56. Simmons, D. T. and J. H. Strauss (1972). *J. Mol. Biol.* **71**, 615–631.
57. Simmons, D. T. and J. H. Strauss (1974). *J. Mol. Biol.* **86**, 397–409.
58. Strauss, E. G. and J. H. Strauss (1983). *Current Topics Microbiol. Immunol.* **105**, 1–98.
59. Strauss, E. G. and J. H. Strauss (1985). *In* "Virus Structure and Assembly" (Ed. S. Casjens), pp. 205–234. Jones and Bartlett Publishers, Portola Valley.
60. Strauss, E. G., C. M. Rice and J. H. Strauss (1983). *Proc. Natl Acad. Sci. USA* **80**, 5271–5275.
61. Strauss, E. G., C. M. Rice and J. H. Strauss (1984). *Virology* **133**, 92–110.
62. Takkinen, K. (1986). *Nuc. Acids Res.* **14**, 5667–5682.
63. Tinoco, I. Jr., P. N. Borer, B. Dengler, M. D. Levine, O. C. Uhlenbeck, D. M. Crothers and J. Gralla (1973). *Nature (New Biol.)* **246**, 40–41.
64. von der Helm, K. (1977). *Proc. Natl Acad. Sci. USA* **74**, 911–915.
65. Wecker, E. (1959). *Virology* **7**, 241–243.
66. Weiner, A. M. and K. Weber (1971). *Nature (New Biol.)* **234**, 206–209.
67. Wengler, G. and G. Wengler (1981). *Virology* **133**, 544–555.
68. Wengler, G., G. Wengler and H. J. Gross (1978). *Virology* **89**, 423–437.
69. Westaway, E. G. (1977). *Virology* **80**, 320–335.
70. Westaway, E. G., G. Speight and L. Endo (1984). *Virus Res.* **1**, 333–350.

7. Replication of Equine Arteritis Virus (EAV): A Comparative Review

Mario F. van Berlo, Willy J. M. Spaan and Marian C. Horzinek

Institute of Virology, Veterinary Faculty, State University, Yalelaan 1, 3506 TD Utrecht, The Netherlands

I. Introduction	105
II. The proteins of EAV	107
III. The RNAs of EAV	108
IV. Strategies of replication	112
References	114

I. INTRODUCTION

Equine arteritis virus (EAV) is a small, enveloped, positive-stranded RNA virus. The virion has a diameter of 60 ± 13 nm and consists of an isometric core of about 35 nm that is surrounded by an envelope carrying ring-like subunits (Horzinek, 1981). On the basis of the morphology, substructure and size of the virion and the properties of the RNA genome, EAV was originally classified as a non-arthropod-borne member of the family Togaviridae, not belonging to one of the existing genera (Horzinek, 1973a,b); this classification has been approved by the ICTV (Porterfield *et al.*, 1978).

The non-arbo togaviruses are structurally similar to their arthropod-borne relatives; examples are rubella virus, bovine viral diarrhoea virus (BVDV) (and the closely related border disease virus of sheep), hog cholera virus, lactic dehydrogenase virus (LDV), EAV and simian haemorrhagic fever virus (SHFV). Further subdivision was made on the basis of serologic relationships and resulted in the establishment of the two additional genera rubivirus and pestivirus (Porterfield 1980).

At that time the strategy of replication had been elucidated primarily for members of the alphavirus genus. Subsequently, additional molecular data on the other genera became available which showed that there exist fundamental differences in replication; these are summarized below, because they provide further criteria for togavirus classification.

Alphaviruses. Alphaviruses produce two mRNAs after infection. One is identical to the virion RNA, and its 5' part is translated into the nonstructural (polymerase) proteins involved in virus replication. The second molecule is a subgenomic mRNA identical to the 3'-terminal third of the genomic RNA; it is translated into the structural proteins of the virion (Strauss and Strauss, 1983).

Flaviviruses. The translation strategy of the flaviviruses is not fully understood. Subgenomic RNA has not been found in infected cells and the general opinion is that the viral genome is the only messenger. Recently, the entire genome of yellow fever virus has been sequenced (Rice et al., 1985); the sequence reveals a single open reading frame (ORF) of 10,233 nucleotides, which could encode a polypeptide of 3411 amino acids. The structural proteins are found within the amino-terminal 780 residues of this polyprotein. The 5' location of the genes encoding the structural proteins, the single long ORF, and the lack of a subgenomic message are characteristics shared with picornaviruses rather than with togaviruses (for a review see Strauss and Strauss, 1983 and this volume, Chapter 6).

Rubiviruses. The replication strategy of rubella virus is very similar to that of the alphaviruses (Oker-Blom et al., 1983; Kalkkinen et al., 1984). A genomic 40S RNA and a subgenomic 24S RNA, identical to the 3' end of the genomic RNA was isolated from infected cells. This subgenomic RNA encodes a precursor (p110) to the structural proteins (Oker-Blom et al., 1984).

Pestiviruses. The replication strategy of the pestiviruses has not been extensively investigated; only BVDV has been studied by Purchio et al. (1983, 1984a,b). In infected cells a single species of virus-specific RNA with a molecular weight of 2.9 MD has been found. Translation *in vitro* of this RNA resulted in the synthesis of several polypeptides, which could be immunoprecipitated using a virus-specific antiserum. None of the polypeptides synthesized *in vitro* appeared to comigrate with authentic viral proteins immunoprecipitated from infected cells. Recently, the genome of BVDV has been cloned; its sequence reveals two non-overlapping ORFs (Renard et al., 1985).

The fundamentally different replication mechanisms have brought the Togavirus Study Group of the International Committee on Taxonomy of Viruses to review the family assignment of its viruses (Westaway et al., 1985). Changes were proposed that resulted in the creation of the new family Flaviviridae (prototype: yellow fever virus) and of a new genus *Arterivirus* (prototype: EAV) in the family Togaviridae (Table 7.1). The Utrecht group has provided most data leading to the establishment of the genus *Arterivirus*; we therefore present a short review of the experimental results which have sparked this taxonomic initiative.

Table 7.1. Current classification of viruses formerly accommodated in the family Togaviridae

Family	Genus	Prototype	Number of viruses
Flaviviridae	*Flavivirus*	Yellow fever virus	64
Togaviridae	*Alphavirus*	Sindbis virus	26
	Rubivirus	Rubella virus	1
	Pestivirus	Mucosal disease/ bovine virus diarrhea	2
	Arterivirus	Equine arteritis virus	1
	not assigned to genera:	Lactic dehydrogenase virus Simian haemorrhagic fever virus	

II. THE PROTEINS OF EAV

Analysis of the protein composition of EAV using radioactively labelled amino acids revealed nine virus-associated proteins, two of which (15 kD and 13 kD) contained most of the label. The latter, an unglycosylated protein associated with the core fraction was thought to represent the nucleoprotein of EAV (Hyllseth, 1973). Zeegers et al. (1976) compared the protein patterns of virus grown in different cell systems. Only three proteins with molecular weights of 12 kD (VP1), 14 kD (VP2) and 21 kD (VP3) were shown to be virus-specific. VP1 is a major phosphorylated core protein, and its molecular weight is close to that calculated by Hyllseth (1973) for the nucleoprotein. The other major 14 kD (VP2) protein, as described by Zeegers et al. (1976), is removed by detergent treatment together with the envelope but does not incorporate glucosamine; this author found only one glycoprotein, VP3 (21 kD), in virus preparations from all three cell systems used. The higher molecular weight glycoproteins observed by Hyllseth (1973) were not well resolved and migrated as a complex within a molecular weight range of 28–40 kD; they are probably cellular contaminants (Fig. 7.1A).

The structural proteins of EAV are of lower molecular weight than those of the other togaviruses. Only LDV, which has been considered as another member of the Togaviridae family on morphological grounds (Horzinek et al., 1975), has polypeptides in the same range (Michaelides and Schlesinger, 1973; Brinton-Darnell and Plagemann, 1975). However, no antigenic relationship at the level of the envelope and nucleocapsid proteins was detected (Van Berlo et al., 1983), confirming that EAV and LDV belong to separate clusters.

Pulse-chase experiments indicated that the viral proteins do not arise by processing from larger precursors. In addition to the viral structural proteins (VP1 and VP2) three polypeptides with molecular weights of 60 kD, 42 kD and 30 kD were found in membrane-containing fractions from infected but not from mock-infected cells (Fig. 7.1B). The glycosylated

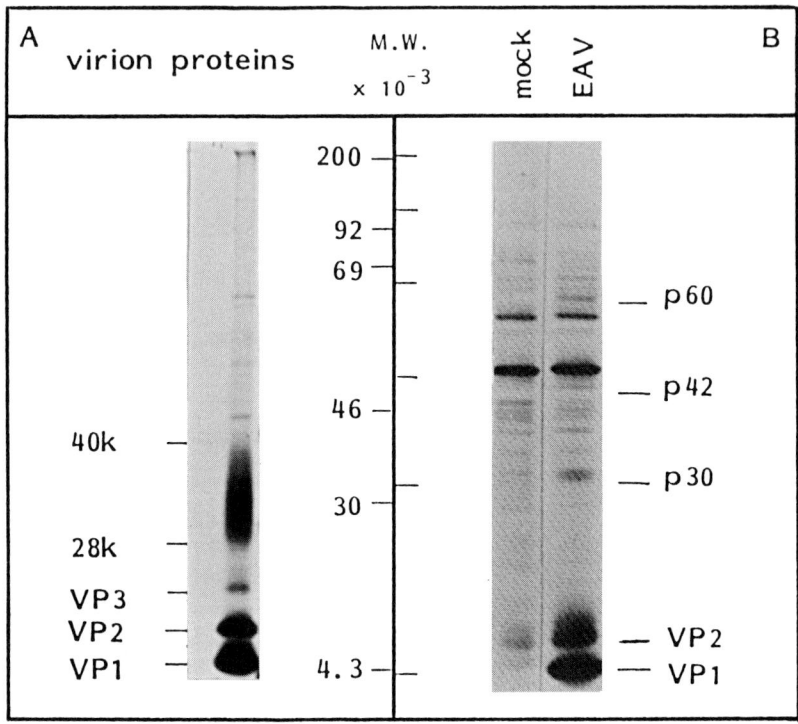

Figure 7.1. Polyacrylamide gel electrophoresis of ^{35}S-methionine labelled polypeptides of equine arteritis virus grown in BHK-21 cells. Virion proteins (A) and intracellular polypeptides (B) after 1 h pulse (7–8 h p.i.) are seen in infected and mock-infected cells, respectively.

envelope protein (VP3) could not be demonstrated directly in infected cells (Van Berlo et al., 1986a).

III. THE RNAs OF EAV

The genome of EAV consists of an infectious, colinear molecule of single stranded RNA (ssRNA), with an $s_{20,w}$ value of 48S and a molecular weight of about 4 MD, as determined by polyacrylamide/agarose gel electrophoresis (Van der Zeijst et al., 1975).

Analysis of the RNAs synthesized in EAV-infected BHK cells revealed the existence of six virus-specific species with molecular weights of 4.3 MD (RNA1, comigrating with the viral genome), 1.3 MD (RNA2), 0.9 MD (RNA3), 0.7 MD (RNA4), 0.3 MD (RNA5) and 0.2 MD (RNA6) (Van Berlo et al., 1982; Fig. 7.2). All RNA species are polyadenylated; translation of the unfractionated polyadenylated RNAs in an mRNA-

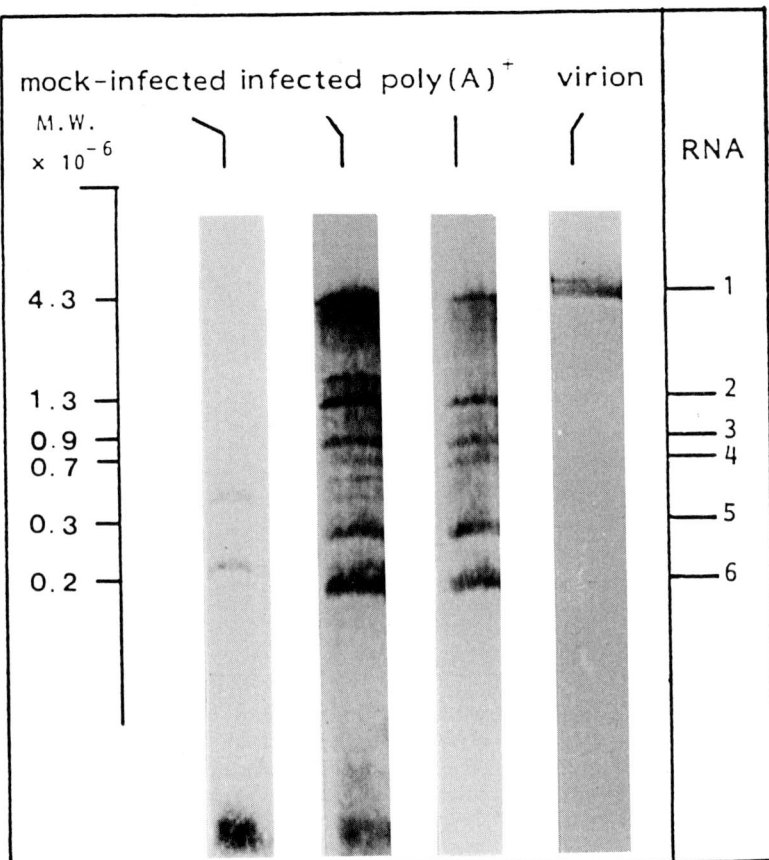

Figure 7.2. Agarose gel electrophoresis of RNAs from equine arteritis virus-infected cells and mock-infected BHK-cells; RNA from purified virions was analysed for comparison. RNA was labelled with ^{3}H-uridine in the presence of actinomycin D. Polyadenylated RNA from EAV-infected cells was purified by oligo(dT)-cellulose chromatography.

dependent reticulocyte cell-free system gave rise to the nucleocapsid protein (VP1) and to a 30 kD protein (Fig. 7.3A).

Of the fractionated molecules, only RNA6 could be translated; its translation product corresponded with the nucleocapsid protein (VP1) (Fig. 7.3B). This observation indicates that the nucleocapsid protein is a primary translation product (Van Berlo et al., 1986a). The sum of the molecular weights of the subgenomic RNAs amounts to 3.4 MD, which is about 20% less than the molecular weight of the genome. The information in the subgenomic RNAs could therefore be adjacent in the genome. On the other hand, the subgenomic RNAs might also contain common

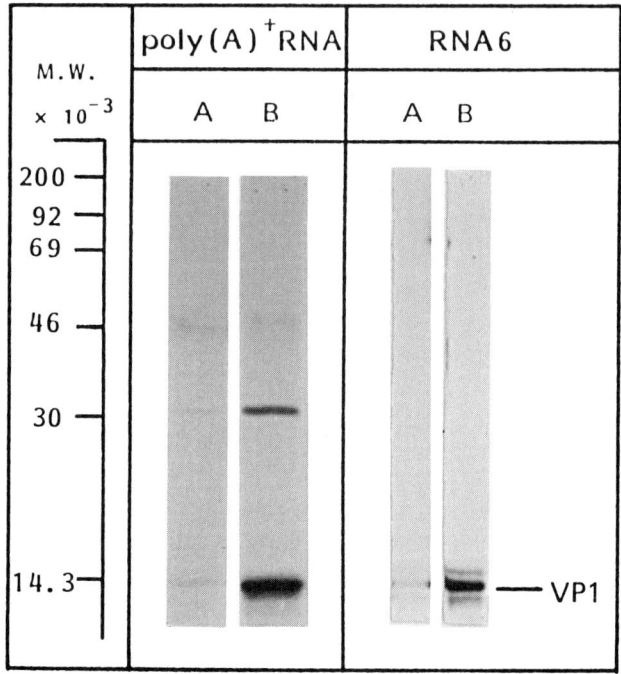

Figure 7.3. Analysis of the translation products *in vitro* of EAV poly(A)-selected RNA and RNA6, respectively. RNA isolated from EAV infected BHK-cells was added to a rabbit reticulocyte system, in the presence of ^{35}S-methionine. Samples of the lysates were immunoprecipitated using hamster sera obtained before (A) and after (B) immunization with EAV, while a rabbit anti-hamster serum was employed for precipitation of the immune complexes.

sequences, as has been found for example in alphaviruses, coronaviruses and several plant viruses (for a review see Strauss and Strauss, 1983).

To obtain a first impression of the sequence relationships between the EAV-specific RNAs, a Northern blot analysis was carried out using a radiolabelled DNA probe complementary to RNA6. It hybridized with all the RNAs indicating that they possess common sequences; the experiment also demonstrates the virus-specificity of the subgenomic RNAs (Van Berlo *et al.*, 1986b).

Using this cDNA probe, homology between the subgenomic RNAs of EAV was established. However, this technique does not allow one to decide whether the positive signals are due to small stretches or extensive regions of homology. We therefore compared the subgenomic molecules by RNase T1 oligonucleotide fingerprinting. ^{32}P-labelled RNA was isolated from infected cells and the molecules were separated by agarose-urea gel electrophoresis after poly(A) selection. The purified EAV RNAs were

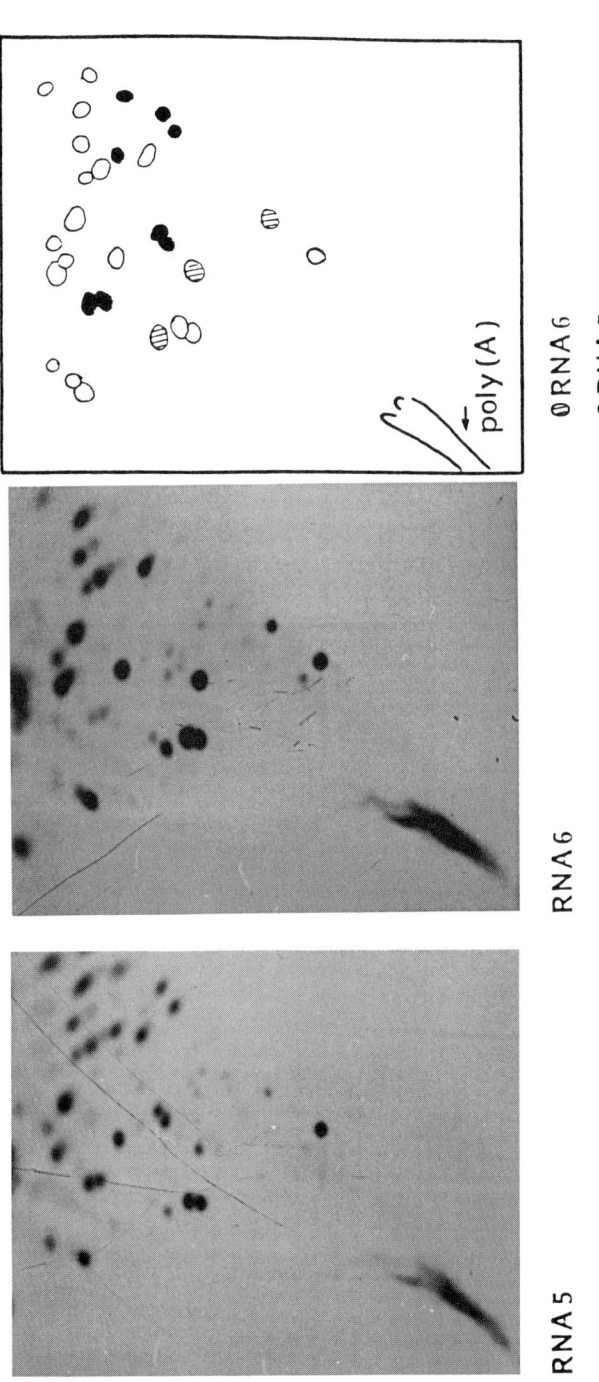

Figure 7.4. RNase T1 oligonucleotide fingerprints of the subgenomic RNAs 5 and 6 of EAV. ^{32}P-labelled intracellular RNAs 5 and 6 were isolated and digested with ribonuclease T1. The products were then subjected to two-dimensional gel electrophoresis and autoradiography. In these fingerprints, electrophoresis was from left to right in the first dimension and from bottom to top in the second dimension. In the diagram, the solid dots indicate the oligonucleotides present in RNA5 that do not occur in RNA6; the hatched spots indicate the oligonucleotides of RNA6 that do not show in RNA5 or show only weakly.

Table 7.2. RNase T1-resistant oligonucleotides in subgenomic RNAs of EAV as compared with the expected frequencies (Beemon, 1978)

	Molecular weight		T1-spots Chain length > 15 nucleotides	
	(MD)	(kb)	Found	Expected
RNA1	4.3	13.2	19	44
RNA2	1.3	4.0	13	13
RNA3	0.9	2.8	9	9
RNA4	0.7	2.2	—	7
RNA5	0.3	0.9	3	3
RNA6	0.2	0.6	4	2

digested with RNAase T1 and the resistant oligonucleotides were separated by electrophoresis in two dimensions. The fingerprints, containing the unique spots of RNA 5 and 6 are shown in Fig. 7.4.; these results support the hypothesis that the intracellular viral RNAs form a nested set. However, some anomalies concerning discrete oligonucleotides were observed; we interpret these as sequence rearrangements, which may occur during the mRNA synthesis (Van Berlo et al., 1986b). The collection of spots found in the fingerprints of the subgenomic RNAs are roughly in agreement with their expected frequency (Beemon, 1978) (Table 7.2). In contrast, RNA1 should have contained about three times the number of oligonucleotides of RNA2. However, instead of about 44 spots only 19 spots were found. This could be due to a high G-content of RNA1, but other explanations (such as the occurrence of polyploid sequences) are possible.

Although the RNAs seem to form a nested set, we do not know yet their positions on the genome. Restriction enzyme analyses of cDNA clones suggest a common 3' terminal end.

IV. STRATEGIES OF REPLICATION

The existence of subgenomic messenger RNAs prompted us to compare the translation and transcription strategy of EAV with that of other positive- and single-stranded RNA viruses whose genes are expressed with the aid of subgenomic RNAs. The subgenomic (26S) RNAs of alphaviruses (Kennedy, 1980) and rubella virus (Oker-Blom et al., 1984) are translated into continuous polypeptide chains, which are further processed to bring forth the viral proteins. The expression of the subgenomic RNAs of coronaviruses (Siddell et al., 1983) and of some plant viruses (Joshi and Haenni, 1984) suggests some similarities with the translation strategy that EAV uses. These subgenomic RNAs form a 3' coterminal nested set and only the 5' terminal parts, which are absent in the next smaller RNAs, are translated. Translation *in vitro* of RNA6 of EAV resulted in the synthesis of a 14 kD protein. As discussed above, RNA5 must be regarded as an extension of RNA6; in that case the unique region of RNA5 would be only 0.3 kb in length which is insufficient to encode any of the known proteins of EAV. These results suggest that overlapping reading frames exist in both RNAs.

Of central importance to the study of gene expression are the mechanisms by which mRNAs are produced. Upon infection a negative-stranded RNA is synthesized from the positive-stranded virion RNA, which serves as the template for novel positive-strand RNA synthesis. At present two mechanisms are known for positive-stranded RNA viruses by which subgenomic RNAs can arise: (i) internal initiation by the RNA-polymerase on the negative strand of genomic RNA (e.g. alphaviruses, Kennedy, 1980; rubella virus, Oker-Blom *et al.*, 1984); (ii) fusion of the leader and body sequences, which are non-contiguous in the genome and are joined in the cytoplasm (e.g. coronaviruses, Spaan *et al.*, 1983; Lai *et al.*, 1984). A number of plant viruses have structural features in common with plus-sense animal viruses and they also share similarities in their replication strategies (Haseloff *et al.*, 1984; Ahlquist *et al.*, 1985). They also use different ways to produce subgenomic messenger RNAs (Joshi and Haenni, 1984), one of them involving premature termination during negative strand synthesis, followed by independent replication of the subgenomic negative strand (Goelet and Karn, 1982). Nucleolytic cleavage has been suggested as another possibility for subgenomic RNA production (Gonda and Symons, 1979).

Further research has focused on the mechanism by which the EAV subgenomic RNAs are synthesized. UV-target sizes for the templates of the EAV specific RNAs were determined and compared to those of mouse hepatitis virus, a coronavirus (Table 7.3). The values for RNAs 2–5 were rather uniform and very close to the physical size of RNA1 (4.3 MD). Only

Table 7.3.

Equine arteritis virus			*Mouse hepatitis virus*	
Target size of template	RNA size	RNA	RNA size	Target size of template
(4.3)	4.3	1	5.6	(5.6)
4.5	1.3	2	4.0	4.0
4.8	0.9	3	3.0	3.0
3.8	0.7	4	1.4	1.4
4.3	0.3	5	1.2	1.2
2.8	0.2	6	0.97	0.9
—	—	7	0.75	0.6

the target size of the template of RNA6 was smaller (2.8 MD), although still much larger than its physical size. The data are consistent with a model in which the individual RNAs are derived from a larger precursor RNA molecule and exclude the possibility that these RNAs are synthesized via independent initiation on one or more template molecules, as described earlier for coronaviral subgenomic RNAs (Jacobs et al., 1981; Stern et al., 1982). In this system the UV target sizes were almost identical with the physical sizes of the RNAs (Table 7.3).

The mechanism of EAV mRNA synthesis is unique among the positive-stranded animal RNA viruses; mRNAs probably arise by nucleolytic cleavage of a genome-sized RNA. Sequence rearrangements can occur if several cleavage products are fused. Although we have no direct evidence for splicing, the anomalies detected in the T1 fingerprints of the individual mRNAs suggest that rearrangements do exist. However, splicing has only been found to occur in the nucleus and in EAV replication nuclear functions have not yet been identified. Although a universal mechanism does not seem to be involved in RNA splicing, highly conserved sequences located at the intron/exon junctions are required for accurate and efficient splicing (Rogers, 1985).

Data on the nucleotide sequence of the genome and of the subgenomic RNAs are needed to unravel the details of the transcription of EAV.

NOTE ADDED IN PROOF

We have shown recently by sequence analysis and hybridization studies that the intracellular RNAs and the genome of EAV share a common 3' region (Spaan et al., manuscript in preparation).

REFERENCES

1. Ahlquist, P., E. G. Strauss, C. M. Rice, J. H. Strauss, J. Haseloff and D. Zimmern (1985). *J. Virol.* **53**, 536–542.
2. Beemon, K. L. (1978). *Current Topics Microbiol. Immunol.* **79**, 73–110.
3. Brinton-Darnell, M. B. and P. G. W. Plagemann (1975). *J. Virol.* **16**, 420–433.
4. Goelet, P. and J. Karn (1982). *J. Mol. Biol.* **154**, 541–550.
5. Gonda, T. J. and R. H. Symons (1979). *J. Gen. Virol.* **45**, 723–736.
6. Haseloff, J., P. Goelet, D. Zimmern, P. Ahlquist, R. Dasgupta and P. Kaesberg (1984). *Proc. Natl Acad. Sci. USA.* **81**, 4358–4362.
7. Horzinek, M. C. (1973a). *Prog. Med. Virol.* **16**, 109–156.
8. Horzinek, M. C. (1973b). *J. Gen. Virol.* **20**, 87–103.
9. Horzinek, M. C. (1981). "Non-Arthropod Borne Togaviruses." Academic Press, London and Orlando.
10. Horzinek, M. C., P. S. van Wielink and D. J. Ellens (1975). *J. Gen. Virol.* **26**, 217–226.
11. Hyllseth, B. (1973). *Arch. Ges. Virusforsch.* **40**, 177–188.
12. Jacobs, L., W. J. M. Spaan, M. C. Horzinek and B. A. M. van der Zeijst (1981). *J. Virol.* **39**, 401–406.
13. Joshi, S. and A. Haenni (1984). *FEBS Lett.* **177**, 163–174.

14. Kalkkinen, N., C. Oker-Blom and R. F. Petterson (1984). *J. Gen. Virol.* **65**, 1549–1557.
15. Kennedy, S. I. T. (1980). In "The Togaviruses." (Ed. R. W. Schlesinger), pp. 351–369. Academic Press, London and Orlando.
16. Lai, M. M., R. S. Baric, P. R. Brayton and S. A. Stohlman (1984). *Proc. Natl Acad. Sci. USA.* **81**, 3626–3630.
17. Michaelides, M. C. and S. Schlesinger (1973). *Virology* **55**, 211–217.
18. Oker-Blom, C., N. Kalkkinen, L. Kääriäinen and R. F. Petterson (1983). *J. Virol.* **46**, 964–973.
19. Oker-Blom, C., I. Ulmanen, L. Kääriäinen and R. F. Petterson (1984). *J. Virol.* **49**, 403–408.
20. Porterfield, J. S. (1980). In "The Togaviruses" (Ed. R. W. Schlesinger), pp. 13–46. Academic Press, London and Orlando.
21. Porterfield, J. S., J. Casals, M. P. Chumakov, S. Ya. Gaidamovich, C. Hannoun, I. H. Holmes, M. C. Horzinek, M. Mussgay, N. Oker-Blom, P. K. Russel and D. W. Trent (1978). *Intervirology* **9**, 129–148.
22. Purchio, A. F., R. Larson and M. S. Collett (1983). *J. Virol.* **48**, 320–324.
23. Purchio, A. F., R. Larson and M. S. Collett (1984a). *J. Virol.* **50**, 666–669.
24. Purchio, A. F., R. Larson, L. L. Torburg and M. S. Collett (1984b). *J. Virol.* **52**, 973–975.
25. Renard, A., C. Guiot, D. Schmetz, L. Dagenais, P. P. Pastoret, D. Dina and J. Martial (1985). *DNA* **4**, 429–438.
26. Rice, C. M., E. M. Lenches, S. R. Eddy, S. J. Shin, R. L. Sheets and J. H. Strauss (1985). *Science* **229**, 726–733.
27. Rogers, J. H. (1985). *Int. Rev. Cytol.* **93**, 187–231.
28. Siddell, S., H. Wege and V. ter Meulen (1983). *J. Gen. Virol.* **64**, 761–776.
29. Spaan, W., H. Delius, M. Skinner, J. Armstrong, P. Rottier, S. Smeekens, B. A. M. van der Zeijst and S. G. Siddell (1983). *EMBO J.* **2**, 1839–1844.
30. Stern, D. F. and B. M. Sefton (1982). *J. Virol.* **42**, 755–759.
31. Strauss, E. G. and J. H. Strauss (1983). *Current Topics Microbiol. Immunol.* **105**, 1–98.
32. Van Berlo, M. F., M. C. Horzinek and B. A. M. van der Zeijst (1982). *Virology* **118**, 345–352.
33. Van Berlo, M. F., J. J. W. Zeegers, M. C. Horzinek and B. A. M. van der Zeijst (1983). *Zbl. Vet. Med. B* **30**, 297–304.
34. Van Berlo, M. F., P. J. M. Rottier, W. J. M. Spaan and M. C. Horzinek (1986a). *J. Gen. Virol.* **67**, 1543–1549.
35. Van Berlo, M. F., P. J. M. Rottier, M. C. Horzinek and B. A. M. van der Zeijst (1986b). *Virology* **152**, 492–496.
36. Van der Zeijst, B. A. M., M. C. Horzinek and V. Moennig (1975). *Virology* **68**, 418–425.
37. Westaway, E. G., M. A. Brinton, S. Ya. Gaidamovich, M. C. Horzinek, A. Igarashi, L. Kääriäinen, D. K. Lvov, J. S. Porterfield, P. K. Russell and D. W. Trent (1985). *Intervirology* **24**, 125–139.
38. Zeegers, J. J. W., B. A. M. van der Zeijst and M. C. Horzinek (1976). *Virology* **73**, 200–205.

8. The Organization and Expression of Coronavirus Genomes

Stuart Siddell

Institute of Virology, Versbacher Straße 7, 8700 Würzburg, Federal Republic of Germany

---

I. Introduction	117
II. Organization and expression	118
A. Structural analysis of viral RNAs	118
B. Translation products of viral mRNAs	119
C. Sequence analysis	120
1. IBV	120
2. MHV	120
III. Regulation	121
References	125

I. INTRODUCTION

The Coronaviridae are a family of enveloped, positive-stranded RNA viruses that infect vertebrates and cause diseases of considerable economic importance. The family is comprised of 11 members of which two, infectious bronchitis virus (IBV) and murine hepatitis virus (MHV), have been used extensively for biochemical studies (Siddell *et al.*, 1983a). In this article, I shall review the available information on the organization and expression of the IBV and MHV genomes. Several comprehensive reviews on coronavirus structure and replication have been published recently (Siddell *et al.*, 1983b; Sturman and Holmes, 1983; Holmes, 1985; Dubois-Dalcq *et al.*, 1984).

The coronavirus genome is a monopartite, positive-stranded RNA of 15–20 kb, which is capped, polyadenylated and infectious (Siddell *et al.*, 1983b). In the virion the genomic RNA is associated with a basic phosphoprotein, N, to form a helical nucleocapsid. This nucleocapsid structure lies within a lipoprotein envelope that contains a largely integral, transmembrane glycoprotein, M, (Armstrong et al., 1984a; Rottier et al., 1984) and a peripheral glycoprotein, S, that assembles to form the surface structures of the virion (Cavanagh, 1983). The replication of coronaviruses is cytoplasmic and can occur in enucleate cells (Brayton et al., 1981; Wilhelmsen et al., 1981) or in cells treated with actinomycin D or α-amanitin (Lai et al., 1981; Mahy et al., 1983; Stern and Kennedy, 1980a; Alonso-Caplen et al., 1984). The duration of the lytic cycle is about 10 h and cytopathic effects such as cell fusions and/or cell lysis develop as the infection proceeds.

II. ORGANIZATION AND EXPRESSION

Our view of the organization and expression of the coronavirus genome is based upon three lines of evidence.

A. Structural Analysis of Viral RNAs

Following infection, the coronavirus genome functions as a mRNA that encodes a membrane-bound RNA polymerase activity. This activity is responsible for the synthesis of genomic-length, negative-stranded RNA (Brayton et al., 1984). As the infection proceeds this RNA serves in replicative-intermediate structures as the template for positive-stranded RNA synthesis (Lai et al., 1982a). The polymerase activities associated with genome and mRNA synthesis are also membrane-bound and can be separated (Brayton et al., 1982; Mahy et al., 1983). As yet, the structural components of these different polymerase activities have not been identified.

In the coronavirus-infected cell it is characteristic that in addition to genomic RNA, multiple subgenomic mRNAs are synthesized. In IBV-infected cells five, and in MHV-infected cells six, subgenomic mRNAs are produced. These mRNAs are capped and polyadenylated and are produced in unequal but constant proportions throughout infection (Stern and Kennedy, 1980a; Spaan et al., 1981; Wege et al., 1981; Lai et al., 1982b; see also Fig. 8.1). The size of these subgenomic mRNAs has been estimated by gel electrophoresis as: *IBV* mRNA A, 2.0 kb; mRNA B, 2.4 kb; mRNA C, 3.7 kb; mRNA D, 4.5 kb and mRNA E, 7.9 kb (Stern and Sefton, 1984; Brown and Boursnell, 1984) and *MHV*, mRNA 7, 1.9 kb; mRNA 6, 2.5 kb; mRNA 5, 3.2 kb; mRNA 4, 3.6 kb; mRNA 3, 7.8 kb and mRNA 2, 10.3 kb (Leibowitz et al., 1981).

The structural relationships of coronavirus mRNAs have been elucidated by RNAse T_1-resistant oligonucleotide fingerprinting (Stern and Kennedy, 1980a, 1980b; Leibowitz et al., 1981; Lai et al., 1981; Spaan et al., 1982; Makino et al., 1984). These data show that the genomic RNA and subgenomic mRNAs form a 3' coterminal overlapping or "nested" set. Each mRNA contains the nucleotide sequence of the next smaller mRNA

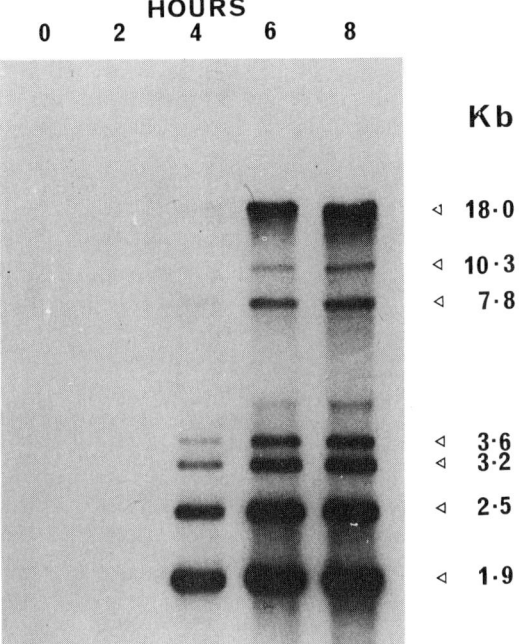

Figure 8.1. Northern blot analysis of intracellular MHV RNA. 5 μg of cytoplasmic RNA, isolated from MHV A59-infected cells (m.o.i. 10) at the times indicated, was electrophoresed in formaldehyde/agarose gels, transferred to nitrocellulose and hybridized to a ^{32}P-labelled cDNA complementary to the coding region of MHV JHM mRNA 7. The filter was washed at low stringency and autoradiographed.

plus additional sequences at the 5' end, which are referred to as unique regions. Northern blot analysis of intracellular mRNAs using cDNA probes complementary to the unique regions of several mRNAs support this conclusion (Cheley *et al.*, 1981; Weiss and Leibowitz, 1983; Budzilowicz *et al.*, 1985; Stern and Sefton, 1984; Skinner and Siddell, 1985).

B. Translation Products of Viral mRNAs

In IBV- or MHV-infected cells the precursors to the N, M and S proteins are readily identified. In MHV-infected cells two additional virus-coded polypeptides of 14 kD and 30–35 kD have also been recognized (Stern and Sefton, 1982a; Alonso-Caplen *et al.*, 1984; Siddell *et al.*, 1981, Siddell, 1982; Cheley and Anderson, 1981; Bond *et al.*, 1984; Rottier *et al.*, 1981a; Holmes *et al.*, 1981).

The viral mRNAs that specify these polypeptides have been identified by the translation *in vitro* of RNA fractions enriched by electrophoresis or velocity sedimentation for each of the subgenomic species (Stern and Sefton, 1984; Siddell *et al.*, 1980; Siddell, 1983; Leibowitz *et al.*, 1982; Rottier *et al.*, 1981b). The coding assignments are, for IBV N (51 kD), mRNA A; M (23 kD), mRNA C and S (110 kD), mRNA E and for *MHV* N (54–60 kD), mRNA 7; M (23 kD), mRNA 6; S (120 kD), mRNA 3; 14 kD, mRNA 4 or 5 and 30–35 kD, mRNA 2 (see Fig. 8.2). Also, genomic RNA from purified MHV virions has been translated *in vitro* to produce a series of structurally related polypeptides of more than 200 kD (Leibowitz *et al.*, 1982). The identity of these products is not known but it seems reasonable to assume that they are related to the components of the RNA polymerase activities.

C. Sequence Analysis

1. IBV

A continuous sequence of about 7.4 kb extending from the 3' end of the IBV (Beaudette) genomic RNA has been determined (Boursnell *et al.*, 1984; Boursnell and Brown, 1984; Boursnell *et al.*, 1985a, 1985b; Binns *et al.*, 1985). Within this sequence the regions encoding the N, M and S protein are readily identified on the basis of the size of the potential open reading frames (ORFs) and the characteristics of their predicted polypeptides. The identification of the *S* gene has also been confirmed by partial amino acid sequencing of the S1 protein (Binns *et al.*, 1985). The position of the *N*, *M* and *S* genes correspond to the unique regions of the subgenomic mRNAs A, C and E. In the genomic sequence corresponding to the unique regions of mRNAs B and D, five large ORFs (as illustrated in Fig. 8.2) are found, but it is not known which of these are expressed.

2. MHV

As with IBV, sequences at the 3' end of the MHV genome have been determined (Armstrong *et al.*, 1983, 1984a, 1984b; Skinner and Siddell, 1983, 1985; Skinner *et al.*, 1985). Again, the genomic regions encoding the N and M proteins have been identified and they correspond to the unique regions of mRNAs 7 and 6. Hybrid-arrested translation has also been used to confirm the position of the *N* gene (Skinner and Siddell, 1983). In the genomic sequence corresponding to the unique region of mRNA 4, a single ORF is found and the predicted polypeptide has a size and charge indicative of the 14 kD polypeptide detected in infected cells. The sequence corresponding to the unique region of mRNA 5 contains two large ORFs (see Fig. 8.2) but translation products for this region have not been identified.

From the preceding information it is possible to formulate a model describing the organization and expression of the IBV and MHV genomes. This model as depicted in Fig. 8.2 is incomplete, but nevertheless some general features can be recognized.

1. The virus structural genes encoding the N, M and S proteins are ordered within the 3' half of the genome. In both IBV and MHV these genes are interspersed by putative non-structural genes.
2. The viral RNA polymerase activity is encoded in the 5' proximal region of the genome. The number and arrangement of ORFs within this region is unknown.
3. The coronavirus genes which have been identified are non-overlapping.
4. The expression of coronavirus genes is mediated by multiple subgenomic mRNAs which form a 3' coterminal nested set.
5. Only the unique regions of the subgenomic mRNAs are translationally active and only a single polypeptide, representing one gene, is translated from each mRNA. The translated polypeptide may, as in the case of the S gene product, undergo post-translational cleavage (Binns et al., 1985). It should be held in mind that the unique regions of several coronavirus mRNAs contain two or more essentially non-overlapping ORFs. The expression of the 3' distal ORFs within these regions would require either the production of novel mRNAs or the internal initiation of protein synthesis, but should be considered a possibility.

The strategy used by coronaviruses has several obvious advantages. Firstly, the genomic location of the polymerase and structural genes, together with the production of a 3' coterminal nested set of mRNAs, facilitates the temporal requirements for these gene products during the infection. Secondly, given the general inability of eukaryotic ribosomes to initiate protein synthesis at internal sites (Kozak, 1983), the production of multiple subgenomic mRNAs allows for the expression of a large viral genome without the synthesis and processing of a correspondingly large primary translation product. Thirdly, this strategy, which is also reflected in the replication of alphaviruses and tobamoviruses (Garoff et al., 1982; Goelet and Karn, 1982) has the major advantage that it permits the independent transcriptional regulation of different regions of the viral genome.

III. REGULATION

The synthesis of subgenomic mRNAs is a key element in the regulation of coronavirus gene expression and it is therefore reasonable to find that the process displays features unique to this group of viruses.

Lai et al. (1983, 1984) were the first to demonstrate that each MHV mRNA possesses at its 5' end a common 70-base leader sequence that is derived from the 5' end of the genome. An analogous leader of 60 bases has also recently been shown for IBV (Brown et al., 1984; T. D. K. Brown, personal communication). During mRNA synthesis this leader RNA is transcribed from the 3' end of the negative-stranded template and is then translocated to specific internal re-initiation sites where it functions as a primer for the transcription of individual mRNAs (Stern and Sefton,

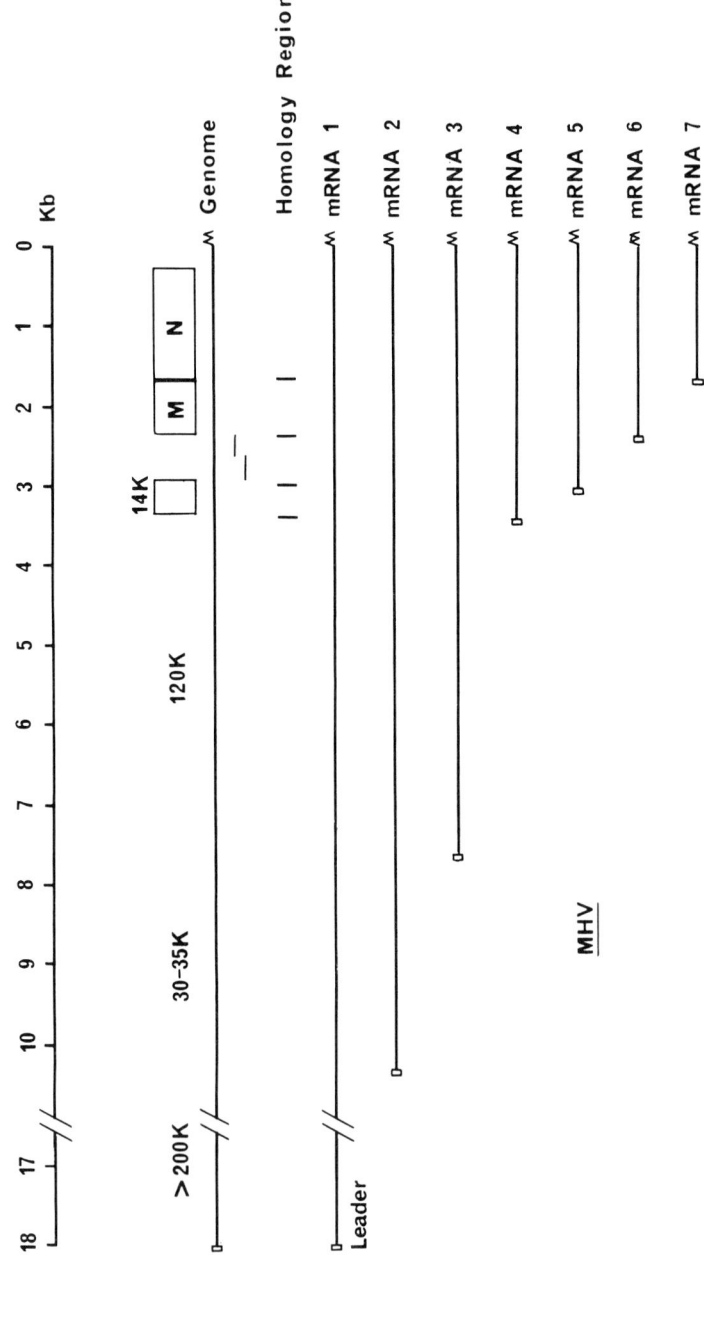

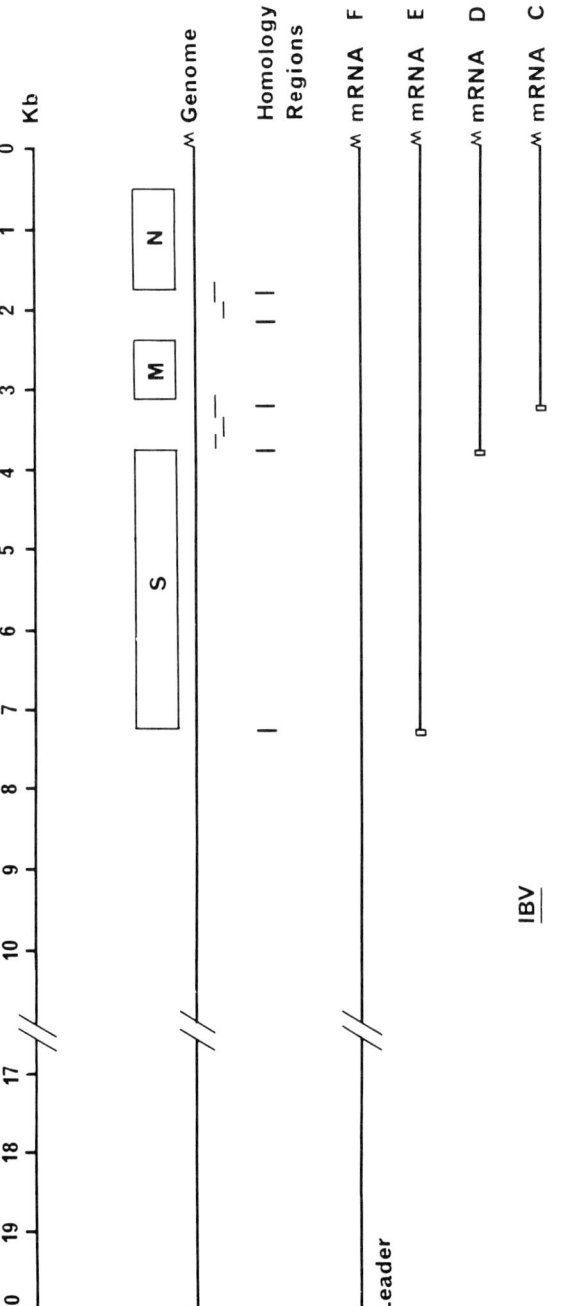

Figure 8.2. The organization and expression of the IBV and MHV genomes. For details see text. (K = kilodaltons.)

1982b; Jacobs et al., 1981; Baric et al., 1983; Lai et al., 1982a; Spaan et al., 1983).

Many aspects of this leader-primed mode of mRNA synthesis are not yet understood. For example, it is not clear whether the leader RNA dissociates from the negative-stranded template or "slides" to the reinitiation sites. However, it is clear that the frequency with which the different sites are used will dictate the relative abundance of the subgenomic mRNAs. The sequences representing these sites are clearly therefore of interest and are listed in Fig. 8.3. These sequences can be recognized by their homology and their position in the genome, which corresponds to the 5' end of the unique region of each mRNA (see Fig. 8.2). As expected, they also represent the positions at which mRNA and genomic RNA sequences diverge (Brown and Boursnell, 1984; Spaan et al., 1983). Evidently, there is sequence homology not only between the sites for each mRNA, but also to a limited extent between both viruses.

It has been speculated that the internal initiation sites are recognized directly by the viral RNA polymerase, or that they are involved in base-pairing with sequences present in the leader RNA (Armstrong et al.,

IBV	A	UUGUGUUUACUUU	CUUAAC	AAAGCAGGACA
IBV	B	UACUUAACAAAAA	CUUAAC	AAAUACGGACG
IBV	C	AUAUGGUAGAAAA	CUUAAC	AAUCCGGAAUU
IBV	D	AACGAUGUGGUAA	CUGAAC	AAUACAGACCU
IBV	E	GUUUAAUUUGAAA	CUGAAC	AAAAGACAGAC

MHV A59	7	UUGAGAAU	CUAAU	CUAAAC	UUUAAG
MHV A59	6	GAUCAUAU	CUAAU	CCAAAC	AUUAUG
MHV A59	5	UUAUGUUA	CUAAU	CUAAAC	CUCAUC

MHV JHM	6	AUGAUAAUAUAA	U	CCAAAC	AUUAUG
MHV JHM	5	UAUAUUACUAGU	U	CUAAAC	CUCAUC
MHV JHM	4	AAAGACAGAAAA	U	CUAAAC	AAUUUA

Figure 8.3. Homologous regions in the genomic RNA of IBV and MHV. The positions of these sequences (shown in Fig. 8.2) correspond to the 5' end of the unique region of the mRNAs indicated. The data is compiled from Boursnell et al. (1984), Boursnell and Brown (1984), Binns et al. (1985), Armstrong et al. (1984a), Skinner and Siddell (1985), Skinner et al. (1985), Spaan et al. (1983) and Budzilowicz et al. (1985).

1984a, Spaan et al., 1983; Budzilowicz et al., 1985). Alternatively, this interaction may involve all three components. At the moment, there is no experimental data to discriminate between these possibilities.

There is still a great deal to be learnt about the organization, expression and regulation of the coronavirus genome. In the foreseeable future the complete nucleotide sequence of IBV and MHV will become available and will provide a major impetus for the study of, in particular, the non-structural genes of these viruses. The function of these genes and their products will be one of the most intensive areas of coronavirus research in the next few years.

ACKNOWLEDGEMENTS

This work was supported financially by the DFG. I would like to thank Helga Kriesinger for typing this manuscript.

REFERENCES

1. Alonso-Caplen, F. V., Y. Matsuoka, G. Wilcox and R. W. Compans (1984). *Virus Res.* **1**, 153–167.
2. Armstrong, J., S. Smeekens and P. Rottier (1983). *Nucl. Acids Res.* **11**, 883–891.
3. Armstrong, J., H. Niemann and G. Warren (1984a). *Nature (London)* **308**, 751–752.
4. Armstrong, J., S. Smeeken, W. Spaan, P. Rottier and B. van der Zeijst (1984b). *Adv. Exp. Med. Biol.* **173**, 155–162.
5. Baric, R. S., S. A. Stohlman and M. M. C. Lai (1983). *J. Virol.* **48**, 633–640.
6. Binns, M. M., M. E. G. Boursnell, D. Cavanagh, D. J. C. Pappin and T. D. K. Brown (1985). *J. Gen. Virol.* **66**, 719–726.
7. Bond, C. W., K. Anderson and J. L. Leibowitz (1984). *Virology* **80**, 333–347.
8. Boursnell, M. E. G. and T. D. K. Brown (1984). *Gene* **29**, 87–92.
9. Boursnell, M. E. G., T. D. K. Brown and M. M. Binns (1984). *Virus Res.* **1**, 303–313.
10. Boursnell, M. E. G., M. M. Binns, I. J. Foulds and T. D. K. Brown (1985a). *J. Gen. Virol.* **66**, 573–580.
11. Boursnell, M. E. G., M. M. Binns and T. D. K. Brown (1985b). *J. Gen. Virol.* **66**, 2253–2258.
12. Brayton, P. R., R. G. Ganges and S. A. Stohlman (1981). *J. Gen. Virol.* **56**, 457–460.
13. Brayton, P. R., M. M. C. Lai, C. D. Patton and S. A. Stohlman (1982). *J. Virol.* **42**, 847–853.
14. Brayton, P. R., S. A. Stohlman and M. M. C. Lai (1984). *Virology* **133**, 197–201.
15. Brown, T. D. K. and M. E. G. Boursnell (1984). *Virus Res.* **1**, 15–24.
16. Brown, T. D. K., M. E. G. Boursnell and M. M. Binns (1984). *J. Gen. Virol.* **65**, 1437–1442.
17. Budzilowicz, C. J., S. P. Wilczynski and S. R. Weiss (1985). *J. Virol.* **53**, 834–840.
18. Cavanagh, D. (1983). *J. Gen. Virol.* **64**, 2577–2583.

19. Cheley, S. and R. Anderson (1981). *J. Gen. Virol.* **54**, 301–311.
20. Cheley, S., R. Anderson, M. Cupples, E. C. M. Lee Chan and V. L. Morris (1981). *Virology* **112**, 596–604.
21. Dubois-Dalcq, M. E., K. V. Holmes and B. Rentier (1984). "Assembly of Enveloped RNA Viruses". Springer Verlag, Vienna.
22. Garoff, H., C. Kondor-Koch and H. Riedel (1982). *Current Topics Microbiol. Immunol.* **99**, 1–50.
23. Goelet, P. and J. Karn (1982). *J. Mol. Biol.* **154**, 541–550.
24. Holmes, K. V. (1985). *In* "Virology" (Ed. B. N. Fields). Raven Press, New York.
25. Holmes, K. V., E. W. Doller and L. S. Sturman (1981). *Virology* **115**, 334–344.
26. Jacobs, L., W. J. M. Spaan, M. C. Horzinek and B. A. M. van der Zeijst (1981). *J. Virol.* **39**, 401–406.
27. Kozak, M. (1983). *Microbiol. Rev.* **47**, 1–45.
28. Lai, M. M. C., P. R. Brayton, R. C. Armen, C. D. Patton, C. Pugh and S. A. Stohlman (1981). *J. Virol.* **39**, 823–834.
29. Lai, M. M. C., C. D. Patton and S. A. Stohlman (1982a). *J. Virol.* **44**, 487–492.
30. Lai, M. M. C., C. D. Patton and S. A. Stohlman (1982b). *J. Virol.* **41**, 557–565.
31. Lai, M. M. C., C. D. Patton, R. S. Baric and S. A. Stohlman (1983). *J. Virol.* **46**, 1027–1033.
32. Lai, M. M. C., R. S. Baric, P. R. Brayton and S. A. Stohlman (1984). *Proc. Natl Acad. Sci. USA* **81**, 3626–3630.
33. Leibowitz, J. L., K. C. Wilhelmsen and C. W. Bond (1981). *Virology* **114**, 39–51.
34. Leibowitz, J. L., S. R. Weiss, E. Paavola and C. W. Bond (1982). *J. Virol.* **43**, 905–913.
35. Mahy, B. W. J., S. Siddell, H. Wege and V. ter Meulen (1983). *J. Gen. Virol.* **64**, 193–211.
36. Makino, S., F. Taguchi, N. Hirano and K. Fujiwara (1984). *Virology* **139**, 138–151.
37. Rottier, P. J. M., M. C. Horzinek and B. A. M. van der Zeijst (1981a). *J. Virol.* **40**, 350–357.
38. Rottier, P. J. M., W. J. M. Spaan, M. C. Horzinek and B. A. M. van der Zeijst (1981b). *J. Virol.* **38**, 20–26.
39. Rottier, P., D. Brandenburg, J. Armstrong, B. van der Zeijst and G. Warren (1984). *Proc. Natl Acad. Sci. USA* **81**, 1421–1425.
40. Siddell, S. G. (1982). *J. Gen. Virol.* **62**, 259–269.
41. Siddell, S. G. (1983). *J. Gen. Virol.* **64**, 113–125.
42. Siddell, S. G., H. Wege, A. Barthel and V. ter Meulen (1980). *J. Virol.* **33**, 10–17.
43. Siddell, S. G., H. Wege, A. Barthel and V. ter Meulen (1981). *J. Gen. Virol.* **53**, 145–155.
44. Siddell, S. G., R. Anderson, D. Cavanagh, K. Fujiwara, H. D. Klenk, M. R. Macnaughton, M. Pensaert, S. A. Stohlman, L. Sturman, and B. A. M. van der Zeijst (1983a). *Intervirology* **20**, 181–189.
45. Siddell, S. G., H. Wege and V. ter Meulen (1983b). *J. Gen. Virol.* **64**, 761–776.
46. Skinner, M. A. and S. G. Siddell (1983). *Nucl. Acids Res.* **11**, 5045–5054.
47. Skinner, M. A. and S. G. Siddell (1985). *J. Gen. Virol.* **66**, 593–596.
48. Skinner, M. A., D. Ebner and S. G. Siddell (1985). *J. Gen. Virol.* **66**, 581–592.

49. Spaan, W. J. M., P. J. M. Rottier, M. C. Horzinek and B. A. M. van der Zeijst (1981). *Virology* **108**, 424–434.
50. Spaan, W. J. M., P. J. M. Rottier, M. C. Horzinek and B. A. M. van der Zeijst (1982). *J. Virol.* **42**, 432–439.
51. Spaan, W., H. Delius, M. Skinner, J. Armstrong, P. Rottier, S. Smeekens, B. van der Zeijst and S. G. Siddell (1983). *EMBO J.* **2**, 1839–1844.
52. Stern, D. F. and S. I. T. Kennedy (1980a). *J. Virol.* **34**, 665–674.
53. Stern, D. F. and S. I. T. Kennedy (1980b). *J. Virol.* **36**, 440–449.
54. Stern, D. F. and B. M. Sefton (1982a). *J. Virol.* **44**, 794–803.
55. Stern, D. F. and B. M. Sefton (1982b). *J. Virol.* **42**, 755–759.
56. Stern, D. F. and B. M. Sefton (1984). *J. Virol.* **50**, 22–29.
57. Sturman, L. S. and K. V. Holmes (1983). *Adv. Virus Res.* **28**, 35–112.
58. Wege, H., S. Siddell, M. Sturm and V. ter Meulen (1981). *J. Gen. Virol.* **54**, 213–217.
59. Weiss, S. R. and J. L. Leibowitz (1983). *J. Gen. Virol.* **64**, 127–133.
60. Wilhelmsen, K. C., J. L. Leibowitz, C. W. Bond and J. A. Robb (1981). *Virology* **110**, 225–230.

9. Genetic Recombination in RNA Viruses

Andrew M. Q. King, Stephen A. Ortlepp, John W. I. Newman and David McCahon

A.F.R.C. Institute for Animal Disease Research, Pirbright Laboratory, Pirbright, Woking, Surrey GU24 ONF, U.K.

I.	Introduction	129
II.	Non-homologous recombination	130
	A. Mechanism of non-homologous recombination	131
III.	Homologous recombination	132
	A. Detection and isolation of recombinants	132
	B. Genetic recombination maps of picornaviruses	133
	C. Biochemical characterization of recombinants	135
	D. Proof of recombination in RNA	138
	E. Mapping recombinant RNA	139
	F. Recombination occurs at many sites in the picornavirus genome	140
	G. Genetic exchanges between distantly related viruses	143
	H. Recombinant proteins	145
	I. Nucleotide sequence homology at sites of recombination	146
	J. Homologous recombination in picornaviruses	148
IV.	Mechanism of recombination	148
V.	Why do RNA viruses recombine?	149
	References	151

I. INTRODUCTION

Genetic recombination in RNA viruses is defined as any process involving the exchange of information between viral RNA molecules. We are not concerned, here, with the reassortment of RNA segments, nor with recombination in retroviruses, which is assumed to involve DNA intermediates.

RNA recombination is of interest for several reasons. First, it offers a means of manipulating RNA genomes (e.g. for genetic mapping) without having to convert them to infectious cDNA. Second, the ability to exchange genetic information may confer a selective advantage on the virus, and thus be a significant factor in its evolution. Third, RNA recombination is a little studied process, whose molecular mechanism has yet to be elucidated.

Recombinational processes are of two kinds, homologous and non-homologous. In the former, the parental RNAs are related to each other, and the location of the genetic cross-over is the same in both sequences, so preserving the reading frame and producing a potentially functional recombinant molecule. In non-homologous recombination neither of these restrictions applies. Of the two processes, homologous recombination has been by far the most extensively studied. Most of this chapter will be devoted to reviewing the progress that has been made towards answering the following questions: can homologous recombination occur anywhere in the genome? To what extent can distantly related RNA viruses exchange genetic information? Must there be homology between the parental nucleotide sequences at the cross-over point? Is there a preferred sequence at which recombination occurs?

II. NON-HOMOLOGOUS RECOMBINATION

There have been two reports of non-homologous recombination in RNA viruses. The first was of a "mosaic" defective interfering DI RNA of influenza virus, consisting of a region of segment 1 RNA flanked by sequences derived from segment 3 (Fields and Winter, 1982). There was little homology between the nucleotide sequences of segments 1 and 3 at the cross-overs, just two bases matching at one cross-over point and none at the other. In this respect, these intermolecular cross-overs are typical of the more common intramolecular rearrangements seen in DI RNAs. Examination of 50 cross-over sequences of influenza DI RNAs reported by Jennings *et al.* (1983) shows that any length of homology, from zero to six bases, is permissible, a match of just one base being the most common.

Second, in the course of determining the RNA sequence of several DI preparations of Sindbis virus, Monroe and Schlesinger (1984) found that two of them had independently acquired part of a cellular RNA molecule, tRNA^{ASP}, at their 5'-termini. Inspection of the sequences in the regions of cross-over between the viral RNA and the tRNA shows only a single matching base (an A residue) in each case. Hybridization studies confirm that tRNA^{ASP} sequences are frequently acquired by DI preparations of Sindbis virus (Tsiang *et al.*, 1985). At first sight, the preference of the Sindbis virus polymerase for a cellular tRNA seems strange, although it should be remembered that tRNA-like structures are found in the genomes of many plant viruses, including tobacco mosaic and brome mosaic viruses

9. Genetic Recombination in RNA Viruses

(see Chapters 10 and 13, this volume), and moreover that both of the latter viruses are related to Sindbis virus in parts of their polymerase genes (Ahlquist *et al.*, 1985; and see Chapter 6, this volume).

To account for homology between families of RNA viruses previously assumed to be entirely unrelated, Haseloff *et al.* (1984) speculated that genes may have been transposed between the ancestors of the present-day plant rhabdoviruses, tobamoviruses and bromoviruses by RNA recombination. Similarly, the tandem triplication of the foot and mouth disease virus (FMDV)VPg gene may have been the result of unequal cross-overs (Forss and Schaller, 1982). These observations suggest that non-homologous recombination may occasionally play an important role in the evolution of RNA viruses, and possibly in the creation of RNA viruses from cellular genes.

A. Mechanism of Non-homologous Recombination

DI RNAs are generally assumed to be produced by a copy-choice mechanism, in which the polymerase dissociates from the template and re-initiates elsewhere, using its growing RNA strand as the primer (Lazzarini *et al.*, 1981). The fact that the 3' end of the tRNA sequence was retained in both Sindbis recombinants, even though the 5' end was not, fits this primer model.

The sequence rearrangements of influenza virus DI RNAs have two characteristic features, both of which are shared, not only by the Sindbis/tRNA recombinants described above, but also by the DI RNAs of vesicular stomatitis virus (Meier *et al.*, 1984): (a) although homology is not essential, the majority of recombining sequences do have a small number of matching bases (typically one) at the cross-over point; and (b) there is a marked preference for a purine residue at the 3' end of any homology region, where present. Assuming the copy-choice model, discussed earlier, this nucleotide would normally represent the 3' terminus of the nascent RNA strand that had been used to prime synthesis on a new region of template. Thus, the non-homologous rearrangements of both Sindbis and influenza DI RNAs are consistent with a model in which re-initiation of synthesis can be directed efficiently by a single base pair, formed between a 3'-terminal A or G of the primer and a complementary base in the template. The absence of any base specificity at the 5' ends of these homology regions implies that most recombination events take place during synthesis of the positive sense strand.

The priming of transcription in influenza virus by 5'-terminal pieces of capped cellular message represents yet another example of non-homologous recombination, albeit not in genomic RNA. The fact that this process shares the same preference for a single homologous purine residue at the 3' end of the primer (Plotch *et al.*, 1981) argues strongly that all the non-homologous rearrangements of influenza virus, and probably Sindbis virus as well, occur by essentially the same copy-choice mechanism.

III. HOMOLOGOUS RECOMBINATION

A. Detection and Isolation of Recombinants

Homologous recombinants may be selected by assaying the progeny of a mixed infection under conditions that restrict the growth of both parental viruses independently (see below). An enhancement in the yield of resistant virus from the mixed infection, compared with singly infected controls, is indicative of genetic recombination. Homologous recombination was first demonstrated in poliovirus by Hirst (1962) and Ledinko (1963). Soon after, FMDV was also shown to have the ability to recombine (Pringle, 1965). Virtually all studies of recombination have been done on these two picornaviruses, although, very recently, good evidence has been obtained for recombination between two strains of mouse hepatitis virus (MHV), a coronavirus (Lai et al., 1985; see Chapter 8, this volume).

The two most commonly used types of selection are shown in Fig. 9.1. Panel A shows two parental genomes with temperature-sensitive (ts) mutations at different locations. Growth at the restrictive temperature selects for recombinants inheriting the wild-type marker from both parents, i.e. for cross-overs occurring between the ts loci. This method was used extensively by Cooper's group working with poliovirus, and has been used in all our own studies with FMDV. In the second method (panel B),

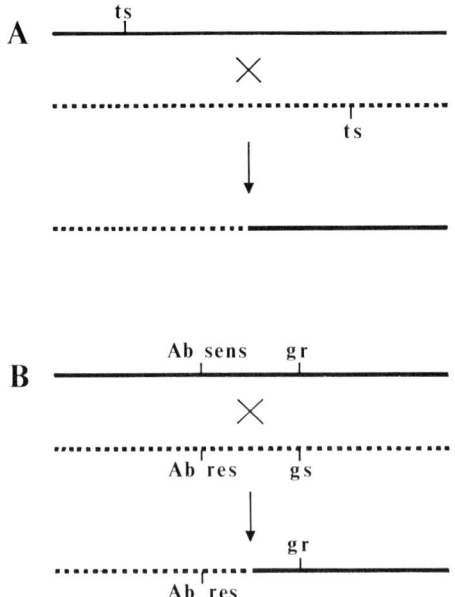

Figure 9.1. Two methods of selecting recombinants.

used by Agol *et al.* (1984) and Emini *et al.* (1984), one of the parents has a mutation *gr* conferring resistance to guanidine, the other being sensitive (*gs*). The *gr* parent is selectively neutralized by an antiserum, so that only recombinants inheriting both *gr* and antibody-resistance (Ab res) markers are able to grow. To cross the Sabin and Mahoney strains of poliovirus type 1, which are closely related serologically, Emini *et al.* (1984) used a specific monoclonal antibody. The disadvantage of method B is that the determinants of both antigenicity and guanidine-resistance have unique loci, and only recombination events within a fixed 2 kb region in the middle of the genome can be selected. Therefore, biological characters like guanidine-resistance can only be mapped to the left, or right, halves of the genome. It is also possible to use a combination of methods A and B, for example selecting with guanidine at high temperature (Pringle, 1965).

The standard procedure for crossing viruses, known as the "yield test", is to grow the mixture of parents under permissive conditions, and assay their progeny under restrictive conditions. A further advantage of temperature selection is that it is possible to begin applying the selective pressure during the initial step. This is the basis of the more sensitive "infectious centre test", introduced by McCahon and Slade (1981). In this method, the mixedly infected cells are resuspended before they have had time to lyse, and assayed at the restrictive temperature. Thus, each ts^+ plaque is derived from an independent cell in which recombination has taken place. In crosses between mutants of the same strain, up to 30% of infected cells can be shown to contain ts^+ recombinants. We have used the infectious centre method in all our recent experiments with different FMDV types and subtypes. A similar method was used for isolating coronavirus recombinants (Lai *et al.*, 1985).

B. Genetic Recombination Maps of Picornaviruses

Frequencies of recombination, as measured by the yield test, tend to be variable, and small in comparison to the background reversion rate. Despite these problems, Cooper (1968) succeeded in constructing a map, based on recombination frequencies between *ts* mutations of a strain of type 1 poliovirus, that was linear and approximately additive. The map featured a single, centrally located guanidine-resistance locus. A similar map was constructed for FMDV (Lake *et al.*, 1975), to be followed by an extended version (McCahon *et al.*, 1977). This version is shown at the top of Fig. 9.2.

These attempts at mapping the picornavirus genome rested on the unproven assumption that in crosses between mutants of the same virus strain recombination frequency is proportional to genomic distance (i.e. that cross-overs are distributed randomly). In fact, the FMDV recombination map does seem to represent the physical genome, orientated with the 5'-end on the left. We have determined the physical locations of many of the FMDV *ts* mutations by correlating their *ts* character with a change in the isoelectric point (pI) of a viral protein. The bottom part of Fig. 9.2

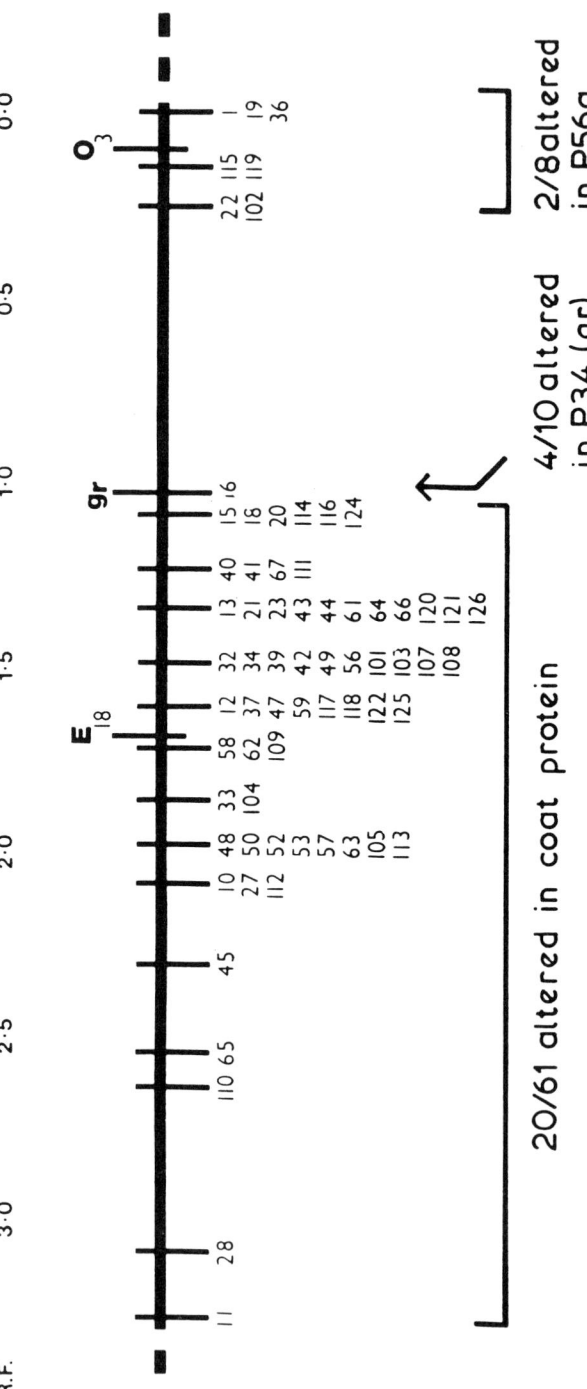

Figure 9.2. The physical distribution of mutations of the FMDV genetic recombination map (McCahon *et al.*, 1977) as determined by electrofocusing. All mutations are *ts* except for those designated *gr*. (Reprinted from King *et al.*, 1983.)

summarizes all the pI changes, detected by gel electrofocusing, that coreverted with a *ts* mutation. Of the mutations that we were able to map physically, all 20 *ts* mutations on the left of the *gr* locus are carried by coat proteins (King *et al.*, 1980) encoded near the 5' end of the genome; the four *gr* mutations are all in the central P34-coding region (Saunders and King, 1982); and the two *ts* mutations at the extreme right of the map are polymerase mutants (Lowe *et al.*, 1981), and therefore near the 3' end of the genome. When the positions of the individual coat protein genes are considered, the relationship between the physical and recombination maps becomes imprecise. The standard error of recombination mapping, estimated from the loci of a mutant that was mapped independently 15 times, is 0.2% map units, a distance roughly equivalent to one of the major coat protein genes (King *et al.*, 1980).

We can use the maps to estimate the fraction of genomes in the virus yield that are derived from recombination events. The standard cross, $gr \times ts03$, included by McCahon *et al.* (1977) in all their recombination experiments, gave a recombination frequency of 0.92%. This was the mean value of 15 replicates. Allowing for the fact that the cross only detected recombination (a) between markers now known to be separated by 1.5–3 kb, (b) in one direction, and (c) between different parents, leads us to conclude that the overall recombination frequency is between 10% and 20%; i.e. in cells infected with a single strain of FMDV, 10–20% of the viral genomes undergo recombination per growth cycle. This estimate is comparable with a figure of about 5% based on the length of the poliovirus map (Cooper, 1968). Both must be regarded as underestimates, since they assume that in every cell, of each cross, the ratio of the parental genomes never fluctuated from the ideal value of 1 : 1.

C. Biochemical characterization of recombinants

The earliest biochemical evidence for recombination in picornaviruses was based on the inheritance of viral proteins, identified either by point mutations that altered pI (King *et al.*, 1982a) or by tryptic peptide fingerprinting (Tolskaya *et al.*, 1983). One-dimensional electrofocusing, used in the former study, is such a simple and sensitive method of distinguishing proteins that we still use it for screening recombinants. Examples of two kinds of recombinant, isolated from a pairwise cross between FMDV subtypes O_1 and O_6, are shown in Fig. 9.3. It is readily seen that both progeny viruses are recombinant, having the coat proteins, VP1, VP0/2, and VP3, of the O_6 subtype together with the P56a of the O_1 subtype. The two recombinants differ in the parental origin of the centrally encoded P34.

Recombinant RNA genomes were analysed in detail by T_1 RNase fingerprinting. This technique has been used by all the laboratories currently working in this field. Figs 9.4, 9.5 and 9.6 illustrate its use with FMDV subtypes O_1 and O_6. Figure 9.4 shows the fingerprints of the

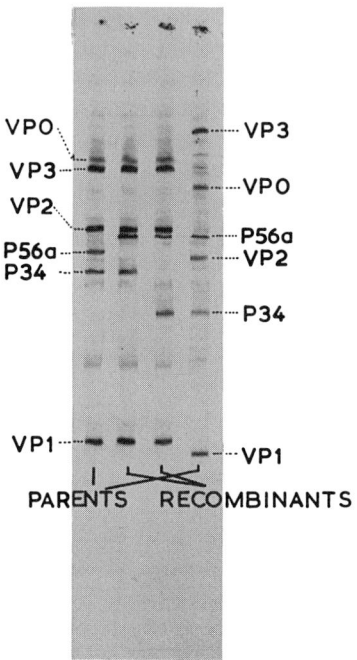

Figure 9.3. Detection of recombinants by electrofocusing virus-induced proteins. Virus-infected BHK cells were labelled with ^{35}S-methionine for 30 min, followed by a 30-min chase. FMDV proteins in cytoplasmic extracts were immunoprecipitated and electrofocused, with the origin (acidic end) at the top, as described previously (King et al., 1981). From left to right, the viruses are: O_6 parent, Rec2, Rec1, O_1 parent.

parental strains; the complexity of the mixture $O_1 + O_6$ shows that almost all the large oligonucleotides are characteristic of their virus strain. The composition of each recombinant was determined by reference to the diagram of this mixture, shown in Fig. 9.5.

Examples of four recombinant RNA fingerprints are shown in Fig. 9.6. The isolation of these recombinants from three genetic crosses is described by Saunders et al. (1985). Each fingerprint is composed of oligonucleotide spots derived from both subtypes O_1 and O_6. The recombinants we have chosen for this illustration happen to fall into two reciprocal pairs, Rec1/Rec3 and Rec2/Rec4, in that all the subtype-specific spots present in Rec1 are missing from Rec3, and vice versa, and likewise for Rec2 and Rec4. A simple way to verify the recombinant nature of these viruses is to superimpose copies of, for example, the Rec1 and Rec3 fingerprints on each other; the resulting mixture is identical to that of the parental mixture shown in Fig. 9.4 (except for one oligonucleotide that was missing in one of the *ts* parents, and therefore also missing from its progeny, Rec1). It

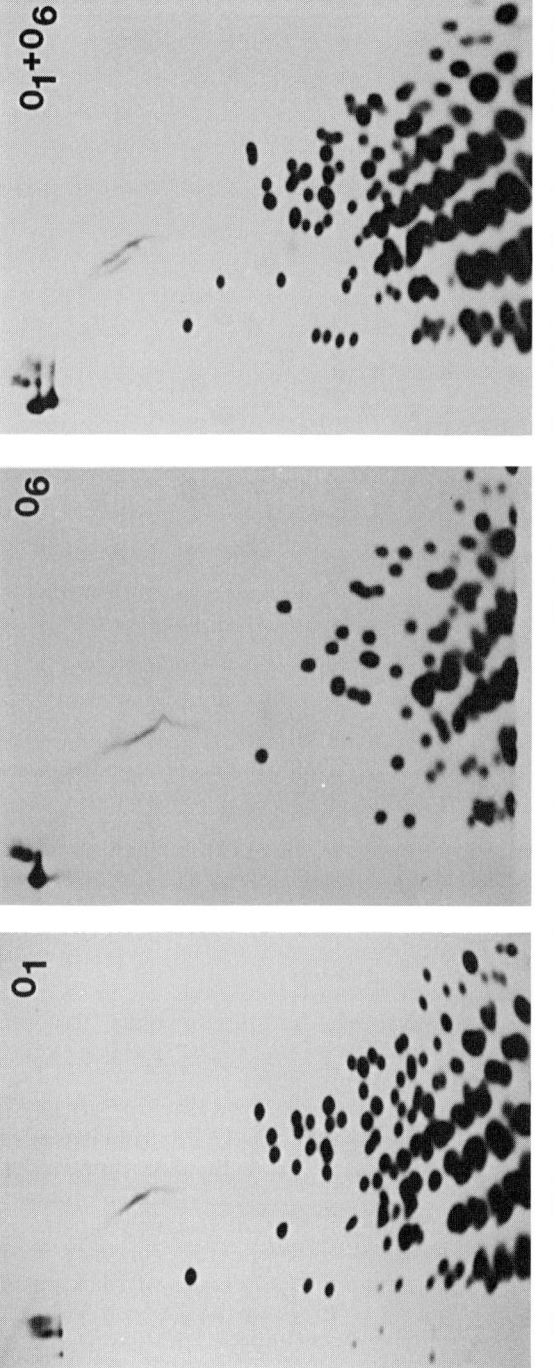

Figure 9.4. RNAase-T_1 fingerprints of the RNA of the two parental FMDV subtypes (O_1 and O_6), and of a mixture of the RNAs ($O_1 + O_6$). The origin of electrophoresis is top left.

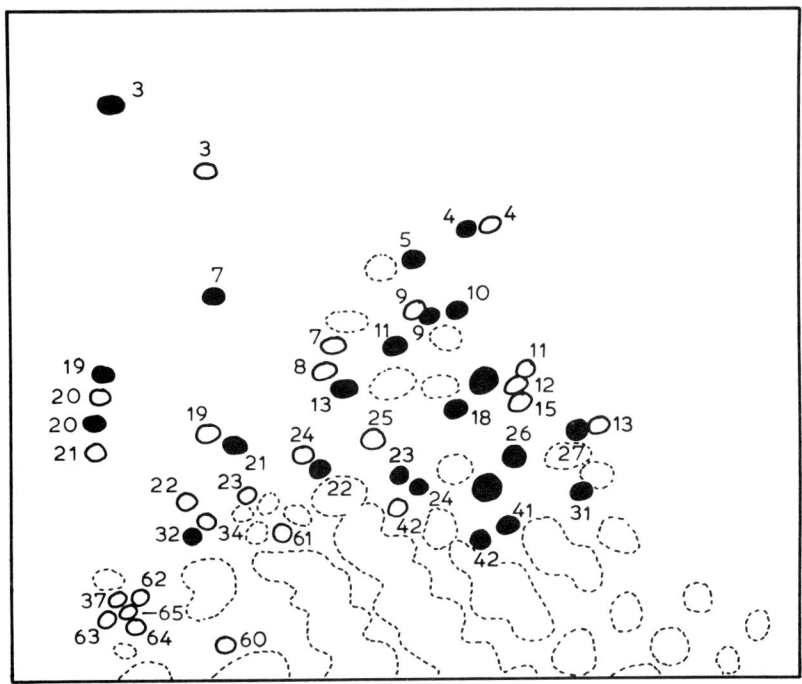

Figure 9.5. Diagram of the $O_1 + O_6$ mixture in Fig. 9.3, showing the oligonucleotides specific to the O_1 strain (filled), and O_6 strain (empty). Regions of the fingerprint in which the two parents could not be distinguished are shown as dotted lines. (Reprinted from King et al., 1985.)

follows that Rec1 and Rec3 must have been formed by genetic cross-overs in the same region of the genome, but in opposite directions.

D. Proof of Recombination in RNA

Publication of the fingerprints of Rec1 and Rec2 (King et al., 1982b) dispelled any doubts about whether picornavirus "recombinants" were really ts^+ revertants; it would need mutations in at least seven oligonucleotides to convert either parent into Rec1.

Is there any other way of explaining the fingerprints in Fig. 9.6 without having to invoke recombination in RNA? Picornaviruses replicate in the cytoplasm, with no requirement for a nuclear function. It is therefore highly unlikely that recombination occurs via a DNA intermediate. Partial heterozygotes, or reassortment of genome fragments, can also be ruled out, since picornaviruses have an unsegmented genome, recombinants are genetically stable and the physical properties of their virus particles and RNA genomes are identical to those of the parents (Cooper et al., 1974). This is also true of the inter-subtype recombinants described here; the

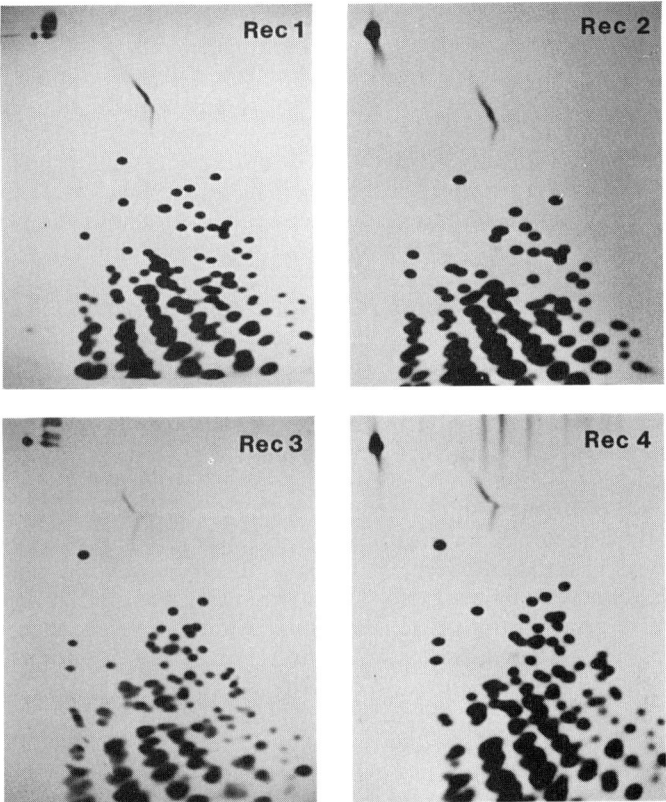

Figure 9.6. Representatives of four types of recombinant RNA fingerprint. (Reprinted from Saunders *et al.*, 1985.)

fingerprints contain no evidence of virus mixtures, and fingerprints of Rec1 and Rec2, essentially identical to those shown here, have been obtained from full-length 35S fractions of viral RNA (King *et al.*, 1982b). Finally, sequencing studies, described later in this paper, prove that genetic information from different picornavirus strains can indeed be present within the same RNA molecule.

E. Mapping Recombinant RNA

To map a recombinant RNA from its fingerprint, we must find out where the oligonucleotides are located in the parental genomes. For recombinants of type 1 poliovirus (Emini et al., 1984; Agol et al., 1984), a complete map of the T_1 oligonucleotides is available (a spin-off of their use as sequencing primers; Kitamura et al., 1981). In the case of the coronavirus recombinants, oligonucleotides were located by fingerprinting a series of size fractions of 3′ coterminal fragments, selected by oligo(dT) cellulose chromatography (Lai et al., 1985). However, this is only a semi-quantitative method of mapping.

We have located the oligonucleotides of both FMDV subtypes shown in Fig. 9.4, and 9.5 by determining their sequences and searching for homologous regions in the published sequences of related FMDV strains (Saunders et al., 1985). Thirty-eight of these sequences (19 of each subtype) could be located with 95% confidence, and their maps are shown at the top of Fig. 9.7. Maps of the four inter-subtype recombinants, in the bottom half of Fig. 9.7, show that each recombinant was formed by a single genetic cross-over. In Rec1 and Rec3, recombination occurred within a centrally located 1 kb region, bounded by oligonucleotides O_1-34 and O_6-61, each parent thus contributing approximately half its genome to each recombinant. In Rec2 and Rec4, the cross-over region was located nearer to the 3′ end.

Figure 9.7 also shows how recombination can be used to map a virus function. Each recombinant in Fig. 9.7 was derived from two parents, one of which was *gr* and the other *gs*. The region of the genome determining the *gr* character was determined by assaying the recombinant progeny for the ability to grow in the presence of guanidine. The results confirm the electrofocusing evidence of Saunders and King (1982), which linked *gr* with changes in P34. They are also consistent with the data of Agol's group (Tolskaya et al., 1983; Agol et al., 1984) and Emini et al. (1984), which place the *gr* locus of poliovirus in the right-hand half of the genome.

F. Recombination Occurs at Many Sites in the Picornavirus Genome

The genetic mapping experiments of Cooper (1968) and McCahon et al. (1977) demonstrated the ability of a large number of independent *ts* mutants to recombine with each other in many different combinations. This is consistent with the notion that the number of potential recombination sites in the picornavirus genome is large.

To investigate biochemically this fundamental aspect of recombination, we crossed mutants of the O_1 and O_6 FMDV subtypes in several other pairwise combinations. To date, we have analysed 43 ts^+ progeny of nine crosses by both electrofocusing and RNA fingerprinting (King et al., 1985). A total of 17 different types of recombinant have been identified, including Rec1, 2, 3 and 4, described above. The results of this study are summarized in Fig. 9.8. The main conclusions are as follows.

1. Most recombinants were produced by a single genetic cross-over, although three, Rec13, 15 and 17, were formed by two cross-overs each. Such a high proportion of double recombinants suggests either that we

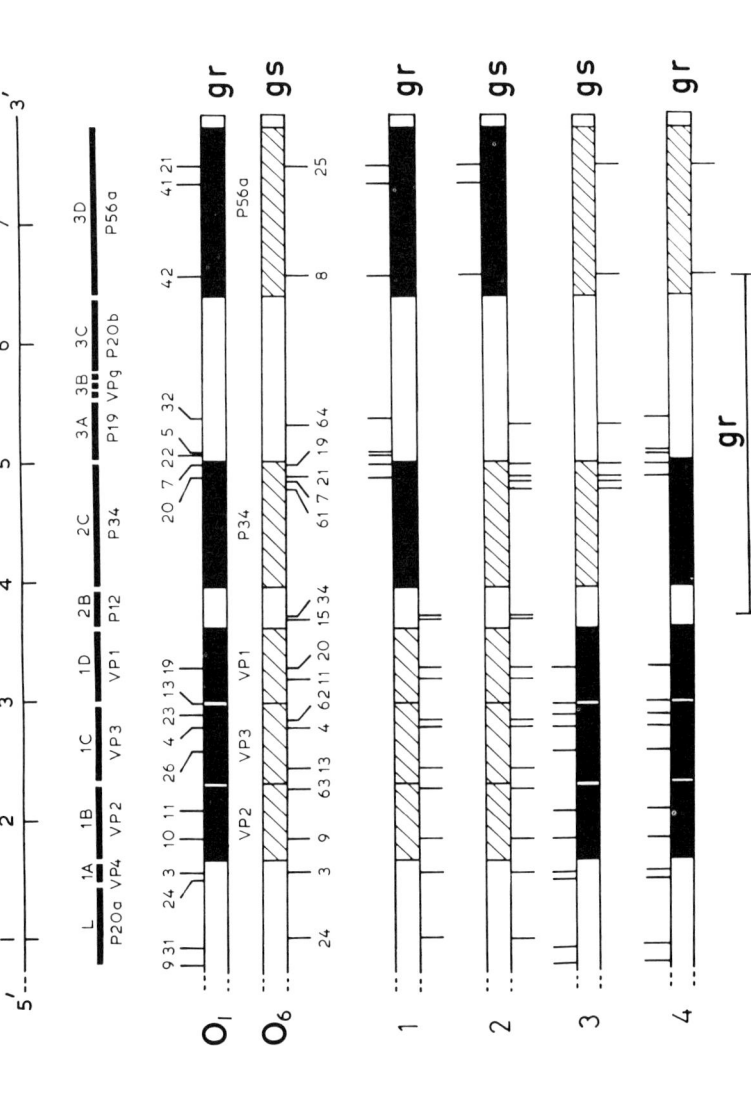

Figure 9.7. Maps of the genomes of the four recombinants, Rec1, 2, 3 and 4. Lines pointing upwards indicate O_1 oligonucleotides, and downwards O_6 oligonucleotides. Filled regions indicate O_1 proteins, and shaded regions O_6 proteins. Coding regions of the viral proteins, shown at the top of the figure, are identified in two ways: names above each gene are based on the unified nomenclature for picornaviruses (Rueckert and Wimmer, 1984), and below using traditional FMDV nomenclature. The bar at the bottom shows the possible locus of gr. (Adapted from Saunders et al., 1985.)

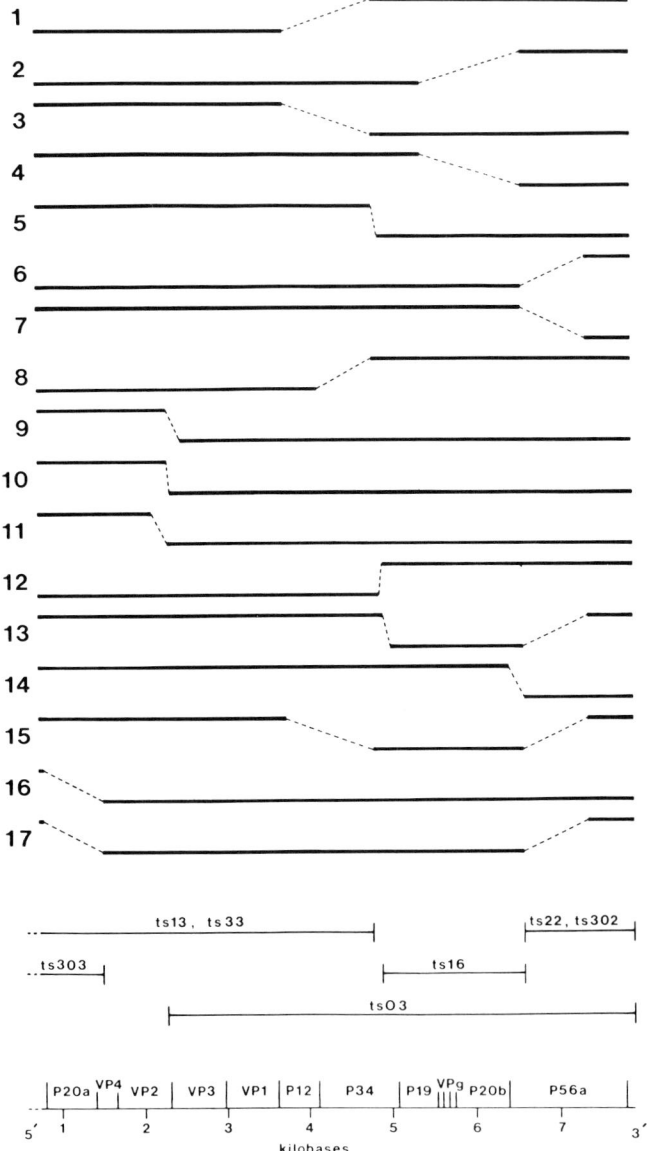

Figure 9.8. Multiple sites of recombination in FMDV. The figure shows maps of the 17 different kinds of inter-subtypic recombinant. Regions of known parentage are represented by solid lines (O_1, upper; O_6, lower); dotted lines linking them indicate cross-over regions. Loci of the parental *ts* mutations, as determined by these recombination studies, are shown at the bottom of the figure. (Reprinted from King *et al.*, 1985.)

have grossly underestimated the recombination frequency, or, more probably, that the occurrence of one recombinational event increases the probability of a second in the same genome.

2. Cross-overs are located in many regions of the genome, at least twelve different sites being needed to explain the data. Given the limited number of genetic markers on which the analysis is based, this result is consistent with a general, rather than a site-specific, mechanism of recombination.

3. The only region of the genome in which recombination is known *not* to have occurred is in the VP3 and adjoining VP1 genes. Since this is the most poorly conserved region of the picornavirus genome (Forss *et al.*, 1984; Stanway *et al.*, 1983), it is possible that recombination in this region was prevented by nucleotide sequence heterogeneity, or that recombination did occur but gave rise to non-functional gene products.

4. All but five of the 43 ts^+ progeny examined proved to be recombinant. The few revertants that did arise were confined to two crosses in which there was no enhancement in the yield of ts^+ progeny over singly infected controls. These results verify that selection at the restrictive temperature does primarily detect recombination, rather than some other process like complementation. This is an important finding since it validates the classical, biological test for recombination on which all the painstaking genetic studies of Cooper, Pringle, McCahon and their associates had been based.

5. Three recombinants exhibited altered, or missing, oligonucleotides. Since all were located far from any known cross-over site, it is assumed that these anomalies were caused by mutations.

G. Genetic Exchanges between Distantly Related Viruses

To assess the potential of recombination for changing viruses, it is important to know the extent to which distantly related strains can exchange genetic information. No attempts at recombining different genera of picornavirus have been reported. (Poliovirus and rhinovirus are closely enough related for a genetic cross to have a fair chance of succeeding.) However, recombinants have been constructed between different serotypes of the same genus.

Agol *et al.* (1984) succeeded in isolating recombinants between poliovirus types 1 and 3 at low but significant frequencies in the range 10^{-5} to 2×10^{-4}. A representative from each of four crosses was analysed. Each possessed type 3 coat proteins and a 3' half of the genome derived from type 1. When one of these recombinants was back-crossed with a type 1 virus, selecting this time with anti-type 3 serum, a double recombinant was formed, entirely type 1 except for a short type 3 stretch in the middle of the genome (Agol *et al.*, 1985).

McCahon *et al.* (1985), in our laboratory, have described a variety of recombinants made between different European serotypes of FMDV (O × A), and also between European and South African serotypes O × SAT2, and A × SAT2; the terms "European" and "South African" refer to the two main evolutionary lines of FMDV (Robson et al., 1977). In the O × A crosses, the proportion of mixedly infected cells yielding recombinants, 1% to 5%, was similar to that of the inter-subtype crosses described above. By contrast, the frequency of recombination between types O and SAT2 was too low to be detected by the infectious centre test, although recombinants did prove to be present among the ts^+ progeny. Figure 9.9 shows the RNA fingerprints of a set of O/SAT2 recombinants from one cross. It can be seen that all the fingerprints resemble that of the type O parent, but with increasing (from left to right) numbers of type O spots replaced by type SAT2 spots. Maps of the recombinant RNAs, based on the known locations of the type O oligonucleotides, show that the recombinants possess various amounts of SAT2 information substituted at the 3' end of the genome (see bottom, Fig. 9.10). Maps of two other O/SAT2 recombinants, Rec E85 and Rec E86, which were isolated from another cross, are also included in Fig. 9.10.

The genetic cross-overs of the two sets of O/SAT2 recombinants shown in Fig. 9.10 differ in orientation, but are all confined to the 3'-terminal one-

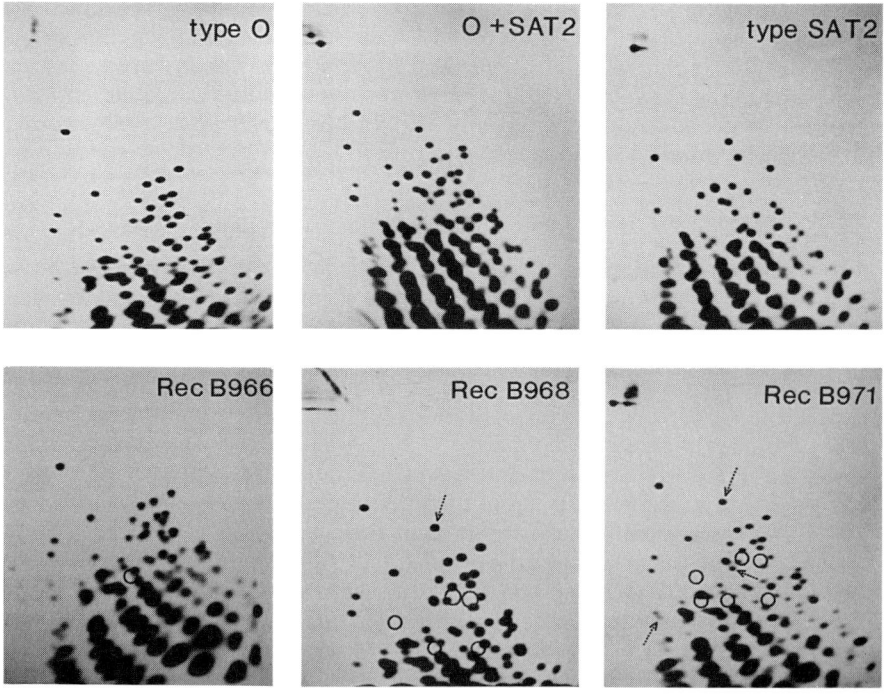

Figure 9.9. RNase-T_1 fingerprints of FMDV strains of type O and type SAT2, and of their recombinants. All three recombinants resemble the type O parent; spots missing are circled, new spots inherited from the SAT2 parent are arrowed. (Reprinted from McCahon et al., 1985.)

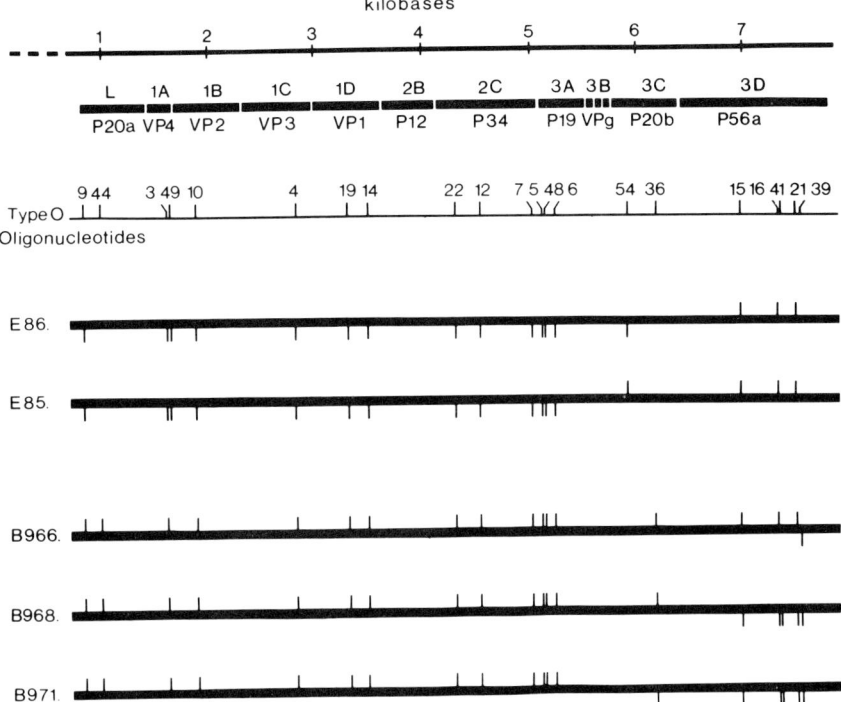

Figure 9.10. Maps of the genomes of type O/type SAT2 recombinants of FMDV, based on the known locations of type O oligonucleotides. Lines pointing upwards indicate type O spots present, and downwards absent. (Reprinted from McCahon et al., 1985.)

third of the genome. By contrast, two European types of FMDV, O and A, were able to recombine over a much wider area of the non-structural coding region (McCahon et al., 1985). To explore whether the most distantly related serotypes can also do this, a cross was performed between a South African strain with a *ts* mutation in the central P34 gene and a European strain with a coat protein mutation. Two recombinant progeny have been characterized, and both exhibit cross-overs in the middle of the genome (B. R. Reavy, unpublished data), similar to the recombinants described by Agol et al. (1984). This result suggests that our ability to generate cross-overs in any region of choice is limited, at least in part, by the availability of selectable markers in the right places.

H. Recombinant Proteins

In addition to fingerprinting the RNAs of the intertypic recombinants, we routinely analysed their induced proteins by gel electrofocusing; examples

O X SAT

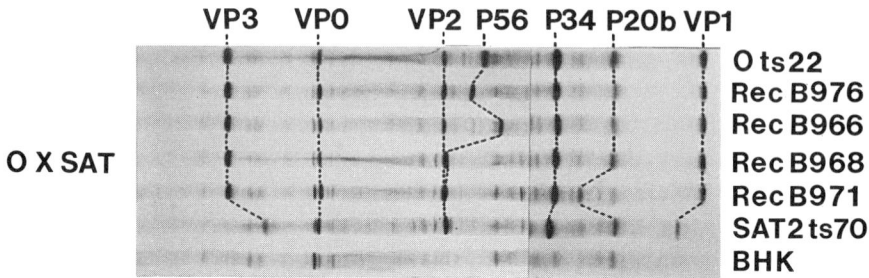

Figure 9.11. Electrofocusing the induced proteins of type O and type SAT2 parents of FMDV, and of their recombinants. Fingerprints of the same viruses are shown in Fig. 9.7. The figure is composed of parts of two electropherograms, joined so that redundant information is omitted. (Reprinted from McCahon et al., 1985.)

of O/SAT2 recombinants are shown in Fig. 9.11. As expected from their RNA compositions (Figs 9.9 and 9.10), these viruses are predominantly O-like. Also as expected, the two recombinants with the most type SAT2 genetic information, Rec B968 and Rec B971, possessed a SAT2-like P56. The interesting point about these analyses is that two proteins, P56 of Rec B968 and P20b of Rec B971, resembled neither parent. Indeed, ten of the eleven O/SAT2 recombinants examined possessed a protein with a novel pI. Since, in each case, that protein is encoded in the vicinity of the cross-over site, we interpret the anomalies as being due to recombination within the protein coding sequence.

I. Nucleotide Sequence Homology at Sites of Recombination

To locate cross-over sites precisely, the parental nucleotide sequences should be as different from each other as possible. To this end, both our research group, and Agol's group in Moscow, are currently engaged in sequencing cross-over regions of intertypic recombinants. As it happens, the first cross-over sequence was reported by Kew and Nottay (1984). The recombinant was recovered from a human contact of a subject who had been vaccinated with trivalent attenuated poliovirus. Primer extension sequencing of the viral RNA showed that a recombination event had occurred near to the 3' end of the P2-X(P2C) gene, the sequence on the 5' side of this site being almost identical to that of the Sabin type 2 strain, and on the 3' side, to Sabin type 1. Between the two, lay a tract of 18 nucleotides — the cross-over region itself — where the parental types were indistinguishable.

We have compared the nucleotide sequence of an FMDV recombinant, Rec B971, with its type O and type SAT2 parents in the region of the cross-over. The sequences of all three viruses were determined from

9. Genetic Recombination in RNA Viruses

cDNA clones that spanned the region between T_1 oligonucleotides O–6 and O–36, these being the limits of the possible cross-over region (see Fig. 9.10). The sequences, part of which are shown in Fig. 9.12, confirm that Rec B971 was formed by a cross-over within the P20b gene, as predicted from the altered pI of P20b (Fig. 9.11). On the 5' side of this point the sequence of Rec B971 is identical to that of the type O parent, and on the 3' side, to the SAT2 parent. With a length of 32 bases, the homology region at the cross-over site is even larger than the one in poliovirus described by Kew and Nottay (1984).

Conclusions drawn from a sample of just two cross-over sequences are necessarily tentative. Yet some comments seem in order.

1. Both the poliovirus and FMDV cross-overs are in conserved genes. For example, the P20b genes of type O and type SAT2 FMDV share 87% nucleotide sequence homology (unpublished data).
2. In neither case could the cross-over be located precisely, because it occurred within a stretch of perfect sequence homology. It is possible, given the high overall homology in the area, that 18- and 32-base homologous stretches were selected by chance. However, the probability of this happening is low, and we strongly suspect that some degree of base pairing is intrinsic to the mechanism of recombination.
3. No sequence longer than a triplet is shared by the two cross-over sites.
4. The surrounding RNA sequences would not be predicted to contain large stem-loop secondary structures.

To summarize, the only obvious feature common to the two cross-over sequences is their large size, and it may be that sequence homology is the sole condition for recombination. However, we need to compare many more cross-over sequences before we can define any more subtle sequence requirements.

```
             5922       5932       5942       5952       5962       5972
Rec     GATCATGCTG GACGGCAGAG CCATGACAGA CAGTGACTAC AGAGTGTTTG AGTTTGAGAT
O       GATCATGCTG GACGGCAGAG CCATGACAGA CAGTGACTAC AGAGTGTTTG AGTTTGAGAT
SAT2    GATCATGATT GACGGCAGGG CCATGACAGA CCGTGATTTC AGAGTGTTTG AGTTTGAGAT

             5982       5992       6002       6012       6022       6032
Rec     CAAAGTAAAA GGACAGGACA TGCTCTCAGA CGCTGCGCTC ATGGTGTTGC ACCGTGGGAA
O       CAAAGTAAAA GGACAGGACA TGCTCTCAGA CGCGGCGCTC ATGGTGCTCC ACCGTGGGAA
SAT2    TAAAGTAAAA GGACAGGACA TGCTCTCAGA CGCTGCGCTC ATGGTGTTGC ACCGTGGGAA

             6042       6052       6062       6072       6082       6092
Rec     CCGCGTGAGA GACATCACGA GACATTCTCG TGATCAAGCA AGAATGAGGA AAGGAACCCC
O       CCGCGTGAGA GACATCACGA AGCACTTTCG TGATACAGCA AGAATGAAGA AAGGCACCCC
SAT2    CCGCGTGAGA GACATCACGA GACATTCTCG TGATCAAGCA AGAATGAGGA AAGGAACCCC
```

Figure 9.12. The nucleotide sequence of a cDNA clone of Rec B971 RNA in the region of the genetic cross-over. Lines indicate identity between the parental and recombinant sequences, the overlap representing the cross-over region. Sequence numbers refer to the positions in the type O genome (Forss et al., 1984.)

J. Homologous Recombination in Coronaviruses

For more than twenty years, picornaviruses appeared to be uniquely endowed with the ability to recombine RNA. Recently however, Lai et al. (1985) have reported the isolation of a recombinant between *ts* mutants of two strains, called A59 and JHM, of mouse hepatitis virus (MHV), a coronavirus. The fingerprint of the genomic RNA revealed that the first 3 kb at the 5' end was inherited from strain JHM, the remainder being like A59. This places the genetic cross-over somewhere in gene *A*, which codes for the viral RNA polymerase. The small mRNAs derived from the 3' end of the genome were, as expected, identical to those of the A59 parent except for those oligonucleotides derived from the leader region, which were like those of strain JHM. This shows that it is possible for the leader RNA of one strain to be fused to the mRNA body of another.

The method that was used to isolate the MHV recombinants, a type of infectious centre method, does not allow a meaningful estimate to be made of recombination frequency. Recently however, Lai and his co-workers (personal communication) have isolated a large variety of recombinants, all with cross-overs in the 5' half of the genome, under conditions that selected against only one of the parents. The results appear to indicate an extremely high recombination frequency in coronaviruses, although it will be necessary to confirm this conclusion by analysing the progeny of a single growth cycle using a standard yield test.

IV. MECHANISM OF RECOMBINATION

From the outset we have distinguished two kinds of recombination in RNA, "homologous" and "non-homologous". That these occur by different mechanisms is suggested by their distinct sequence requirements. The 18- and 32-base homology regions, seen in the two picornavirus cross-over sequences, appear to indicate that a much longer sequence match is required for homologous, than for non-homologous, recombination. In addition, it is hard to envisage how such a high proportion ($>5\%$) of viable recombinants with cross-overs at so many (>12) possible locations, could be produced without invoking some mechanism for aligning the parental RNAs.

Different families of RNA virus appear to vary in their ability to recombine. Pfefferkorn (1977) has reviewed the compelling evidence that Sindbis virus undergoes homologous recombination infrequently, if at all. Such variation implies that RNA recombination is a function of the viral polymerase, rather than the host cell, and therefore that some form of copy-choice mechanism may be responsible. This hypothesis was originally proposed by Cooper *et al.* (1974). Since the poliovirus replicative intermediate is single-stranded *in vivo* (Richards *et al.*, 1984), the growing RNA strand must unwind from its template as it is synthesized. We speculate that, occasionally, the growing strand dissociates from the template, and primes synthesis of a recombinant RNA molecule on a second template. The site of re-initiation is determined by base-pairing between the template and the 3′ end of the growing strand.

The essential difference between this mechanism and the one proposed above for non-homologous recombination lies in the minimum length of base pairing that is required for a primer chain to be elongated. This parameter is likely to be an intrinsic property of the RNA polymerase. As we saw earlier, that length is optimally just one for influenza and Sindbis viruses. We postulate that the polymerase of picornaviruses differs in that it is only able to extend a primer that is bound to the template by a longer stretch of base pairing. We do not yet know how long this stretch must be; as few as six base pairs would ensure that a fair proportion of re-initiations would take place at the same, or homologous, sites.

The mechanism by which coronaviruses synthesize their messenger RNAs provides a model for homologous recombination. Several lines of evidence suggest that the mRNA leader sequence is synthesized independently and then primes transcription at a set of discrete sites on the negative strand template (Baric et al., 1985). All the initiation sites that have been sequenced possess extensive complementarity with the 3′ end of the leader RNA, suggesting that recognition of these sites depends, at least in part, on base pairing between template and primer. The number of base pairs involved, 6 to 8 (Budzilowicz et al., 1985), is sufficient to specify a limited number of genetic loci, and therefore fulfils the basic requirement for homologous recombination stated above.

It thus appears that RNA viruses can be classified into two groups: (a) those viruses, represented by influenza, Sindbis and vesicular stomatitis, that require little homology at the cross-over site, and are prone to making DI RNAs; and (b) viruses like picornaviruses and coronaviruses, which exchange genetic information by homologous recombination but which have only a limited capacity to make DI RNAs (Makino et al., 1984). The extensive homology seen in the cross-over sites of picornaviruses is consistent with the observed requirement of their polymerases for a large pre-formed primer. We speculate that influenza virus may have solved the problem of how to copy the first nucleotide in its genome by having a polymerase able to use a complementary mononucleotide as "primer". This would be consistent with the fact that the 5′-terminal nucleotide of genomic RNA is always a purine.

V. WHY DO RNA VIRUSES RECOMBINE?

Many RNA viruses exchange information by either genetic recombination or reassortment of genome segments, but the reason for their doing so has never been established conclusively. Here are four possible explanations. The last three are not mutually exclusive.

Recombination is a side-effect of replication. The selective advantages of recombination are far from clear. Some viruses, like Sindbis, seem to get along quite well without being able to undergo homologous recombination. Earlier, we blamed that inability on the replication mechanism adopted by those viruses, and it is only consistent to argue that the ability of picornaviruses to recombine is likewise incidental to their replication mechanism.

Recombination generates new virus variants. We have demonstrated that different serotypes of FMDV can recombine and generate novel recombinant proteins. Up to three different serotypes have been recovered from a single animal in the field (Hedger, 1972), showing that conditions do exist in nature for creating new recombinant viruses. The flaw in this theory is that it does not explain why such variants should have a selective advantage; in general, the reverse will be true. It is true that an advantage would be conferred on FMDV by a genetic cross-over in the VP1 gene, that altered antigenicity, but cross-overs within the coat protein genes are highly restricted, and none has ever been observed in VP1, even between closely related strains. Thus, although new recombinant virus strains probably arise occasionally, any advantage will be exceptional. Whether a rapidly mutating virus would be able to maintain the capacity for recombination, solely to take advantage of such rare events, is doubtful.

Recombination is an RNA repair mechanism. According to this theory, recombination between overlapping fragments of viral RNA might generate complete, viable genomes. This could explain the rescue of UV-inactivated poliovirus (Drake, 1958) and FMDV (Pringle, 1965) at high multiplicity of infection. However, the mechanism of RNA repair must be regarded as speculative, since it has yet to be shown at a molecular level that sub-genomic fragments can take part in recombination.

Recombination eliminates deleterious mutations. Much has been written about the high mutation rate of RNA viruses. Reanny (1984) has suggested that the infidelity of RNA polymerases imposes a limit on the maximum size of RNA genomes and that reassortment of genome segments (and hence recombination also) may serve to reduce this "noise". If there has to be a reason for viruses to recombine, then this is the most plausible on several counts. (i) The process would normally involve recombination between isogenic viruses, which occurs at high frequency, throughout the genome. (ii) Mis-sense mutations are constantly arising in RNA virus populations, and are much more likely to be deleterious than advantageous. (iii) The selective advantage of recombination between defective mutants is experimentally demonstrable, both in tissue culture, and in poliovirus vaccinees (Kew and Nottay, 1984; P. Minor, personal communication). (iv) The theory suggests a possible reason why not all unsegmented RNA viruses have the ability to recombine. Sindbis virus does not need to recombine because complementation enables mixtures of defective mutants to grow efficiently. By this means, the virus population can accumulate many deleterious mutations. However, complementation is inefficient in picornaviruses (McCahon, 1981). To the extent that their replicative genes function only in *cis*, deleterious mutations will be propagated poorly. (This explains why the covariant reversion test, as used in Fig. 9.2, is so reliable with FMDV mutants.) In these circumstances, the only way in which picornaviruses can utilize their non-defective genes is by recombination. Thus, picornaviruses have adopted a distinctive replicative strategy, in which the ability to recombine and the inability to complement are two crucial elements ensuring high-fidelity reproduction during passage at high multiplicity.

ACKNOWLEDGEMENT

Work in our laboratory was supported by the Biomolecular Engineering Programme of the Commission of the European Communities, under research contract no. GB1-2-010-UK to D. McCahon.

REFERENCES

1. Agol, V. I., V. P. Grachev, S. G. Orozdov, M. S. Kolesnikova, V. G. Kozlov, N. M. Ralph, L. I. Romonova, E. A. Tolskaya, A. V. Tyufanov and E. G. Viktorova (1984). *Virology* **136**, 41–55.
2. Agol, V. I., S. G. Drozdov, V. P. Grachev, M. S. Kolesnikova, V. G. Kozlov, N. M. Ralph, L. I. Romanova, E. A. Tolskaya, A. V. Tyufanov and E. G. Viktorova (1985). *Virology* **143**, 467–477.
3. Ahlquist, P., E. G. Strauss, C. M. Rice, J. H. Strauss, J. Haseloff and D. Zimmern (1985). *J. Virol.* **53**, 536–542.
4. Baric, R. S., S. A. Stohlman, M. K. Razavi and M. M. C. Lai (1985). *Virus Res.* **3**, 19–33.
5. Budzilowicz, C. J., S. P. Wilczynski and S. R. Weiss (1985). *J. Virol.* **53**, 834–840.
6. Cooper P. D. (1968). *Virology* **35**, 584–596.
7. Cooper P. D., A. Steiner-Pryor, P. D. Scotti and D. Delong (1974). *J. Gen. Virol.* **23**, 41–49.
8. Drake J. W. (1958). *Virology* **6**, 244–264.
9. Emini, E. A., J.Leibowitz, D. C. Diamond, J. Bonin and E. Wimmer (1984). *Virology* **137**, 74–85.
10. Fields, S. and G. Winter. (1982). *Cell* **28**, 303–313.
11. Forss, S and H. Schaller (1982). *Nucl. Acids Res.* **10**, 6441–6450.
12. Forss, S., K. Strebel, E. Beck and H. Schaller (1984). *Nucl. Acids Res.* **12**, 6587–6601.
13. Haseloff, J., P. Goelet, D. Zimmern, P. Ahlquist, R. Dasgupta and P. Kaesberg (1984). *Proc. Natl. Acad. Sci. USA* **81**, 4358–4362.
14. Hedger, R. S. (1972). *J. Comp. Pathol.* **82**, 19–28.
15. Hirst, G. K. (1962). *Cold Spring Harbor Symp. Quant. Biol.* **27**, 303–308.
16. Jennings, P. A., J. T. Finch, G. Winter and J. S. Robertson (1983). *Cell* **34**, 619–627.
17. Kew, O. M. and B. K. Nottay (1984). *In* "Modern Approaches to Vaccines" (Eds R. Chanock and R. Lerner), pp. 375–362. Cold Spring Harbor, New York.
18. King, A. M. Q., W. R. Slade, J. W. I. Newman and D. McCahon (1980). *J. Virol.* **34**, 67–72.

19. King, A. M. Q., B. O. Underwood, D. McCahon, J. W. I. Newman and F. Brown (1981). *Nature (London)* **293**, 479–480.
20. King, A. M. Q., D. McCahon, W. R. Slade and J. W. I. Newman (1982a). *J. Virol.* **41**, 66–77.
21. King. A. M. Q., D. McCahon, W. R. Slade and J. W. I. Newman (1982b). *Cell* **29**, 921–928.
22. King A. M. Q., D. McCahon, J. W. I. Newman, J. R. Crowther and W. C. Carpenter (1983). *Current Topics Microbiol. Immunol.* **104**, 219–233.
23. King, A. M. Q., D. McCahon, K. Saunders, J. W. I. Newman and W. R. Slade (1985). *Virus Res.* **3**, 373–384.
24. Kitamura, N., B. L. Semler, P. G. Rothberg, G. R. Larsen, C. J. Adler, A. J. Dorner, E. A. Emini, R. Hanecak, J. J. Lee, S. van der Werf, C. W. Anderson and E. Wimmer (1981). *Nature (London)* **291**, 547–553.
25. Lai, M. M. C., R. S. Baric, S. Makino, J. G. Keck, J. Egbert, J. L. Leibowitz and S. A. Stohlman (1985). *J. Virol.* **56**, 449–456.
26. Lake J. R., R. A. J. Priston and W. R. Slade (1975). *J. Gen. Virol.* **27**, 355–367.
27. Lazzarini, R. A., J. D. Keene and M. Schubert (1981). *Cell* **26**, 145–154.
28. Ledinko, N. (1963). *Virology* **20**, 107–119.
29. Lowe, P. A., A. M. Q. King, D. McCahon, F. Brown and J. W. I. Newman (1982). *Proc. Natl. Acad. Sci. USA* **78**, 4448–4452.
30. Makino, S., F. Taguchi and K. Fujiwara (1984). *Virology* **133**, 9–17.
31. McCahon, D. (1981). *Intervirology* **69**, 1–23.
32. McCahon, D and W. R. Slade (1981). *J. Gen. Virol.* **53**, 333–342.
33. McCahon, D., W. R. Slade, R. A. J. Priston and J. R. Lake (1977). *J. Gen. Virol.* **35**, 555–565.
34. McCahon, D., A. M. Q. King, D. S. Roe, W. R. Slade, J. W. I. Newman and A. M. Cleary (1985). *Virus Res.* **3**, 87–100.
35. Meier, E., G. G. Harmison, J. D. Keene and M. Schubert (1984). *J. Virol.* **51**, 515–521.
36. Monroe, S. S. and S. Schlesinger (1984). *J. Virol.* **49**, 865–872.
37. Pfefferkorn, E. R. (1977). In "Comprehensive Virology" (Eds H. Fraenkel-Conrat and R. R. Wagner), vol. 9, pp. 209–238. Plenum Press, New York.
38. Plotch, S. J., M. Bouloy, I. Ulmanen and R. M. Krug (1981). *Cell* **23**, 847–858.
39. Pringle, C. R. (1965). *Virology* **25**, 48–54.
40. Reanney, D. (1984). In "The Microbe: Part I, Viruses" (Eds B. W. J. Mahy and J. R. Pattison), pp. 175–196. Cambridge University Press, Cambridge.
41. Richards, O. C., S. C. Martin, H. G. Jense and E. Ehrenfeld (1984). *J. Mol. Biol.* **173**, 325–340.
42. Robson, K. J. R., T. J. R. Harris and F. Brown (1977). *J. Gen. Virol.* **37**, 271–276.
43. Rueckert, R. R. and E. Wimmer (1984). *J. Virol.* **50**, 957–959.
44. Saunders, K. and A. M. Q. King (1982). *J. Virol.* **42**, 389–394.
45. Saunders, K., A. M. Q. King, D. McCahon, J. W. I. Newman, W. R. Slade and S. Forss (1985). *J. Virol.* **56**, 921–929.
46. Stanway, G., A. J. Cann, R. Hauptmann, P. Hughes, L. D. Clarke, R. C. Mountford, P. D. Minor, G. C. Schild and J. W. Almond (1983). *Nucl. Acids Res.* **11**, 5629–5643.
47. Tolskaya, E. A., L. A. Romanova, M. S. Kolesnikova and V. I. Agol (1983). *Virology* **124**, 121–132.
48. Tsiang, M., S. S. Monroe and S. Schlesinger (1985). *J. Virol*, **54**, 38–44.

10. Organization of Plant Virus Genomes that Comprise a Single RNA Molecule

Marie-Dominique Morch and Anne-Lise Haenni

Institut Jacques Monod, C.N.R.S. and Université Paris VII, 2 Place Jussieu – Tour 43, 75251 Paris Cédex 05, France

I.	Introduction	154
II.	Strategies of expression	156
III.	Potyviruses	160
IV.	Tobamoviruses	162
V.	Tymoviruses	166
VI.	Satellites	168
	A. Satellite of the TNV group	169
	B. Satellites of the Tombusvirus group	170
	C. Satellites of the Sobemovirus group	170
VII.	Conclusions and perspectives	171
	References	172

Abbreviations: AlMV, alfalfa mosaic virus; BMV, brome mosaic virus; BYDV, barley yellow dwarf virus; CarMV, carnation mottle virus; CcTMV, cowpea strain of tobacco mosaic virus; CGMMV, cucumber green mottle mosaic virus; CLV, carnation latent virus; LTSV, lucerne transient streak virus; MCDV, maize chlorotic dwarf virus; NMV, narcissus mosaic virus; PeMV, pepper mottle virus; PMV, papaya mosaic virus; PRSV, papaya ringspot virus; PVX, potato virus X; PVY, potato virus Y; SBMV, southern bean mosaic virus; SBYV, sugar beet yellows virus; SCMoV, subterranean clover mottle virus; SMV, soybean mosaic virus; SNMV, solanum nodiflorum mottle virus; S-TCV, satellite of turnip crinkle virus; STNV, satellite tobacco necrosis virus; TBSV, tomato bushy stunt virus; TCV, turnip crinkle virus; TEV, tobacco etch virus; TMV, tobacco mosaic virus; TMV-L, tomato strain of tobacco mosaic virus; TNV, tobacco necrosis virus; TRosV, turnip rosette virus; TVMV, tobacco vein mottling virus; TYMV, turnip yellow mosaic virus; VTMoV, velvet tobacco mottle virus.

THE MOLECULAR BIOLOGY OF THE
POSITIVE STRAND RNA VIRUSES   ISBN 0-12-599930-5

Copyright © 1987 Academic Press Inc. (London) Ltd.
All rights of reproduction in any form reserved

I. INTRODUCTION

In the plant kingdom, viruses whose genomes are made up of RNA are by far in the majority. This contrasts with viruses of higher animals, arthropods and bacteria, where both RNA and DNA types are abundant. The situation among plant viruses is strikingly demonstrated by the fact that until fairly recently, only RNA viruses were known to infect plants. The reasons for this bias in favour of RNA genomes, and whether it corresponds to a selective advantage for the viruses are still unknown.

Viruses with an RNA genome make up about 94% of all plant viruses described (Hull and Davies, 1983). They are classified into groups or families on the basis of their morphology and structure, their serological relationships, and certain elements of their genomic organization. The members thus classified in a given group generally possess other common properties such as stability, mode of transmission and translational strategies. The vast majority ($\sim 82\%$) of plant RNA viruses contain as their genome, single-stranded RNA of positive polarity (i.e. of messenger sense). They contain either a monopartite, a bipartite or a tripartite genome.

This review is concerned solely with plant viruses possessing a monopartite RNA genome of positive polarity, i.e. a single RNA molecule that is infectious and thus contains all the information required for the replication of the virus. After a general presentation of these viruses, the recent developments in understanding of the translational strategies they use will be considered with the help of a few well-studied examples. The reader is referred to other recent reviews for additional information (Atabekov and Morozov, 1979; Davies and Hull, 1982; Van Vloten-Doting and Neeleman, 1982; Joshi and Haenni, 1984).

Plant viruses with a single-stranded RNA genome of positive polarity have been classified into 11 groups (Matthews et al., 1982). Here, carnation mottle virus (CarMV) has been added to this list as a twelfth group; a taxonomic designation is however still pending for this virus as well as for a number of other viruses. Table 10.1 lists the 12 virus groups, gives the name of the type member of each group, the morphology of the virus particle, the molecular weight of its RNA and the mode of transmission of the virus.

Mechanical transmission is common to most plant RNA viruses. In the field, arthropods (aphids, leafhoppers and beetles) are the main vectors of viruses. Two modes of transmission by aphids can be distinguished. (1) In the circulative (persistent) mode of transmission, the virus is acquired via the food canal, absorbed and transported to the salivary glands. It is designated propagative if the virus multiplies in the aphid, or non-propagative if it does not. (2) The non-circulative (including non-persistent and semipersistent subcategories), possibly stylet-borne, mode of transmission is characterized by the loss of vector transmission through

Table 10.1. Certain characteristics of plant viruses with monopartite positive stranded RNA genomes

Group and/or type member	Morphology	Mol.wt. of genomic RNA (MD)	Transmission
Carlavirus, carnation latent virus (CLV)	flexuous rods	2.7	mechanically, sometimes by aphids non-persistently
Carnation mottle virus (CarMV)	isometric	1.3–1.4	
Closterovirus, sugar beet yellows virus (SBYV)	flexuous rods	2.2–4.7	mechanically (with difficulty), some by aphids semi-persistently
Luteovirus, barley yellow dwarf virus (BYDV)	isometric	2.0	aphids persistently
Maize chlorotic dwarf virus (MCDV)	isometric	3.2	leafhopper semi-persistently
Potexvirus, potato virus X (PVX)	flexuous rods	2.1	mechanically, by contact between plants, no known vectors
Potyvirus, potato virus Y (PVY)	flexuous rods	3.0–3.5	mechanically, aphids non-persistently, some by whiteflies, mites, fungi
Sobemovirus, southern bean mosaic virus (SBMV)	isometric	1.4	mechanically, beetles, seed
Tobacco necrosis virus (TNV)	isometric	1.3–1.6	mechanically, through soil by chytrid fungus
Tobamovirus, tobacco mosaic virus (TMV)	rigid rods	2.0	mechanically, some by seed
Tombusvirus, tomato bushy stunt virus (TBSV)	isometric	1.5	mechanically, soil
Tymovirus, turnip yellow mosaic virus (TYMV)	isometric	2.0	mechanically, beetles

Adapted from Matthews *et al.* (1982) except for CarMV, where indications are from Tremaine (1970), Hull (1977), Waterworth and Kaper (1972) and Nelson and Tremaine (1975). The international name of the group is italicized where reported.

moulting and absence of virus from the salivary glands. Some aphids transmit only a particular virus, usually in a persistent manner, whereas others can transmit over 70 different viruses, usually in a non-persistent or semipersistent manner. Further information can be found in the review of Harris (1983).

The viruses considered here are all simple and non-enveloped. The RNA is contained within a protein capsid for which it displays more or less affinity depending on the virus. The capsid is composed of only one molecular species of small molecular weight (generally 20–35 kD), the virus-coded coat protein. A large number of such coat protein molecules assemble to form the virus capsid. Consequently, the virus must preferentially synthesize this structural protein, since many copies are required to encapsidate a single RNA molecule.

The molecular weight of the RNA rarely exceeds 3 MD ($\sim 10\,000$ nucleotides). In addition to the coat protein, the RNA also directs the synthesis of non-structural proteins most likely involved in functions such as viral RNA replication, the formation of inclusion bodies, cell-to-cell migration of the virus, its transmission by insects, or other as yet unknown functions.

At their 5' end, the viral RNAs generally have either a cap structure or a small virus-coded protein termed VPg. The 5' terminal non-coding segment of the viral RNAs varies in length from 10 nucleotides in one of the tobacco mosaic virus (TMV) subgenomic RNAs (Guilley et al., 1979) to 95 nucleotides in the turnip yellow mosaic virus (TYMV) genomic RNA (Briand et al., 1978); in the latter case, the second AUG from the 5' terminus is probably the initiation triplet (Briand et al., 1978). The 3' noncoding region of the RNA adopts one of 3 features: the RNA can be aminoacylated at its 3' end by a specific amino acid (Haenni et al., 1982), it can be terminated by a poly(A) stretch, or it may possess neither feature.

II. STRATEGIES OF EXPRESSION

In spite of the large amount of information that has accumulated on these viral RNAs, the strategies of expression used by viruses of only eight groups have been examined in any detail. These are listed in Table 10.2. The strategies assigned to these viruses are mainly based on translation studies *in vitro*. However, in certain cases, experiments *in vivo* and recently nucleotide sequence data, make it possible to verify the strategies proposed on the basis of experiments *in vitro*. Such comparisons have already been possible with the tobamoviruses.

It is generally agreed that cellular and viral messenger RNAs of eukaryotic origin are functionally monocistronic even though in the case of many plant and animal viruses, the RNAs are structurally polygenic, i.e. they contain information for the synthesis of several proteins. In the genomic RNAs of such viruses however, only the 5' proximal gene can be expressed as protein, the other genes being silent. Since the single RNA species of viruses with a monopartite genome is infectious, it must contain the information for the synthesis of the non-structural as well as the structural (capsid) proteins. Consequently such viruses have resorted to several strategies to produce their different proteins.

A number of strategies employed by these viruses are schematized in Fig. 10.1.

1. Premature termination at the level of a stop signal different from a termination codon leads to the formation of two proteins that differ in size,

Virus group	Type and other members	Structure 5'	Structure 3'	Additional encapsidated RNA	PreT	RT	PTcl	MRF	SG[a]	Int.Init.	Reference
Carnation mottle virus	CarMV	?	?	SG		+			+		[17, 34]
Potexvirus	PVX	cap	poly(A)	0		(+)					[71, 96, 107]
	PMV	(cap)	?	0					+		[11]
	NMV	?	?	SG					+		[94]
Potyvirus	PVY, PRSV, TEV	VPg	poly(A)	0	(+)	(+)	+		(+)		[2, 22, 36, 58, 108]
	PeMV, SMV	VPg	?	0	(+)	(+)	+		(+)		[22, 99]
	TVMV	VPg	?	0		(+)	(+)		(+)		[42, 95]
Sobemovirus	SBMV	VPg	pXOH	SG				+	+		[27, 28, 89, 90]
	TRosV	VPg	(pXOH)	SG					+		[59, 72]
	SNMV, LTSV	?	?	SG, Sat		+	+[c]		+		[48, 49, 59]
	VTMoV	?	?	SG, (Sat)		+	+		+		[59]
Tobacco necrosis virus	TNV	ppA	pXOH	(SG), Sat[b]					(+)	+	[18, 105, 106]
Tobamovirus	TMV	cap	tRNA[His]	SG		+		+	+		[7, 16, 30, 57, 77, 81, 112]
	TMV-L	cap	(tRNA[His])[d]	(SG)		(+)[d]		(+)[d]	(+)[d]		[66, 79, 97]
	CcTMV	(cap)	tRNA[Val]	SG				(+)[d]	(+)[d]		[6, 65]
	CGMMV	(cap)	tRNA[His]	(SG)				(+)[d]	(+)[d]		[50, 67]
Tombusvirus	TCV[e]	?	?	SG, Sat	+	+		+	+		[3, 23]
Tymovirus	TYMV	cap	tRNA[Val]	SG				+	+		[15, 60, 68–70, 83, 85, 87]

PreT = premature termination; RT = read-through; PTcl = post-translational cleavage; MRF = multiple reading frames; SG = subgenomic; Int.Init. = internal initiation; ? = unknown; ( ) = to be confirmed; 0 = none. [a]SG could result from RNA fragmentation. [b]The satellite is encapsidated in separate capsids composed of its own coat protein molecules. [c]This strategy is not used by LTSV(ref. 59). [d]Deduced from cDNA sequence data. [e]The strategy of expression of the type member, TBSV, is not known; TCV, although serologically unrelated to TBSV, is a tentative member of the tombusvirus group.

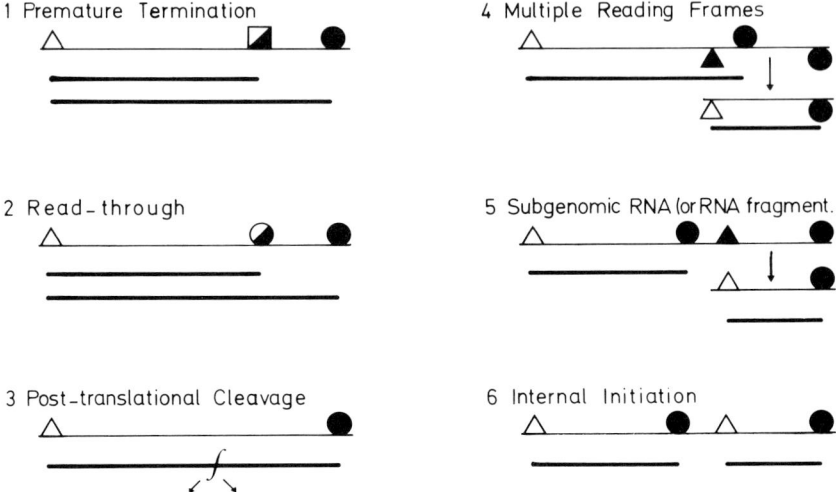

Figure 10.1. Translation strategies of plant viruses with a monopartite RNA genome. Viral RNAs and corresponding translation products are schematized by thin and thick lines respectively. △(▲), accessible (inaccessible) initiation codon; ●, termination codon; ◐, leaky termination codon; ◢, signal for alternative arrest of translation (premature termination); ∫, site of proteolytic processing; "fragment.", fragmentation. Translation signals are placed above or below the RNA to specify the utilization of different reading frames.

but whose amino acid sequences are identical over the total length of the shorter protein. The nature of these stop signals is unknown.

2. Read-through or suppression of a termination codon also leads to the appearance of two proteins. As in the preceding strategy, these two proteins are identical over the total length of the shorter protein; the longer one results from the elongation of the shorter protein beyond the termination codon mediated by a suppressor tRNA. One can easily envisage that in this and in the preceding strategy these proteins have different functions depending on whether or not they are elongated. Examples of such a situation will be illustrated in the following sections.

3. Post-translational cleavage of a polyprotein that yields mature functional proteins.

4. Multiple reading frames: this leads to the synthesis of distinct proteins. In the cases described to date, this strategy has involved the formation of a second messenger RNA termed subgenomic RNA.

5. Whether or not multiple reading frames are used, most viruses produce, in addition to the genomic RNA, one or more subgenomic RNAs that derive from the genomic RNA by an unknown mechanism sometimes postulated to be RNA fragmentation. These subgenomic RNAs are not

required for infectivity. The cistrons that in the genomic RNA occupied internal positions and could not be expressed are now 5' proximal in the various subgenomic RNAs and can be translated. This strategy is commonly used by numerous plant viruses for the synthesis of their coat protein. Whether or not the subgenomic RNA is encapsidated depends on the virus (see Table 10.2).

6. Internal initiation, similar to a strategy commonly used by prokaryotic systems, has been proposed in some cases. This model has been abandoned in the case of CarMV, but subsists for the moment for other viruses (see Table 10.2).

All these strategies not only allow the virus to synthesize more than one protein from a unique RNA, but also permit the virus to modulate the relative abundance of the viral proteins. In the case of premature termination (strategy 1) or read-through (strategy 2), the relative amount of the two proteins depends on the efficiency with which the termination signal is overcome, for example on the amount of a suppressor tRNA and on the nucleotide sequence surrounding the termination codon. The simultaneous presence of several RNAs (strategy 5) also leads to a possible regulation of the expression of these RNAs, since some have a much greater affinity for the translational machinery than others. In particular this is the case for the subgenomic RNAs that code for the viral coat protein which are preferentially translated.

Most viruses make use of more than one strategy for the synthesis of their proteins. Table 10.2 summarizes the data presently available concerning the situations encountered in the eight most studied virus groups. The structure at the 5' and 3' end of the RNA is given whenever it has been determined. The table also indicates whether additional RNAs other than the genomic RNA are encapsidated. For each family the members have been grouped when they possess similar characteristics and adopt similar strategies. In some cases the results or strategies are still not well established and this is indicated by parentheses. Among the potexviruses, narcissus mosaic virus (NMV) uses a subgenomic RNA to produce the coat protein, whereas the type member, potato virus X (PVX) and papaya mosaic virus (PMV) do not. Certain recently isolated viruses with characteristics of sobemoviruses also differ from the type member, southern bean mosaic virus (SBMV), by the presence of an encapsidated satellite RNA (see Section VI). In the case of the tobamoviruses, RNA from one member, the cowpea strain of TMV (CcTMV) = sunn-hemp mosaic virus can be aminoacylated with valine, whereas RNA of common strain TMV and cucumber green mottle mosaic virus (CGMMV) are aminoacylated with histidine. Moreover, in CcTMV (and presumably also in CGMMV) the subgenomic RNA coding for the coat protein can be encapsidated, whereas that of common TMV (or presumably in the tomato strain of TMV (TMV-L)) is not. Finally, TYMV is exceptional since it uses five translational strategies; these have been established on the basis of experiments *in vitro*.

A few examples of viruses for which the strategies of expression have been extensively studied and are now well documented have been selected and are illustrated in the following sections. These are the potyviruses, the tobamoviruses and the tymoviruses. A section is also devoted to satellites that accompany certain monopartite RNA viruses, since the diversity in the nature of satellites, in the expression of their RNA and in the mechanisms whereby they alter the symptoms produced by viruses is being actively pursued.

III. POTYVIRUSES

The potato virus Y group or potyvirus group is one of the largest groups of plant viruses, containing over 60 members. Potyviruses are flexuous rods (filaments) 680–900 nm long and 11 nm wide. They are unusual because they induce in the host cells the production of large amounts of non-structural virus-coded proteins that aggregate to form inclusion bodies. These include cytoplasmic (cylindrical, and in the case of papaya ringspot virus (PRSV) also amorphous (De Mejia et al., 1985)) inclusion bodies that contain at least one virus-specific protein of 68–79 kD. Some potyviruses such as tobacco etch virus (TEV) also induce the formation of nuclear inclusion bodies composed primarily of two other virus-specific proteins of 49 and 54 kD (Dougherty and Hiebert, 1980). In addition, the RNA of potyviruses also codes for a helper component of 53 kD needed for aphid transmission (Pirone and Thornbury, 1984). The RNA molecules of at least some potyviruses such as TEV, potato virus Y (PVY) and tobacco vein mottling virus (TVMV) appear to carry a 22 kD protein at their 5' ends termed VPg (Hari, 1981; Khaddam, 1985; Siaw et al., 1985). Although not yet demonstrated, it seems likely that VPg is coded by the virus. The genomic RNA (3.5 MD) of at least some potyviruses carries a poly(A) stretch at the 3' end (Hari et al., 1979; Allison et al., 1985; Khaddam, 1985).

The genomic organization and the translational strategies used by potyviruses have been the subject of several investigations. In these studies, the RNA of PRSV, TEV, TVMV, pepper mottle virus (PeMV) and soybean mosaic virus (SMV) have been the most frequently used. The data obtained originally led to the proposal of several strategies of expression for this virus group, including read-through, post-translational cleavage, and even RNA fragmentation and/or internal initiation although these are slowly being abandoned. Considerable controversy still exists concerning the strategies used in spite of the availability of antibodies against the various viral proteins. Such antibodies have served to identify the corresponding proteins from among the translation products and thereby locate the genes on the RNA. Although this led to the assignment of functions for several of the non-structural proteins in addition to the coat protein, the interpretation of the results was hampered by the fact that a given antiserum immunoprecipitates many virus-coded proteins.

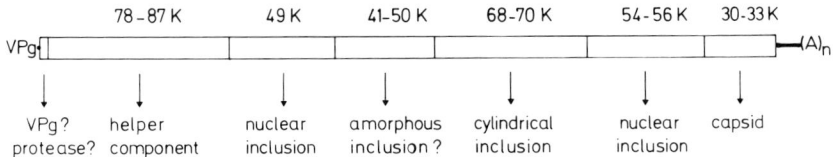

Figure 10.2. Proposed combined genetic map of potyvirus RNA (adapted from Dougherty and Hiebert, 1980). Boxes on the RNA represent coding regions. The approximate molecular weight of the translation products (K = kilodalton) is indicated above each box. A function (below the box) has been proposed for each protein (see text). Question mark signifies that no evidence is as yet available for this assignment.

The genetic map proposed in Fig. 10.2 is based on a compilation of the results of translation experiments *in vitro* and immunoprecipitation studies with various potyvirus RNAs. At least seven (6 + VPg) genes are most likely contained in the genomic RNA. The gene coding for VPg could be 5' proximal (Hellmann et al., 1985); the second gene from the 5' end could code for the helper component (Hellmann et al., 1985); the third and sixth for the nuclear inclusion proteins (Dougherty and Hiebert, 1980); the fifth for the cylindrical inclusion protein (Dougherty and Hiebert, 1980) and the seventh for the coat protein (Dougherty and Hiebert, 1980; Allison et al., 1985). This would leave the fourth gene as possibly encoding the amorphous inclusion protein. Still to be located on the genomic map is a putative protease that might be involved in post-translational cleavage; Hellmann et al. (1985) have suggested that its gene is a candidate for the 5' position on the genome.

Sequence determination of the 3' terminal approximately 2300 nucleotides of TEV (Allison et al., 1985) has now unambiguously located the coat protein gene of this virus in the 3' region of the RNA. It is positioned upstream of the poly(A) stretch and of a 189 nucleotide-long untranslated region, and is produced by processing of a precursor polypeptide. The 85 kD precursor, which is synthesized *in vitro*, is immunoprecipitated with antisera to either the coat protein or the 54 kD nuclear inclusion bodies. On the other hand, by hybrid-arrested translation, Hellmann et al. (1985) have mapped the position of the helper component gene near the 5' terminus of TVMV RNA.

Although a large body of information suggested that the translation products of potyvirus RNAs might undergo proteolytic cleavage (Hellmann et al., 1983; Dougherty, 1983; Vance and Beachy, 1984), clear evidence for the formation of a polyprotein and for a direct precursor–product relationship between the translation products has only recently been provided. Using the RNA of PRSV, Yeh and Gonsalves (1985) have demonstrated the synthesis *in vitro* of a 330 kD polyprotein that corresponds to nearly the total coding capacity of the viral RNA, and have established a relationship between this polyprotein and subsequent cleavage products. Based on an observation made by Pelham (1979) that dithiothreitol (DTT) can favour post-translational cleavage, incubations were carried out in the absence of DTT. In these conditions, all the translation products shifted to high molecular weight proteins and a 330 kD protein was the major product formed. This polypeptide is immunologically related to the cylindrical inclusion bodies, to the coat protein and also to the amorphous inclusion bodies characteristic of PRSV. Two shorter polypeptides of 112 kD and 220 kD that derive from the 330kD polyprotein react with antibodies against the amorphous inclusions, and against both the cylindrical inclusions and the coat protein respectively. These experiments will undoubtedly open the way to a clearer understanding of the mechanisms involved in the synthesis of the potyvirus products.

The results of Hellman et al. (1985), however, raise new questions concerning the translational strategies of potyviruses. Sequencing of DNA probes complementary to nucleotides contained within the 5′ terminal third of the genome of TVMV reveals that the region encoding the C-terminus of the helper component is terminated by two UAG triplets in tandem. Although this observation could exclude a single translation product for TVMV, the authors propose that *in vivo* the overcoming of the UAG codons might regulate the level of production of the different virus proteins, either by suppression or by frame-shifting.

IV. TOBAMOVIRUSES

Of monopartite RNA plant viruses, tobacco mosaic virus (TMV) has certainly been the most studied. Its contribution to virology and molecular biology is impressive. It was the first virus to be purified, the first one for which the genetic material was shown to be RNA, the first to be reassembled and crystallized. It was one of the first whose genome was used for translation studies *in vitro*, and it was the first plant virus whose genome was entirely sequenced. The structure of TMV is reviewed in Chapter 14 of this volume.

Tobamoviruses consist of rigid rod-like nucleoprotein particles. Over 2000 copies of viral coat protein encapsidate a single genomic RNA of 2 MD. The genomic RNA is capped (Zimmern, 1975) and that of the common (wild-type or *vulgare*) TMV strain can be specifically aminoacylated at its 3′ end with histidine (Öberg and Philipson, 1972). In addition to the genomic RNA, 3′-coterminal subgenomic RNAs of which a major species of 0.68 MD designated I_2 RNA are present in the viral RNA preparations; the latter RNA is encapsidated in intermediate-length rods (Beachy and Zaitlin, 1977) and appears not to be capped (Joshi et al., 1983b; Hunter et al., 1983).

In TMV-infected tobacco leaves, at least eight 3′-coterminal RNAs are formed. These RNAs include the genomic RNA, the I_2 RNA and the coat protein mRNA as well as additional subgenomic RNAs whose translation products have not been characterized *in vivo* but which have mRNA properties *in vitro* (Goelet and Karn, 1982).

The genetic map of TMV RNA is schematized in Fig. 10.3. When introduced into a cell-free system *in vitro*, the RNA isolated from TMV directs the synthesis of several non-structural polypeptides, among them a major 126 kD protein and a less abundant 183 kD protein. The strategies adopted by this virus for the synthesis of these two high molecular weight proteins was first elucidated by Pelham (1978). The 126 and 183 kD proteins are initiated on the same AUG and have overlapping amino acid sequences. The 126 kD protein gene is terminated by a UAG codon. The leakiness of this codon leads to the synthesis in small amounts of the read-through protein of 183 kD. The synthesis of the latter protein can be considerably increased by addition to the system *in vitro* of a yeast tRNA suppressor of the UAG codon.

In TMV-infected cells, high molecular weight proteins appear, comigrating with the 126 kD and 183 kD proteins synthesized *in vitro* (Paterson and Knight, 1975; Beier *et al.*, 1980). The production of the 183 kD protein *in vivo* has led to the suggestion that tobacco plants and possibly other eukaryotic systems contain a natural UAG suppressor tRNA. Such a suppressor tRNA has indeed been isolated and sequenced from tobacco leaves (Beier *et al.*, 1984a), wheat leaves and wheat germ (Beier *et al.*, 1984b). A UAG suppressor tRNA has also been isolated from *Drosophila melanogaster* (Bienz and Kubli, 1981). In all cases, the suppressor tRNA is a species acceptor of tyrosine with the anticodon GΨA. Another major tRNA species acceptor of tyrosine that does not behave as a UAG suppressor has the anticodon QΨA. Thus the replacement of the G in the first position of the anticodon by queuosine (Q) restricts recognition of the tyrosyl-tRNA to the tyrosine codons and prevents the tRNA from reading the UAG termination triplet.

In addition to the two high molecular proteins of 126 and 183 kD, a 30 kD protein is also produced *in vitro*. It has no peptides in common with the high molecular weight proteins and is synthesized from the subge-

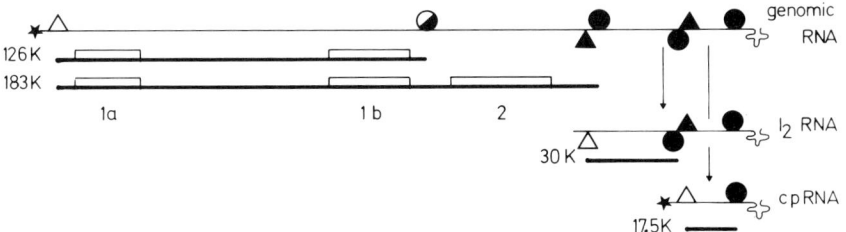

Figure 10.3. Genetic map of tobacco mosaic virus RNA (based on *vulgare* strain). Symbols are as in Figure 1; cp, coat protein; ★, cap structure; ⇌, tRNA-like structure. Boxes 1a, 1b and 2 correspond to regions of homology with proteins encoded by other viruses (see text). (K = kilodalton.)

nomic I$_2$ RNA (Beachy and Zaitlin, 1977). It is also produced in infected tobacco leaves (Joshi et al., 1983b) and tobacco protoplasts (Beier et al., 1980; Watanabe et al., 1984). It appears early after the onset of infection, before the synthesis of the high molecular weight proteins (Joshi et al., 1983b).

The RNA extracted from the common strain of TMV does not direct the synthesis *in vitro* of the viral coat protein of 17.5 kD, even though it contains information for the synthesis of this protein. From infected tobacco plants, however, where synthesis of the coat protein is intense, Hunter et al. (1976) isolated from polysomes a short RNA species active in the synthesis of the coat protein. This subgenomic RNA is capped at its 5′ end (Guilley et al., 1979) and is 3′ coterminal with the genomic and I$_2$ RNAs.

The entire nucleotide sequence of the genomic RNA of the *vulgare* TMV strain (Goelet et al., 1982) and of the TMV-L strain (Ohno et al., 1984), and large parts of the RNA sequence of the other tobamoviruses CcTMV (Meshi et al., 1981, 1982) and CGMMV (Meshi et al., 1983a) have been established. The sequence of the *vulgare* TMV RNA and of the TMV-L RNA has enabled a comparison of the open reading frames present in the virus genome with the polypeptides that appear in the systems *in vitro* or in infected plant material, and has confirmed the genome organization proposed for tobamoviruses (Fig. 10.4). Three open reading frames are detected. In both strains the longest one contains the 183 kD protein gene interrupted by the leaky UAG codon. The C-terminal end of this protein overlaps by five codons an open reading frame in a

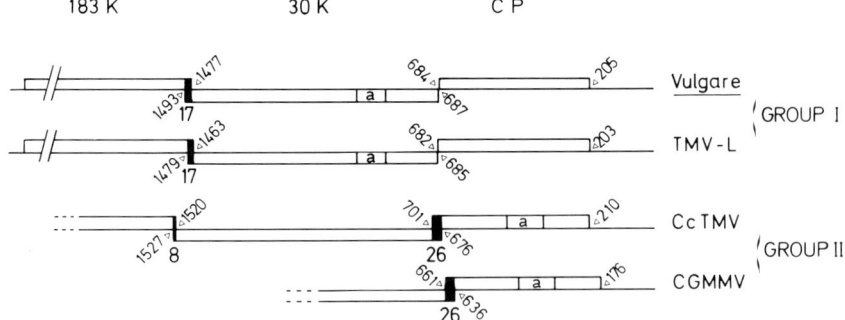

Figure 10.4. Genome organization in the 3′ region of four TMV strains (adapted from Takamatsu et al., 1983). The regions encoding the coat, the 30 kD and the C-terminal part of the 183 kD proteins are boxed above or below the RNA depending on the reading frame. Nucleotides are numbered starting from the 3′ end. The number of the last nucleotide of each cistron (including termination codon) as well as of the first nucleotide of the 30 kD and coat protein cistrons are indicated when determined. Overlapping coding regions are shown solid and their size in nucleotides is written below. Boxed "a" represents the assembly origin for encapsidation. (K = kilodalton.)

second phase that encodes the 30 kD protein. The gene for the 30 kD protein is followed two untranslated nucleotides downstream by the coat protein gene. Thus the 30 kD protein is read in a different frame from the one used for the 183 kD protein and the coat protein. In CcTMV the 30 kD protein is 15 residues longer than in the *vulgare* and TMV-L strains and overlaps in its 5' and 3' regions with the 183 kD and the coat protein genes respectively.

The assembly origin for encapsidation of *vulgare* TMV and TMV-L is contained within the 30 kD protein gene. This results in the encapsidation of the genomic and the I_2 RNAs, but not of the coat protein mRNA. In CcTMV and CGMMV, the position of the assembly origin is shifted to the coat protein gene, which explains why the coat protein mRNA of CcTMV at least is encapsidated. On the basis of the location of the assembly origin and of the genome structure of several tobamoviruses, Fukuda *et al.* (1981) and Takamatsu *et al.* (1983) have classified these viruses into two groups (Fig. 10.4). In the RNAs of group I (*vulgare* and L strains) two nucleotides separate the 30 kD and the coat protein genes, and the assembly origin is located in the 30 kD protein gene. In the RNAs of group II (CcTMV and CGMMV), the 30 kD and the coat protein genes overlap, and the assembly origin is located in the coat protein gene.

The search for the role of the non-structural proteins of TMV has been facilitated by the availability of the nucleotide sequence of the viral RNA. Indeed, homologies are striking in the amino acid sequences of the non-structural proteins deduced from the nucleotide sequence of plant RNA viruses differing in their genomic organization and of certain animal RNA viruses. Thus, the two regions of the 126 kD protein of TMV designated 1a and 1b in Fig. 10.3 show sequence homologies with two corresponding regions in the proteins coded by the RNAs 1 of alfalfa mosaic virus (AlMV) and brome mosaic virus (BMV), both tripartite plant RNA viruses (Cornelissen and Bol, 1984; Haseloff *et al.*, 1984). Furthermore, a large portion (designated 2 in Fig. 10.3) of the read-through region of the 183 kD protein bears important similarities with the protein coded by the RNA 2 of AlMV and BMV (Cornelissen and Bol, 1984; Haseloff *et al.*, 1984), and with a read-through domain in the non-structural polyprotein of Sindbis virus and Middelburg virus, both animal alphaviruses (Haseloff *et al.*, 1984). In addition, the read-through portion of the 183 kD protein of TMV also exhibits distant amino acid sequence homology to the RNA-dependent RNA polymerase of poliovirus (Kamer and Argos, 1984). The products encoded by AlMV and BMV RNAs 1 and 2 are involved in viral RNA replication (discussed by Haseloff *et al.*, 1984) as is also the read-through polyprotein of Sindbis virus (Schlesinger, 1980; see also Chapters 6 and 11 of this volume for discussion of these similarities).

Although it has been suggested that the 126 kD protein could participate in viral RNA replication, this protein has so far not been detected in preparations of RNA-dependent RNA polymerase complexes from TMV-infected tissues. It has been reported that the 126 kD protein of TMV is associated with host chromatin in infected plants, suggesting that this protein might directly interfere with host gene expression, leading to pathogenesis and production of symptoms (Van Telgen et al., 1985). One can postulate that viruses coding for similar polypeptides might cause similar diseases. Hence, if the proteins coded by the RNAs 1 of AlMV and BMV can associate with host chromatin as does the 126 kD protein of TMV, such association could constitute a more general strategy of plant RNA viruses to affect host cells and cause disease. Recent sequencing of an attenuated strain of TMV-L reveals that three amino acid substitutions in the 126 kD protein are probably sufficient to abolish production of symptoms in tomato leaves (Nishiguchi et al., 1985).

The search for a possible role for the viral proteins of TMV has also been facilitated by the availability of mutants. Among these, some are altered in the function of systemic spreading from one cell to another. The examination of a temperature-sensitive strain of TMV led Zaitlin and Leonard (1982) to propose that the 30 kD protein might be implicated in cell-to-cell migration of the virus. Ohno et al. (1983) have compared the nucleotide sequence of the 30 kD protein gene and the amino acid sequence of the 30 kD protein itself between wild-type and a temperature-sensitive mutant defective in cell-to-cell movement at non-permissive temperatures. In the mutant, a base substitution in the first position of the codon for Pro153 results in the replacement of this amino acid by Ser in the 30 kD protein. This change occurs in a highly conserved region of the 30 kD protein, supporting the notion that this protein is somehow involved in virus transport. This in turn could determine the host range of the virus and could also affect virus multiplication in a given host. Moreover, the most striking differences between common TMV and TMV-L lie in the C-terminal part of the 30 kD protein. As discussed by Ohno et al. (1984), although TML-V infects both tomato and tobacco plants, it multiplies much more in tomato plants, and this could be due at least in part to the 30 kD protein. However, nothing is known about the properties and function of the transport protein, and it still remains to be demonstrated that the 30 kD protein is indeed the transport protein.

V. TYMOVIRUSES

Turnip yellow mosaic virus (TYMV) is the type member of this virus group. The genetic map showing the strategies used by this icosahedral virus is schematized in Fig. 10.5. In addition to the genomic RNA of 2 MD, TYMV also encapsidates a major subgenomic RNA species which codes *in vitro* for the coat protein. The sequence of the coat protein subgenomic RNA of 0.24 MD has been established (Guilley and Briand, 1978). It corresponds to the 3' terminal portion of the genomic RNA and like the genomic RNA can be valylated *in vitro* and *in vivo* at its 3' end (Haenni et al., 1982).

When translation systems *in vitro* are programmed with unfractionated

10. Plant Virus Genomes of a Single RNA Molecule

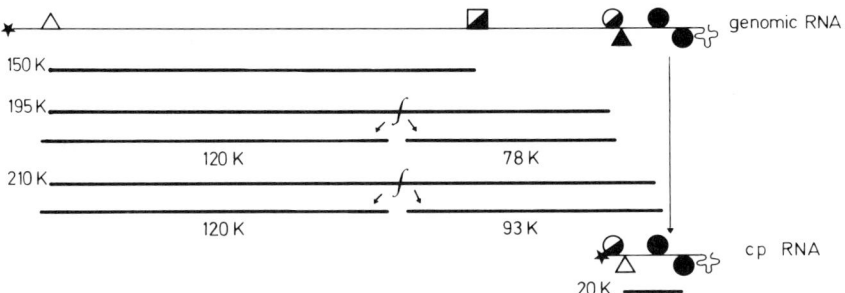

Figure 10.5. Genetic map of turnip yellow mosaic virus RNA. Symbols are as in Figures 1 and 3. The molecular weight of the products of proteolytic processing is indicated below each fragment. (K = kilodalton.)

TYMV RNA, the major products to appear are two high molecular weight proteins of 150 kD and 195 kD in a ratio approaching 1 : 1, and the coat protein (Benicourt et al., 1978). Synthesis of the coat protein is programmed by the subgenomic RNA (Pleij et al., 1976; Klein et al., 1976; Ricard et al., 1977) whilst synthesis of the high molecular weight proteins is directed by the genomic RNA. The two high molecular weight proteins are initiated at the same AUG (Benicourt and Haenni, 1978) and the 150 kD protein results from the arrest of elongation at the level of a stop signal on the RNA. The nature of the stop signal is still unknown. It does not correspond to a termination codon since yeast tRNAs suppressors of the codons UAG, UGA or UAA are incapable of enhancing the synthesis of the 195 kD protein to the detriment of the 150 kD protein (Morch et al., 1982a).

In the presence of the yeast tRNA suppressor of UAG, a 210 kD protein appears, resulting from the elongation of the 195 kD protein beyond the UAG triplet terminating the 195 kD protein gene (Morch and Benicourt, 1980). The C-terminal end of the 210 kD protein is coded by part of the region of the viral RNA that contains the silent coat protein gene. However, different reading frames are used for the 210 kD and for the coat protein (Morch et al., 1982a).

Another strategy utilized by TYMV is post-translational cleavage. Kinetic experiments have shown that the 195 kD protein undergoes a specific cleavage leading to the appearance at late times during the incubations of two polypeptides of 120 kD and 78 kD (Morch and Benicourt, 1980; Morch et al., 1982a). They derive respectively from the N- and C-terminus of the 195 kD protein. The read-through protein of 210 kD undergoes a similar cleavage. Surprisingly, the 150 kD protein is not cleaved in spite of the fact that it contains the cleavage site. This suggests that the two major high molecular weight proteins of 150 kD and 195 kD could have different functions even though they are identical in three-quarters of the length of the 195 kD protein.

The fact that the same high molecular weight proteins and the same cleavage pattern of the 195 kD protein are observed in three different translation systems *in vitro* suggests that cleavage is encoded by the viral RNA (Zagorski *et al.*, 1983). This point was further examined by extensive dilution of the translation samples and subsequent incubation. Analyses of the resulting translation products indicated that cleavage is at least partially insensitive to dilution. This probably reflects the capacity for self-cleavage of the 195 kD protein, and it has been proposed that maturation of the 195 kD protein occurs by internal catalysis (Morch *et al.*, 1982b).

To investigate whether the presence of two mRNAs, genomic and subgenomic, allows the virus to modulate the synthesis of its various proteins, increasing concentrations of unfractionated TYMV RNA were added to a translation system *in vitro* (Benicourt and Haenni, 1978). At low RNA concentrations, the high molecular weight proteins are produced as well as the coat protein. However, as the RNA concentration increases, synthesis of the coat protein increases to the detriment of the high molecular weight proteins. Thus the coat protein mRNA possesses a much higher affinity for the ribosomal machinery than the genomic RNA. In accordance with this model, it has been proposed that *in vivo*, early after infection, when the host cells contain only small amounts of the viral genomic and subgenomic RNAs, both the non-structural and the structural proteins are synthesized. Later in infection, when large amounts of the viral RNAs are present, essentially coat protein is produced.

Little is known about the role of the non-structural proteins of TYMV. It is conceivable that one of them participates in viral RNA replication. Support for this comes from experiments by Mouchès *et al.* (1984) who have purified from TYMV-infected Chinese cabbage leaves an RNA-dependent RNA replicase involved in TYMV RNA synthesis. This enzyme contains two major subunits of 115 and 45 kD. Using antibodies raised against the replicase, it could be shown that the 45 kD subunit is host-coded and is also present in uninfected leaves, whereas the 115 kD protein is virus-coded. The anti-RNA replicase antibodies interact with a translation product *in vitro* of 115 kD. If the 115 kD product in the replicase complex is identical to the 120 kD protein resulting from cleavage of the 195 kD protein, then it will be possible to conclude that the 5' part of the genome codes for a subunit of the replicase. Sequence data (Morch and co-workers, unpublished) should soon allow protein sequence comparisons to be made.

Another non-structural protein presumably contains the protease activity. Since cleavage appears to be intramolecular, and since the 150 kD protein is not cleaved, the protease activity could be contained within the *C*-terminal part of the 195 kD protein (Morch *et al.*, 1982b).

VI. SATELLITES

Satellite viruses were first described by Kassanis and Nixon (1960, 1961) and by Kassanis (1962) in connection with a small virus that accompanies tobacco necrosis virus (TNV). Since then, satellites have been described in many plant viruses with monopartite (TNV group, tombusviruses, sobemoviruses), bipartite (nepoviruses) and tripartite (cucumoviruses) genomes. They designate a virus (satellite virus) or a nucleic acid (satellite RNA) that depends on a specific "helper" virus for its multiplication, is not required for multiplication of the helper virus, and has no significant sequence homology with the helper genome. The RNAs of satellites are also characterized by their unusual stability: they retain potential infectivity for several days in inoculated leaves, and for several hours when incubated in extracts of healthy plants. Satellites have recently been reviewed by Murant and Mayo (1982).

A. Satellite of the TNV Group

TNV particles are 30 nm in diameter and, in some isolates, are accompanied by smaller particles 17 nm in diameter. These two types of particles are serologically unrelated, indicating that their capsid proteins are distinct. The satellite particles designated satellite tobacco necrosis virus (STNV) require the presence of TNV, the helper, for their multiplication, whereas the TNV particles can be propagated in the total absence of STNV. The satellite interferes with TNV multiplication and thus leads to a diminution of the size and number of necrotic lesions produced by TNV (Kassanis, 1962).

The strategy of expression of TNV is not entirely elucidated, although it is known that the coat protein of 30 kD is synthesized *via* a subgenomic RNA of 0.47 MD that is 3' coterminal with the genomic RNA of 1.28 MD (Condit and Fraenkel-Conrat, 1979). The 5' end of the genomic RNA carries the sequence ppAGU (Lesnaw and Reichmann, 1970); this same sequence is found at the 5' end of STNV RNA (Wimmer *et al.*, 1968; Wimmer and Reichmann, 1968). The 3' end of these two RNAs does not seem to possess any particular feature.

The complete nucleotide sequence of STNV RNA has been established (Ysebaert *et al.*, 1980). It reveals one long open reading frame from which the amino acid sequence of the coat protein can be deduced. The coat protein gene is followed by a 3' untranslated region 622 nucleotides long (out of a total of 1239 nucleotides). Translation studies *in vitro* have indicated that STNV RNA codes for its coat protein both in eukaryotic and in prokaryotic translation systems (Klein *et al.*, 1972; Van Emmelo *et al.*, 1984). The reason why faithful synthesis of the coat protein can be achieved using an *Escherichia coli* translation extract probably stems from the fact that the sequence AGGA, which is part of the Shine and Dalgarno (1974) sequence, is found upstream of the initiating AUG.

STNV is characterized by the fact that it codes for its own coat protein and that it forms particles that are morphologically and serologically distinct from the helper TNV.

B. Satellites of the Tombusvirus Group

Tomato bushy stunt virus (TBSV) encapsidates a small satellite-like RNA of 0.15 MD (Hillman and Morris, 1982) that is as yet poorly characterized.

Turnip crinkle virus (TCV), a possible member of the tombusvirus group, contains an RNA genome of 1.4 MD that codes *in vitro* for several proteins; TCV particles contain a heterogeneous population of smaller RNAs including a major subgenomic RNA species of 0.15 MD that codes for the coat protein of 38 kD. Studies *in vitro* have led to the suggestion that the genomic RNA codes for two overlapping non-structural proteins of 25 and 80 kD, the latter being the read-through product of the former. Likewise, translation of a subgenomic RNA leads to the synthesis of a 36 kD protein and of its read-through product, the non-structural 56 kD protein. The C-termini of the 80 and the 56 kD proteins seem to coincide (Dougherty and Kaesberg, 1981).

TCV is capable of supporting the replication of a satellite RNA, S-TCV. The satellite RNA (~ 500 nucleotides) is packaged in the mature virion (Altenbach and Howell, 1981); it requires coinfection with the genomic RNA to replicate in plants, and it intensifies the symptoms produced by the helper virus. It does not have mRNA activity in wheat germ or reticulocyte extracts. Interestingly, some homology seems to exist between S-TCV RNA and the DNA of the host turnip plants (Altenbach and Howell, 1984).

C. Satellites of the Sobemovirus Group

Sobemoviruses contain a genomic RNA that carries a VPg protein at its 5' end. A subgenomic RNA 3'-coterminal with the genomic RNA that also bears a VPg protein in the type member SBMV (Ghosh *et al.*, 1981) directs the synthesis of the virus coat protein.

Solanum nodiflorum mottle virus (SNMV), lucerne transient streak virus (LTSV), velvet tobacco mottle virus (VTMoV) and subterranean clover mottle virus (SCMoV) are recently described viruses that originate from Australasia. They have been compared to sobemoviruses (Gerber, 1981) on the basis of their virion structure and messenger functions *in vitro*; in addition to the genomic RNA of about 4500 nucleotides, they contain a subgenomic RNA that codes for the coat protein. However, unlike other sobemoviruses, they also encapsidate a circular (or linear), single-stranded RNA of 0.13 MD.

The nucleotide sequences of the circular RNA of SNMV and VTMoV have been determined (Haseloff and Symons, 1982). These two RNAs are very closely related to each other, having over 90% sequence homology. They adopt a very strongly basepaired structure similar to the one adopted by viroids, and for this reason are often designated virusoids (Haseloff *et al.*, 1982). Their viroid-like properties extend to the fact that they cannot direct the synthesis of proteins *in vitro*, and that they do not contain long open reading frames.

The virusoids of SNMV and LTSV can be classified as satellites since cultures of the helper virus devoid of virusoids are infective (Jones and Mayo, 1984). The virusoids alone are incapable of autonomous replication, but rely on the helper virus for their propagation. Virusoids are thus characterized by the fact that they are covalently closed circular RNA molecules, but that as opposed to viroids, they are encapsidated with a helper virus.

VII. CONCLUSIONS AND PERSPECTIVES

Having a monopartite genome for a virus has some advantages, but also inconveniences. It certainly leads to a greater efficiency in the infection process, since only one RNA molecule needs to penetrate into the cell. However, monopartite viruses have developed RNA genomes longer than multipartite viruses. Thus, in the former case, the error frequency during replication is higher than in the latter case, and should a lethal error be made, the total information is lost.

Monopartite genomes enable RNA viruses to minimize the non-coding regions and to store their information in a very compact manner. However, the major problem then becomes the expression of such a genome. The variety of strategies described in the present paper show how these viruses circumvent this inconvenience and can produce several proteins with different yields, starting from a single RNA molecule.

It should be stressed that our understanding of these strategies of expression has greatly benefited from the availability of reliable eukaryotic translation systems *in vitro*. This has been particularly important since the development of systems *in vivo* has lagged far behind.

Indeed, as opposed to animal viruses, plant viruses do not cause significant shut-off of protein synthesis in the host cells they infect. Thus the synthesis of virus-coded proteins, and in particular of non-structural virus proteins, is often difficult to detect over a high background of endogenous host-cell protein synthesis. A further complication related to the study of virus development in the host plant has been the difficulty of obtaining synchronous infections.

In recent years, however, the availability of plant protoplasts amenable to synchronous infection has facilitated the interpretation of results *in vivo*. Likewise, the synthesis of virus proteins in protoplasts or leaf tissue has become easier to follow by treatment of the host with UV light or chloramphenicol and actinomycin D. Since at least some of the virus non-structural proteins are synthesized in relatively low amounts, or only transiently, during the virus cycle, these improvements are proving particularly valuable for studies *in vivo*. Moreover, the accumulation of sequence data on plant RNA viruses will allow the production of antibodies raised against synthetic peptides deduced from these sequences. These antibodies will in turn help detect the virus-coded proteins synthesized *in vivo*.

Knowledge of plant RNA virus replication will further benefit from the genetic engineering techniques used for sequencing. Mutations and/or deletions can be envisaged in a full-length DNA copy of the viral genome at precise locations. The DNA can then be transcribed *in vitro* into an RNA that can be tested for infectivity. Such techniques have already been developed for the animal counterpart of a monopartite plant RNA virus, poliovirus (Racaniello and Baltimore, 1981; Semler *et al.*, 1984), as well as for a tripartite plant RNA virus, BMV (Ahlquist *et al.*, 1984; and see Chapter 13, this volume). Finally, the determination of the functions of the non-structural virus proteins as well as the mechanisms of virus–host interaction should enable us to understand how virus development can be controlled and curtailed.

NOTE ADDED IN PROOF

Recently, the total nucleotide sequence of the RNA of TVMV, a potyvirus, has been established (Domier, L. L. *et al.* (1986). *Nucl. Acids Res.* **14**, 5417–5430). It reveals one long open reading frame which can encode a polyprotein of 340 kD, and consequently rules out the previous assumption (see page 162) that internal termination codons are present within this reading frame.

ACKNOWLEDGEMENTS

This work was supported in part by a grant from the "ATP: Interactions entre plantes et microorganismes", Centre National de la Recherche Scientifique.

REFERENCES

1. Ahlquist, P., R. French, M. Janda and L. S. Loesch-Fries (1984). *Proc. Natl Acad. Sci. USA* **81**, 7066–7070.
2. Allison, R. F., J. C. Sorenson, M. E. Kelly, F. B. Armstrong and W. G. Dougherty (1985). *Proc. Natl Acad. Sci. USA* **82**, 3969–3972.
3. Altenbach, S. B. and S. H. Howell (1981). *Virology* **112**, 25–33.
4. Altenbach, S. B. and S. H. Howell (1984). *Virology* **134**, 72–77.
5. Atabekov, J. G. and Yu. Morozov (1979). *Adv. Virus Res.* **25**, 1–91.
6. Beachy, R. N., M. Zaitlin, G. Bruening and H. W. Israel (1976). *Virology* **73**, 498–507.
7. Beachy, R. N. and M. Zaitlin (1977). *Virology* **81**, 160–169.
8. Beier, H., K. W. Mundry and O. G. Issinger, (1980). *Intervirology* **14**, 292–299.
9. Beier, H., M. Barciszewska, G. Krupp, R. Mitnacht and H. J. Gross (1984a). *EMBO J.* **3**, 351–356.
10. Beier, H., M. Barciszewska and H. D. Sickinger (1984b). *EMBO J.* **3**, 1091–1096.
11. Bendena, W. G., M. Aboudhaidar and G. A. Mackie (1985). *Virology* **140**, 257–268.
12. Benicourt, C. and A. L. Haenni (1978). *Biochem. Biophys. Res. Commun.* **84**, 831–839.
13. Benicourt, C., J. P. Péré and A. L. Haenni (1978). *FEBS Lett.* **86**, 268–272.

14. Bienz, M. and E. Kubli (1981). *Nature (London)* **294**, 188–190.
15. Briand, J. P., G. Keith and H. Guilley (1978). *Proc. Natl Acad. Sci. USA* **75**, 3168–3172.
16. Bruening. G., R. N. Beachy, R. Scalla and M. Zaitlin (1976). *Virology* **71**, 498–517.
17. Carrington, J. C. and T. J. Morris (1985). *Virology* **144**, 1–10.
18. Condit, C. and H. Fraenkel-Conrat (1979). *Virology* **97**, 122–130.
19. Cornelissen, B. J. C. and J. F. Bol (1984). *Plant Mol. Biol.* **3**, 379–384.
20. Davies, J. W. and R. Hull (1982). *J. Gen. Virol.* **61**, 1–14.
21. De Mejia, M. V. G., E. Hiebert and D. E. Purcifull (1985). *Virology* **142**, 24–33.
22. Dougherty, W. G. and E. Hiebert (1980). *Virology* **104**, 174–182.
23. Dougherty, W. G. and P. Kaesberg (1981). *Virology* **115**, 45–56.
24. Dougherty, W. G. (1983). *Virology* **131**, 473–481.
25. Fukuda, M., T. Meshi, Y. Okada, Y. Otsuki and I. Takebe (1981). *Proc. Natl Acad. Sci. USA* **78**, 4231–4235.
26. Gerber, R. S. (1981). *Aust. J. Biol. Sci.* **34**, 369–378.
27. Ghosh, A., R. Dasgupta, T. Salerno-Rife, T. Rutgers and P. Kaesberg (1979). *Nucl. Acids Res.* **7**, 2137–2146.
28. Ghosh, A., T. Rutgers, M. Ke-Qiang and P. Kaesberg (1981). *J. Virol.* **39**, 87–92.
29. Goelet, P. and J. Karn (1982). *J. Mol. Biol.* **154**, 541–550.
30. Goelet, P., G. P. Lomonosoff, P. J. G. Butler, M. E. Akam, M. J. Gait and J. Karn (1982). *Proc. Natl Acad. Sci. USA* **79**, 5818–5822.
31. Guilley, H. and J. P. Briand (1978). *Cell* **15**, 113–122.
32. Guilley, H., G. Jonard, B. Kukla and K. E. Richards (1979). *Nucl. Acids Res.* **6**, 1287–1308.
33. Haenni, A. L., S. Joshi and F. Chapeville (1982). *Prog. Nucl. Acid Res. Mol. Biol.* **27**, 85–104.
34. Harbison, S. A., T. M. A. Wilson and J. W. Davies (1984). *Biosci. Rep.* **4**, 949–956.
35. Hari, V., A. Siegel, C. Rozek and W. E. Timberlake (1979). *Virology* **92**, 568–571.
36. Hari, V. (1981). *Virology* **112**, 391–399.
37. Harris, K. F. (1983). *Adv. Virus Res.* **28**, 113–140.
38. Haseloff, J. and R. H. Symons (1982). *Nucl. Acids Res.* **10**, 3681–3691.
39. Haseloff, J., N. A. Mohamed and R. H. Symons (1982). *Nature (London)* **299**, 316–320.
40. Haseloff, J., P. Goelet, D. Zimmern, P. Ahlquist, R. Dasgupta and P. Kaesberg (1984). *Proc. Natl Acad. Sci. USA* **81**, 4358–4362.
41. Hellmann, G. M., D. W. Thornbury, E. Hiebert, J. G. Shaw, T. P. Pirone and R. E. Rhoads (1983). *Virology* **124**, 434–444.
42. Hellmann, G. M., J. G. Shaw and R. E. Rhoads (1985). *Virology* **143**, 23–34.
43. Hillman, B. and T. J. Morris (1982). *Phytopathology* **72**. 952–953.
44. Hull, R. (1977). *J. Gen. Virol.* **36**, 289–295.
45. Hull, R. and J. W. Davies (1983). *Adv. Virus Res.* **28**, 1–33.
46. Hunter, T. R., T. Hunt, J. Knowland and D. Zimmern (1976). *Nature (London)* **260**, 759–764.
47. Hunter, T., R. Jackson and D. Zimmern (1983). *Nucl. Acids Res.* **11**, 801–821.

48. Jones, A. T., M. A. Mayo and G. H. Duncan (1983). *J. Gen. Virol.* **64**, 1167–1173.
49. Jones, A. T. and M. A. Mayo (1984). *J. Gen. Virol.* **65**, 1713–1721.
50. Joshi, S., R. L. Joshi, A. L. Haenni and F. Chapeville (1983a). *Trends Biochem. Sci.* **8**, 402–404.
51. Joshi, S., C. W. A. Pleij, A. L. Haenni, F. Chapeville and L. Bosch (1983b). *Virology* **127**, 100–111.
52. Joshi, S. and A. L. Haenni (1984). *FEBS Lett.* **177**, 163–174.
53. Kamer, G. and P. Argos (1984). *Nucl. Acids Res.* **12**, 7269–7282.
54. Kassanis, B. and H. L. Nixon (1960). *Nature (London)* **187**, 713–714.
55. Kassanis, B. and H. L. Nixon (1961). *J. Gen. Microbiol.* **25**, 459–471.
56. Kassanis, B. (1962). *J. Gen. Microbiol.* **27**, 477–488.
57. Keith, J. and H. Fraenkel-Conrat (1975). *FEBS Lett.* **57**, 31–33.
58. Khaddam, A. (1985). Thèse de Doctorat d'Etat, Université Paris 6.
59. Kiberstis, P. A. and D. Zimmern (1984). *Nucl. Acids Res.* **12**, 933–943.
60. Klein, C., C. Fritsch, J. P. Briand, K. E. Richards, G. Jonard, and L. Hirth (1976). *Nucl. Acids Res.* **3**, 3043–3061.
61. Klein, W. H., C. Nolan, J. M. Lazar and J. M. Clark (1972). *Biochemistry* **11**, 2009–2014.
62. Lesnaw, J. A. and M. E. Reichmann (1970). *Proc. Natl Acad. Sci. USA* **66**, 140–145.
63. Matthews, R. E. F. (1982). *Intervirology* **17**, 1–199.
64. Meshi, T., T. Ohno, H. Iba and Y. Okada (1981). *Mol. Gen. Genet.* **184**, 20–25.
65. Meshi, T., T. Ohno and Y. Okada (1982). *Nucl. Acids Res.* **10**, 6111–6117.
66. Meshi, T., M. Ishikawa, N. Takamatsu, T. Ohno and Y. Okada (1983a). *FEBS Lett.* **162**, 282–285.
67. Meshi, T., R. Kiyama, T. Ohno and Y. Okada. (1983b). *Virology* **127**, 54–64.
68. Morch, M. D. and C. Benicourt (1980). *J. Virol.* **34**, 85–94.
69. Morch, M. D., G. Drugeon and C. Benicourt (1982a). *Virology* **119**, 193–198.
70. Morch, M. D., W. Zagorski and A. L. Haenni (1982b). *Eur. J. Biochem.* **127**, 259–265.
71. Morozov, S. Yu., V. M. Zakhariev, B. K. Chernov, V. S. Prasolov, Yu. V. Kozlov, J. G. Atabekov and K. G. Skryabin (1983). *Dokl. Akad. Nauk SSSR* **271**, 211–215.
72. Morris-Krsinich, B. A. M. and R. Hull (1981). *Virology* **114**, 98–112.
73. Mouchès, C., T. Candresse and J. M. Bové (1984). *Virology* **134**, 78–90.
74. Murant, A. F. and M. A. Mayo (1982). *Annu. Rev. Phytopathol.* **20**, 49–70.
75. Nelson, M. R. and J. H. Tremaine (1975). *Virology* **65**, 309–319.
76. Nishiguchi, M., S. Kikuchi, Y. Kiho, T. Ohno, T. Meshi and Y. Okada (1985). *Nucl. Acids Res.* **13**, 5585–5590.
77. Öberg, B. and L. Philipson (1972). *Biochem. Biophys. Res. Commun.* **48**, 927–932.
78. Ohno, T., N. Takamatsu, T. Meshi, M. Nishiguchi and Y. Kiho (1983). *Virology* **131**, 255–258.
79. Ohno, T., M. Aoyagi, Y. Yamanashi, H. Saito, S. Ikawa, T. Meshi and Y. Okada (1984). *J. Biochem. (Japan)* **96**, 1915–1923.
80. Paterson, R. and C. A. Knight (1975). *Virology* **64**, 10–22.
81. Pelham, H. R. B. (1978). *Nature (London)* **272**, 469–471.
82. Pelham, H. R. B. (1979). *Virology* **96**, 463–477.

83. Pinck, M., P. Yot, F. Chapeville and H. Duranton (1970). *Nature (London)* **226**, 954–956.
84. Pirone, T. P. and D. W. Thornbury (1984). *Trends Microbiol. Sci.* **1**, 191–193.
85. Pleij, C. W. A., A. Neeleman, L. Van Vloten-Doting and L. Bosch (1976). *Proc. Natl Acad. Sci. USA* **73**, 4437–4441.
86. Racaniello, V. R. and D. Baltimore (1981). *Science* **214**, 916–919.
87. Ricard, B., C. Barreau, H. Renaudin, C. Mouchès and J. M. Bové, (1977). *Virology* **79**, 231–235.
88. Rutgers, T., T. Salerno-Rife and P. Kaesberg (1980). *Virology* **104**, 506–509.
89. Salerno-Rife, T., T. Rutgers and P. Kaesberg (1980). *J. Virol.* **34**, 51–58.
90. Schlesinger, R. W. (Ed.) (1984). "The Togaviruses". Academic Press, London and Orlando.
91. Semler, B. L., A. Dorner and E. Wimmer (1984). *Nucl. Acids Res.* **12**, 5123–5140.
92. Shine, J. and L. Dalgarno (1974). *Proc. Natl Acad. Sci. USA* **71**, 1342–1346.
93. Short, M. N. and J. W. Davies (1983). *Biosci. Rep.* **3**, 837–846.
94. Siaw, M. F. E., M. Shahabuddin, S. Ballard, J. G. Shaw and R. E. Rhoads (1985). *Virology* **142**, 134–143.
95. Sonenberg, N., A. J. Shatkin, R. P. Ricciardi, M. Rubin and R. M. Goodman (1978). *Nucl. Acids Res.* **5**, 2501–2512.
96. Takamatsu, N., T. Ohno, T. Meshi and Y. Okada (1983). *Nucl. Acids Res.* **11**, 3767–3778.
97. Tremaine, J. H. (1970). *Virology* **42**, 611–620.
98. Vance, V. B. and R. N. Beachy (1984). *Virology* **132**, 271–281.
99. Van Emmelo, J., P. Ameloot, G. Plaetinck and W. Fiers (1984). *Virology* **136**, 32–40.
100. Van Telgen, H. J., R. W. Goldbach and L. C. Van Loon (1985). *Virology* **43**, 612–616.
101. Van Vloten-Doting, L. and L. Neeleman (1982). *In:* "Encyclopedia of Plant Physiology New Series, Nucleic Acids and Proteins in Plants II" (Eds D. Boulter and B. Parthier), vol. **14B**, pp. 337–367. Springer Verlag, Berlin.
102. Watanabe, Y., Y. Emori, I. Ooshika, T. Meshi, T. Ohno and Y. Okada (1984). *Virology* **133**, 18–24.
103. Waterworth, H. E. and J. M. Kaper (1972). *Phytopathology* **62**, 959–962.
104. Wimmer, E. and M. E. Reichmann (1968). *Science* **160**, 1452–1454.
105. Wimmer, E., A. Y. Chang, J. M. Clark and M. E. Reichmann (1968). *J. Mol. Biol.* **38**, 59–73.
106. Wodnar-Filipowicz, A., L. J. Skrzeczkowski and W. Filipowicz (1980). *FEBS Lett.* **109**, 151–155.
107. Yeh, S. D. and D. Gonsalves (1985). *Virology* **143**, 260–271.
108. Ysebaert, M., J. Van Emmelo and W. Fiers (1980). *J. Mol. Biol.* **143**, 273–287.
109. Zagorski, W., M. D. Morch and A. L. Haenni (1983). *Biochimie* **65**, 127–133.
110. Zaitlin, D. A. and M. Leonard (1982). *Virology* **117**, 416–424.
111. Zimmern, D. (1975). *Nucl. Acids Res.* **2**, 1189–1201.

11. A Comparison of the Translation Strategies used by Bipartite Genome, RNA Plant Viruses

M. A. Mayo

Scottish Crop Research Institute, Invergowrie, Dundee, U.K.

I. Introduction	178
II. Comoviruses	179
A. General properties	179
B. Properties of RNA species	180
C. Translation products	180
D. CPMV-coded proteases	183
E. Sequence comparisons with other viruses	184
III. Tobraviruses	185
A. General properties	185
B. Properties of RNA species	185
C. Translation products	185
D. Results from sequence determinations	186
E. Sequence comparisons with other viruses	188
IV. Nepoviruses	188
A. General properties	188
B. Properties of RNA species	189
C. Translation products	190
D. Results from sequence determinations	192
E. Sequence comparisons with other viruses	192
V. Fungus-transmitted rod-shaped viruses	193
VI. Dianthoviruses	196
VII. Pea enation mosaic virus	197
VIII. Discussion	198
References	201

THE MOLECULAR BIOLOGY OF THE
POSITIVE STRAND RNA VIRUSES   ISBN 0-12-599930-5

Copyright © 1987 Academic Press Inc. (London) Ltd.
All rights of reproduction in any form reserved

Abbreviations: BNYVV, beet necrotic yellow vein virus; CLRV, cherry leafroll virus; CPMV, cowpea mosaic virus; CPSMV, cowpea severe mosaic virus; CRSV, carnation ringspot virus; GFLV, grapevine fanleaf virus; PCV, peanut clump virus; PEBV, pea-early browning virus; PEMV, pea enation mosaic virus; PRV, pepper ringspot virus; RCNMV, red clover necrotic mosaic virus; RRV, raspberry ringspot virus; SBWMV, soil-borne wheat mosaic virus; SqMV, squash mosaic virus; TBRV, tomato black ring virus; TMV, tobacco mosaic virus; TobRV, tobacco ringspot virus; TRV, tobacco rattle virus.

I. INTRODUCTION

A bipartite genome is one in which the genetic information necessary for virus multiplication is divided between two RNA species. This arrangement is relatively common among plant viruses. Of the 278 viruses described in the series of Descriptions of Plant Viruses published by CMI/AAB, 53 have or probably have this sort of genome and of the 21 groups of ssRNA (single-stranded RNA) plant viruses listed by Matthews (1982), five have a bipartite genome as one of their distinguishing characters.

However, such an apparently simple definition needs careful interpretation. It has been known for a number of years that, although tobraviruses have a bipartite genome, the larger part of their genomes will, independently of the smaller genome part, infect plants, multiply and induce symptoms (Harrison and Robinson, 1978). No virus particles are made, because the coat protein gene is contained in the smaller genome part. Recently, other viruses with bipartite genomes have been found to behave in a similar fashion. Thus the larger genome parts of cowpea mosaic (Goldbach *et al.*, 1980) and tomato black ring (Robinson *et al.*, 1980) viruses (comoviruses and nepoviruses respectively) multiply in inoculated protoplasts independently of the smaller genome part.

Some bipartite genome viruses are associated in nature with satellite RNAs that are specific to the viruses but are not essential for virus multiplication (Murant and Mayo, 1982; Francki, 1985) and there would seem to be rather a small separation between a dispensable satellite molecule and a genome part that is dispensable for some functions. Indeed, Harrison and Robinson (1978) have suggested that one possible origin for the bipartite genome of tobraviruses could be evolution from a monopartite genome resembling RNA-1 (such as TMV) and an associated satellite molecule. Apart from its intrinsic fascination, one interest in an examination of the translation strategies used by bipartite genome viruses is that it might contribute clues to discussions on the origins and advantages of a divided genome.

The groups of viruses, and ungrouped viruses, that have bipartite genomes are listed in Table 11.1 together with some other virus properties. It is clear from this list that quite dissimilar viruses share this characteristic. Except for broad bean wilt (Taylor and Stubbs, 1972; Doel, 1975) and pelargonium zonate spot (Gallitelli *et al.*, 1983) viruses, about which

11. Strategies of Bipartite Genome, Plant Viruses

Table 11.1. Some properties of plant viruses with bipartite RNA genomes

Virus group[a]	Particle morphology	Genome size (kb)	Principal type of vector
Comovirus	isometric	9.4	beetles
Tobravirus	rod-shaped	8.3–10.5	nematodes
Nepovirus	isometric	11.3–14.1	nematodes
"Furovirus"	rod-shaped	9.7–11.3	fungi
Dianthovirus	isometric	5.5	?
Pea enation mosaic	isometric	7.7–9.1	aphids
Broad bean wilt	isometric	10	aphids
Pelargonium zonate spot	isometric	6.1	?

[a]Groups according to Matthews (1982), except "furovirus" which was proposed by Shirako and Brakke (1984).

little has been reported, the translation products of the RNA species of the viruses will be reviewed in turn, together with as much detail as is available about the genome RNA molecules. Most studies have been by translation *in vitro*, although where possible the results of studies of protein synthesis *in vivo* are also summarized. In general terms these have largely confirmed the results of translation *in vitro*. No attempt will be made to cover the topics of RNA transcription or RNA replication.

II. COMOVIRUSES

A. General Properties

Comoviruses are a group of viruses with isometric particles 28 nm in diameter that are usually transmitted by beetles. Virus particles are shells, made from 60 copies each of two proteins with molecular weights of about 25 kD and 40 kD, that contain either no RNA or single molecules of RNA with molecular weights of about 1.4 MD (M-RNA in M component) or about 2.1 MD (B-RNA in B component) (Bruening, 1978). Infectivity for plants depends on both nucleoprotein components or RNA species being present and pseudo-recombinants have been made with heterologous mixtures of RNA species from different strains. Experiments with pseudo-recombinants showed that M- and B-RNA can each specify the symptoms induced in infected plants (Table 11.5) and that M-RNA determines at least one of the coat proteins (Bruening, 1977). However, although inoculation with B component of cowpea mosaic virus (CPMV) has no effect on plants (van Kammen, 1968), B-RNA multiplies in isolated protoplasts inoculated with B component alone (Goldbach et al., 1980); no coat protein and no particles are formed in such inoculated protoplasts.

B. Properties of RNA Species

Comovirus RNA species are polyadenylated (Bruening, 1978); the mean lengths of poly(A) in RNA of CPMV are 167 residues in M-RNA and 87 residues in B-RNA (Ahlquist and Kaesberg, 1979). A small protein (VPg) is covalently attached to the 5'-end of each RNA molecule (Stanley et al., 1978; Daubert and Bruening, 1979); CPMV VPg is 28 residues long (Wellink et al., 1986). Protease treatment of RNA degrades VPg but does not affect the infectivity or translation behaviour of the RNA (Stanley et al., 1978). The sequences of CPMV B- and M-RNA have been determined (Lomonossoff and Shanks, 1983; van Wezenbeek et al., 1983) and show that open reading frames occupy more than 90% of each RNA molecule. The non-coding regions of each RNA species are similar in being rich in U and poor in G (Lomonossoff et al., 1982). M- and B-RNA have similar 5' and 3' non-coding sequences (Najarian and Bruening, 1980; Lomonossoff et al., 1982; Davies et al., 1979) and the 5' non-coding sequences contain an oligonucleotide complementary to sequences in the 3' non-coding regions (Najarian and Bruening, 1980).

C. Translation Products

Translation *in vitro* of RNA of several comoviruses yields polypeptides that correspond to almost the full length of each RNA species (Table 11.2); a variety of smaller translation products are also formed. A striking feature of the translation of M-RNA is that two major products are made, the smaller having a molecular weight about 6–10 kD less than that of the larger (Goldbach and Krijt, 1982). By far the majority of work on the

Table 11.2. Translation products *in vitro* of comovirus RNA

Virus	Mol. wt (kD) of translation products of		Reference
	RNA-1	RNA-2	
Cowpea mosaic	200, 170, 110, 87, 84, 60, 32, 28	105, 95, 60[a], 58[a], 48[a]	[41, 82]
Bean pod mottle	181, 158, 30	107, 97, 58, 49, 40	[32]
Squash mosaic	190, 51, 32	112, 105, 90, 64, 22	[53]
Red clover mottle	—	102, 96	[38]
Cowpea severe mosaic	200, 125[b], 42[b]	108, 98	[7, 38]

[a]Made when RNA-1 translation products are present.
[b]Assignment to RNA-1 deduced by subtracting RNA-2 products [38] from the translation products of unfractionated RNA[7].

11. Strategies of Bipartite Genome, Plant Viruses

translation of comovirus RNA has been with CPMV and this will be described in some detail prior to a brief section about the translation of RNA of other comoviruses.

The molecular weight of CPMV-coded polypeptides have recently been calculated from nucleic acid and protein sequence data (Wellink et al., 1986) and these values are slightly different from those used in many previous publications (see Fig. 11.1). For comparability with published work the old values are used throughout the text except in Fig. 11.1, which shows both values for each polypeptide.

In reticulocyte lysate, CPMV B-RNA is translated into a 200 kD polypeptide which is then cleaved into 32 and 170 kD polypeptides, the 32 kD being the N-terminal part of the 200 kD precursor (Pelham, 1979a). The cleavage reaction occurs on nascent 200 kD chains once these have exceeded about 110 kD in size (Franssen et al., 1984a). Incubation of these translation products in the translation mixture results in slow conversion to 110, 87, 84, 60 and 28 kD polypeptides. These products have been identified in protoplasts infected with B component particles and there are alternative cleavages of the 170 kD to give 110 kD + 60 kD or 87 kD + 84 kD (Goldbach and Rezelman, 1983).

The relationships among the polypeptides were established from comparisons of their proteolytic peptide patterns (Rezelman et al., 1980) and by their reaction with antiserum to VPg (Zabel et al., 1982). For example, the 60 kD product was found to be the immediate precursor to VPg (Goldbach et al., 1982). Wellink et al. (1987) have recently shown that an antiserum against a peptide, synthesized to mimic part of the putative 24 kD protein released when the 110 kD or 84 kD products are processed, reacted with a 24 kD protein in protoplast extracts, thus confirming the processing scheme proposed by Rezelman et al. (1980). This is shown in Fig. 11.1.

The precise locations of the cleavage sites in the 200 kD polyprotein have been established by Wellink et al. (1986) who determined the N-terminal sequences of each product by Edman degradation. Cleavage of the dipeptides glutamine-serine (twice), glutamine-methionine and glutamine-glycine accounts for the complete processing scheme.

Translation of CPMV M-RNA yields 105 kD and 95 kD polypeptides as primary translation products in both reticulocyte lysate and wheat germ extracts. The proteins have different N-termini (Franssen et al., 1982) and result from translation commencing at initiation sites either at position 161 or position 512 (Vos et al., 1984). The first initiation site has U and not G at position +4 and may therefore be weak and allow ribosomes to pass without initiating (Vos et al., 1984).

When B-RNA and M-RNA are translated together, both 110 kD and 95 kD polypeptides are cleaved by a protease coded for by B-RNA (Pelham, 1979a; Franssen et al., 1982). Extracts of cowpea protoplasts infected with B component also contain this protease activity (Franssen et al., 1982). The products are 60 kD, which contains the two coat protein

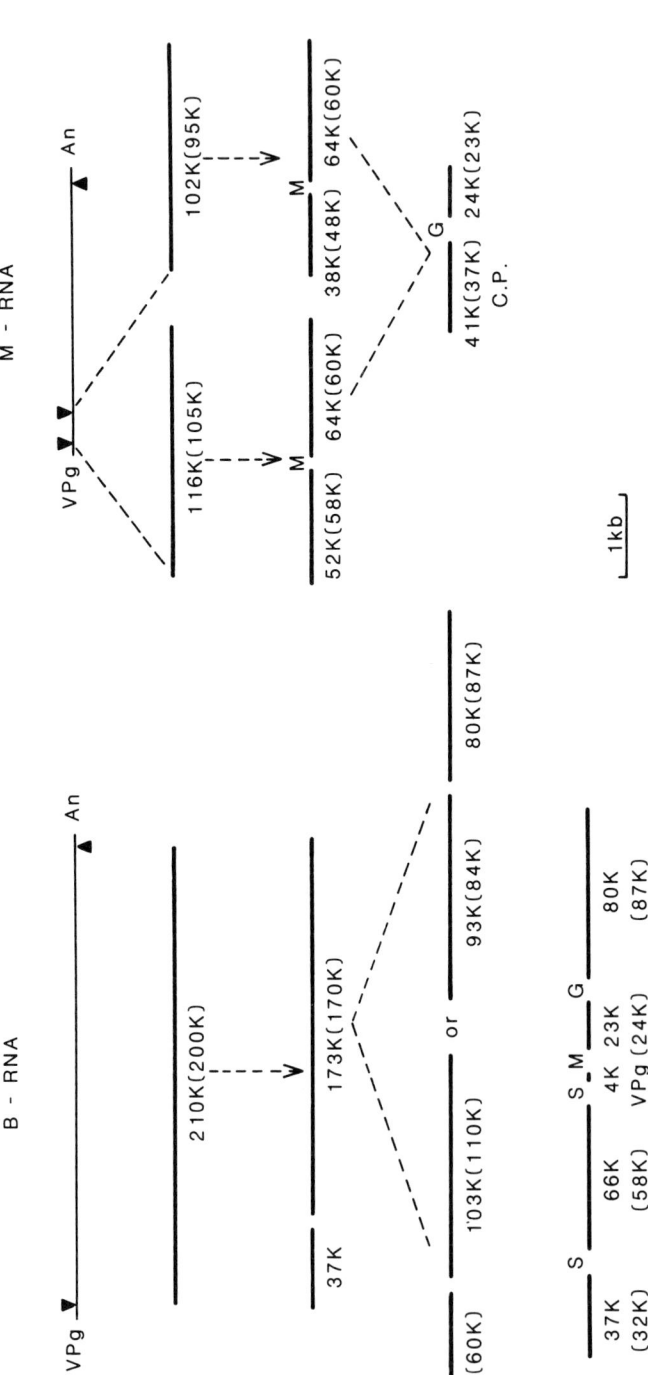

Figure 11.1. Diagram of RNA and translation products of CPMV. The molecules are drawn to scale. Narrow horizontal lines represent RNA molecules, wide horizontal lines represent polypeptide translation products. Broken arrows represent the derivation of polypeptides by proteolytic processing of larger precursor molecules. Dipeptides cleaved in this processing are indicated as S (Gln-Ser), M (Gln-Met) and G (Gln-Gly). Other symbols are: VPg, genome-linked protein; A_n, poly(A) at the 3' end of the RNA; ▼, initiation codon; ◀, termination codon; C.P., coat proteins. The molecular weights of the translation products are those deduced by Wellink et al. (1986a) from sequence data; those in parentheses are values used in previous work and in the text. (K=kilodalton.)

The alternative cleavages of the 173 kD product of B-RNA (left) are shown and the bottom line shows the order and relative sizes of the smallest cleavage products of the 210 kD precursor. The alternative products of M-RNA (right) arise from initiation at different sites.

polypeptides, 37 kD and 23 kD, and 58 kD or 48 kD (Fig. 11.1). The dipeptide cleaved to separate the 60 kD protein is thought to be glutamine-methionine (Franssen, 1984) and that to separate the two coat proteins is glutamine-glycine (van Wezenbeek et al., 1983).

The protease activity coded for by CPMV B component did not cleave the polyprotein precursor translation products of M-RNA of the comoviruses squash mosaic (SqMV), red clover mottle, or cowpea severe mosaic (CPSMV) or of RNA-2 of tomato black ring virus (Goldbach and Krijt, 1982).

Translation of SqMV RNA yields high molecular weight products, and smaller proteins (Hiebert and Purcifull, 1981). However, in this work all products of M-RNA reacted with antiserum and the authors concluded that coat protein genes were located at the 5' end of M-RNA and that the larger products arose by read-through and/or premature termination. More work is needed to test the surprising possibility of quite different translation strategies being used by viruses in the same group.

In results more akin to those with CPMV, Gabriel et al. (1982) found that the large translation products of bean pod mottle virus RNA were cleaved by a virus-coded protease activity which, like that of CPMV, was inhibited by Zn^{2+} and required dithiothreitol for activity. However this protease did not cleave CPMV M-RNA translation products.

In a comparison of translation of CPSMV RNA in vitro and in vivo, Beier et al. (1981) found 125, 98, 86, 65, 39 and 22 kD proteins in infected cells, of which 125 and 86 kD proteins were found in protoplasts infected only with B component, and 39 and 22 kD proteins were coat proteins. The largest translation product in vitro of M-RNA was 108 kD and Beier et al. (1981) suggested that the 125 kD polypeptide is coded for by B-RNA and is analogous to the 170 kD product of the first cleavage of the 200 kD primary product of CPMV B-RNA.

D. CPMV-coded Proteases

Three dipeptides are cleaved during the processing of CPMV RNA translation products, glutamine-methionine, glutamine-serine and glutamine-glycine. Antiserum to the 32 kD protein from the N-terminus of the 200 kD polypeptide, but not antisera to other B-RNA coded proteins, inhibits the cleavage of 105 and 95 kD proteins to release the 60 kD coat protein precursor (Franssen et al., 1984c). This cleavage is thought to occur at a glutamine-methionine bond (Franssen, 1984) and this enzyme might therefore cleave the other such bond between the 60 and 110 kD proteins (Fig. 11.1).

The other protease activity has been assigned to the 24 kD protein by experiments in which a restriction enzyme fragment of cloned DNA, complementary to B-RNA that contained the 24 kD sequence, was inserted into cloned DNA complementary to M-RNA. When transcripts of this DNA were translated, a 23 kD protein was released from the fusion protein produced (P. Vos and R. Goldbach, personal communication), whereas none was released from the translation product of the transcript of unmodified cloned cDNA. Furthermore deletion of 29 amino acids from the 24 kD sequence abolished protease activity associated with the fusion protein and when the deletion was introduced into the B-RNA clone, the cleavage at the glutamine-serine bond between 170 kD and 32 kD proteins (P. Vos and R. Goldbach, personal communication) did not occur.

E. Sequence Comparisons with Other Viruses

Striking comparisons have been published between the sequence of the CPMV B-RNA-coded 200 kD polyprotein and those of picornavirus polyproteins (Argos *et al.*, 1984; Franssen *et al.*, 1984b). Four regions of similarity were discovered. These were in the 58 kD, VPg, 24 kD and 87 kD proteins of CPMV.

The centre of the 58 kD protein has a region of about 143 amino acids that shows about 30% homology to a region of polypeptide 2C of picornaviruses, a protein that may be involved in RNA replication (Argos *et al.*, 1984).

VPg molecules are too small to allow statistical matching, but Argos *et al.* (1984) suggest a consensus sequence between CPMV and several picornaviruses. Moreover, they point out that VPg of encephalomyocarditis virus (EMC) and VPg of CPMV are more similar to each other than either is to VPg of poliovirus.

There is a region of about 60 residues in the *C*-terminal sequences of the 24 kD protease of CPMV and the 3C *pro*-proteinase of picornaviruses which shows marked homology. In this region CPMV was nearly as well related to EMC, as EMC was to poliovirus; the sequences show that CPMV and picornavirus proteases are of the cysteine type (Argos *et al.*, 1984).

The fourth region of similarity is between the 87 kD protein of CPMV and the RNA polymerase 3D *pol* region of picornavirus polyproteins. For example the homology between 87 kD and poliovirus P3–4b was 20.9%, which is equivalent to 113 identical residues out of 541 (Franssen *et al.*, 1984b).

In addition to these regions of homology the arrangement of the gene products in the polyproteins of CPMV and picornaviruses are strikingly similar. The four regions of homology occur in the same order in the respective polyproteins and the 200 kD CPMV polyprotein resembles the *C*-terminal half of the picornavirus polyprotein (Argos *et al.*, 1984; Franssen *et al.*, 1984b). The *N*-terminal half of this polyprotein contains the genes for the particle proteins, and is therefore functionally analogous to M-component 105 kD polyprotein (Franssen *et al.*, 1984b). These comparisons have led to the suggestion that comoviruses and picornaviruses might be evolutionarily related (Goldbach, 1986).

III. TOBRAVIRUSES

A. General Properties

This group comprises tobacco rattle virus (TRV), pepper ringspot virus (PRV) and pea-early browning virus (PEBV). PRV was described as the CAM strain of TRV in earlier papers (Harrison and Robinson, 1978), but is now thought to be a separate virus (Robinson and Harrison, 1985a). Tobraviruses are transmitted by trichodorid nematodes and have rod-shaped particles that are about 23 nm in diameter and of two or more lengths, the longest being about 200 nm and the most abundant between 46 nm and 105 nm in length (Harrison and Robinson, 1981). The coat protein is about 23 kD (Mayo and Robinson, 1975) and the larger estimates obtained for some tobraviruses by gel electrophoresis were incorrect (Ghabrial and Lister, 1973; Mayo and Robinson, 1975). Tobravirus RNA sizes reflect the particle sizes; the molecular weight of RNA-1 is about 2.4 MD and that of RNA-2 is 0.6–1.4 MD depending on the virus and its strain. One feature that distinguishes tobraviruses is that RNA-1 is infective and causes a spreading infection in individual host plants; symptoms induced by infection with RNA-1 alone are usually more severe than those induced by infection with the complete virus genome and such plants contain no virus particles because RNA-2 codes for coat protein (Harrison and Robinson, 1981).

B. Properties of RNA Species

PRV RNA-2 molecules have 7-methylguanosine triphosphate (m^7GTP) caps at the 5' ends (Abou Haidar and Hirth, 1977; Bergh et al., 1985) and translation of RNA-1 and RNA-2 of TRV was inhibited by m^7GTP, suggesting that both are capped (Pelham, 1979b). None of the RNA species are polyadenylated but similar 5'-terminal sequences (of 5–10 nt) and similar 3'-terminal sequences (of 497 nt) are present in the RNA species of TRV strain PSG (Cornelissen et al., 1986). The 3' terminal 14 nt of RNA-1 and RNA-2 of PEBV are identical and are the same as those of RNA-1 and RNA-2 of TRV PSG (G. Hughes, K. R. Wood, and J. W. Davies, personal communication).

Amongst strains of TRV (Robinson, 1983; Robinson and Harrison, 1985a) and PEBV (Robinson and Harrison, 1985b) there are considerable degrees of sequence homology between RNA-1 molecules but there is much less between molecules of RNA-2.

C. Translation Products

Translation in vitro of RNA-1 of PRV (Mayo et al., 1976), TRV (Fritsch et al., 1977; Pelham, 1979b) or PEBV (G. Hughes, K. R. Wood and J. W. Davies, personal communication) gives prominent products of

165–170 kD and 120–140 kD molecular weight. The relative amounts of the two translation products of TRV RNA-1 was affected by magnesium concentration and the relative yield of 170 kD polypeptides was increased by adding yeast suppressor tRNA (Pelham, 1979b). Sequence studies have shown that read-through of a UGA termination codon in RNA-1 of TRV SYM accounts for the translation products (W. Hamilton and D. Baulcombe, personal communication). The proportion of 170 kD polypeptide made by translation of PRV RNA-1 was enhanced by the addition of yeast opal suppressor tRNA or by tRNA from tobacco tissue (H. Beier, personal communication). Thus it is likely that read-through of a termination codon gives rise to the larger product. Polypeptides induced by infection of tobacco protoplasts with PRV or PRV RNA-1 comigrated with the products of translation *in vitro* (Mayo, 1982).

Translation of RNA-2 of PRV (Ball *et al.*, 1973; Mayo *et al.*, 1976), TRV strain PRN (Fritsch *et al.*, 1977; Pelham, 1979b) and PEBV (G. Hughes, K. R. Wood and J. W. Davies, personal communication), gives coat protein. However, TRV PRN RNA-2 preparations often contain a slightly smaller RNA species (Ramirez-Baudrit, 1981; Robinson *et al.*, 1983) and this may be the active mRNA, because RNA-2 of TRV strain SYM (mol. wt 1.4 MD) did not act as a coat protein mRNA, whereas a subgenomic RNA from it (RNA-3; mol. wt 0.6 MD) did (Robinson *et al.*, 1983).

A subgenomic fragment of RNA-1 has been found in RNA from particles of TRV strains Lisse, SYM and PRN (Pelham, 1979b; Ramirez-Baudrit, 1981; Robinson *et al.*, 1983) that translates to give a 30 kD protein. Unlike that to give 170 and 120 kD, translation to give 30 kD protein was not inhibited by the addition of cap analogue to translation mixtures (Pelham, 1979b).

Figure 11.2 shows a diagram of the translation products of TRV and their relation to the genome RNA species.

D. Results from Sequence Determinations

The sequences of the 3′-terminal 2 kb of RNA-1 of TRV strains SYM (Boccara *et al.*, 1986) and PSG (Cornelissen *et al.*, 1986) show the termination codons of the 170 kD proteins and two other open reading frames in the same (TRV SYM) or a different phase (TRV PSG). The putative translation products are 29 kD and 16 kD. The 29 kD product is almost certainly the same as the 30 kD translation product of TRV SYM RNA-4 and this species comprises the 3′-terminal 1.6 kb of RNA-1. A 16 kD translation product has not been identified in translation experiments, although it might well have been obscured by fragments of larger products. Cornelissen *et al.* (1986) have detected a 0.7 kb RNA (RNA-5) subgenomic RNA in RNA preparations. A rod containing such a molecule would be about 20 nm long, which might result in its preferential loss from preparations of virus particles, and hence RNA preparations; alternatively

11. Strategies of Bipartite Genome, Plant Viruses

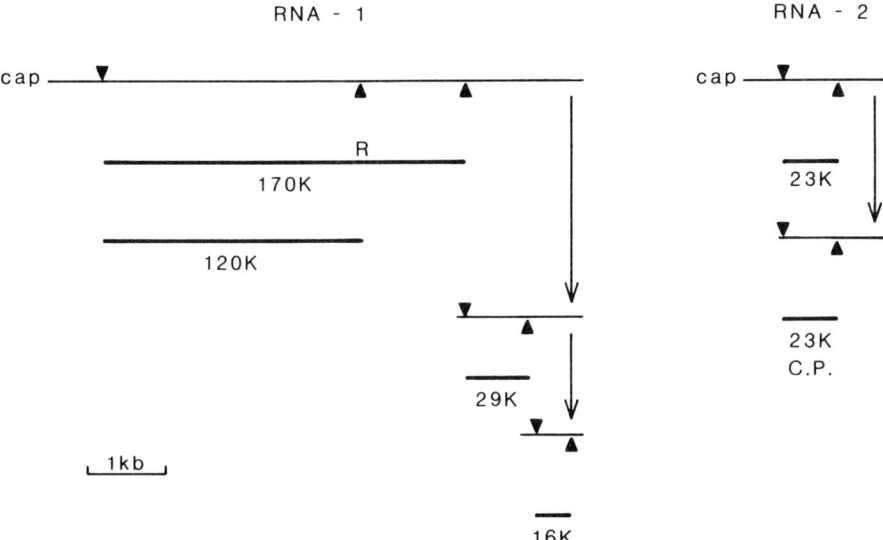

Figure 11.2. Diagram of RNA and translation products of TRV strain PSG. Details are as in Fig. 11.1, otherwise vertical solid arrows represent the derivation of subgenomic RNA molecules and other symbols are: cap, 7-methyl G cap at the 5' end of RNA molecules; R, position of read-through of a termination codon. The positioning of codons in RNA-1 is speculative. (K = kilodalton.)

RNA-5 might lack the putative recognition site for particle assembly (Boccara et al., 1986).

The sequences of RNA-2 of PRV (Bergh et al., 1985) and of TRV strain PSG (Cornelissen et al., 1986) show a single open reading frame within very large non-coding regions of 573 and 554 nt or 570 and 708 nt respectively. The 5' non-coding region in PRV RNA-2 contained direct repeats but that of TRV PSG RNA-2 did not. Cornelissen et al. (1986) detected a subgenomic RNA-4 component in their RNA preparations from particles of TRV PSG that was the 3' terminal 1431 nt of RNA-2 (1905 nt) and speculate that this is the active mRNA for coat protein. The same authors found subgenomic species of RNA-2 about 400–500 nt shorter than RNA-2 in RNA from preparations of several tobraviruses and suggest that the 5' non-coding sequence of these molecules is needed for replication but impedes translation. The results of translation of TRV SYM RNA (Robinson et al., 1983) support this suggestion. Bisaro and Siegel (1980) also found an RNA species of molecular weight about 0.5 MD that was a subgenomic fragment of RNA-2 of PRV. Although these authors found an apparent translation product for this RNA that migrated more slowly than coat protein, this RNA species may be the same as that found by Cornelissen et al. (1986) for TRV CAM strain (=PRV). Although there is little nucleotide sequence homology between RNA-2 of different tobraviruses (Robinson and Harrison 1985a) the amino acid sequences of the coat proteins of PRV and TRV PSG are 40% homologous (Cornelissen et al., 1986). Recent work (W. Hamilton and D. Baulcombe, personal communication) has shown that some strains of TRV have open reading frames in RNA-2 other than that for the coat protein.

E. Sequence Comparisons with Other Viruses

The amino acid sequence of the C-terminal portions of the 170 kD proteins of TRV SYM and TRV PSG share extensive homology with parts of the 183 kD protein of tobacco mosaic virus (TMV) and the translation products of RNA-2 of brome mosaic and alfalfa mosaic viruses (Boccara et al., 1986; Cornelissen et al., 1986). These polypeptides share similar homologies with proteins of other plant and animal viruses, and these common sequences are thought to represent the "core" sequence of RNA-dependent RNA polymerases (Kamer and Argos, 1984).

The sequence of the 29 kD protein also has considerable homology with a TMV gene product, in this case the 30 kD protein (Boccara et al., 1986; Cornelissen et al., 1986), which overlaps and extends to the 3' side of the 183 kD protein gene (Goelet et al., 1982). Indeed, in two regions of about 90 residues in the TRV SYM 29K protein about as many amino acids match those in the 30 kD protein of TMV as match between the same regions of the 30 kD proteins of TMV and sunn-hemp mosaic virus (=cowpea strain of TMV) (Boccara et al., 1986). This protein has been thought to play a role in cell-to-cell transport of TMV in infected plants (Ohno et al., 1983; Zimmern and Hunter, 1983). Infections by TRV RNA-1 alone spread in infected tissue, and if a tobravirus transport gene exists it should be located on RNA-1.

The sequence of the 16 kD protein did not reveal homologies with those proteins tested (Boccara et al., 1986; Cornelissen et al., 1986) but Boccara et al. (1986) speculate that because Harrison and Robinson (1986) consider that tobravirus coat protein is not the determinant for nematode transmission, and no other product is coded for by RNA-2, the 16 kD protein may play this role.

IV. NEPOVIRUSES

A. General Properties

Nepoviruses are transmitted by dorylaimid nematodes and have isometric particles about 28 nm in diameter that in most cases are made from 60 subunits with molecular weights of about 55 kD (Harrison and Murant, 1977a). The RNA species have molecular weights of about 2.8 MD (RNA-1) and 1.3–2.3 MD (RNA-2), depending on the virus (Murant et al., 1981). Virus particles can be protein shells (T component) or can contain

RNA-2 (M component) or either RNA-1 or, if RNA-2 is of molecular weight 1.4 MD or less, two molecules of RNA-2 (B component) (Harrison and Murant, 1977a).

Maximum infectivity of nepovirus RNA for plants depends on both species of RNA being present in inocula (e.g. Harrison et al., 1972), and by mixing RNA species in heterologous combinations it was possible to make pseudo-recombinant isolates that possessed some properties of either virus "parent". For example, for raspberry ringspot virus (RRV) and/or tomato black ring virus (TBRV), RNA-2 carries determinants for some symptom reactions, serological specificity and nematode transmissibility, and RNA-1 carries determinants for host range, seed transmission and other kinds of symptom (Harrison et al., 1974; Hanada and Harrison, 1977; Harrison and Murant, 1977b). It was therefore concluded that nepovirus genomes comprise RNA-1 and RNA-2 (Harrison and Murant, 1977a).

However, although RNA-1 of TBRV did not multiply detectably in inoculated leaves (Randles et al., 1977), it did in inoculated protoplasts, but without the synthesis of coat protein (Robinson et al., 1980). Thus RNA-1 can replicate independently and RNA-2 contributes a coat protein gene and the ability of an infection to spread in plants.

B. Properties of RNA Species

Both species of nepovirus RNA are polyadenylated (Mayo et al., 1979) and both are covalently bound to a small virus-specific protein (VPg) of about 5 kD (Mayo et al., 1982), presumably at the 5' end of the molecules. The partial destruction of VPg by protease treatment decreases the infectivity of nepovirus RNA, depending on the virus, to between 50% and less than 1% of that of the untreated RNA (Mayo et al., 1982). The infectivity of RRV RNA is relatively resistant to protease treatment, being decreased by about 50% by Proteinase K and about 70% either by Pronase, or a mixture of the two proteases. However, the infectivity of protease-treated RNA can be restored somewhat by adding greater than normal amounts of bentonite to inocula (Barker and Mayo, 1982). Thus the effect of VPg on infectivity varies with the type of protease used and may be related to protecting RNA from nuclease attack. Protease treatment did not alter the messenger properties in vitro of RNA of tobacco ringspot virus (TobRV) or TBRV (Chu et al., 1981; Koenig and Fritsch, 1982). Because VPg on TBRV RNA is essential for infectivity, the synthesis of infective RNA-1 in protoplasts not containing RNA-2 shows that RNA-1 codes for VPg (Robinson et al., 1980).

RNA-1 and RNA-2 of nepoviruses show little (TobRV; Rezaian and Francki, 1974) or no (TBRV; Robinson et al., 1980) cross-hybridization although the 126 nucleotides at the 3' end (preceding the poly(A)) of the RNA species of TBRV are almost identical (Dodd, 1983). The extent of hybridization between RNA from different strains of TBRV reflects the degree of serological relatedness among them (Dodd and Robinson, 1984).

C. Translation Products

When nepovirus RNA is translated in reticulocyte lysates several polypeptide species are made. The largest are usually a 220 kD polypeptide from RNA-1 and a product usually equivalent to about 80% of the coding capacity of RNA-2 (Table 11.3). Other similar polypeptide species are found, but usually little or no product is made which has the size of the coat protein. Several RNA-2 translation products react with antiserum to virus particles and it is thought that functional gene products arise by proteolytic cleavage of the large translation products (Fritsch et al., 1980; Morris-Krsinich et al., 1983b; Forster and Morris-Krsinich, 1985; Jobling and Wood, 1985). Incorporation of amino acid analogues into the translation products of RNA of grapevine fanleaf virus (GFLV) and TobRV inhibited the appearance of smaller products, whereas longer than normal incubation or other modifications favoured their production. It was also shown that a translation product of RNA-1 of GFLV or TobRV caused the proteolytic cleavage of the respective translation products of RNA-2 (Morris-Krsinich et al., 1983b; Forster and Morris-Krsinich, 1985).

The large translation products detected for RNA-1 and RNA-2 leave about 1.8 kilobases unaccounted for in RNA-1 of several nepoviruses and also in RNA-2 of cherry leafroll virus (CLRV). Although such large noncoding regions are known with other systems, it remains possible that another coding region exists in these large molecules. Fig. 11.3 shows a

Table 11.3. Translation products and RNA species of nepoviruses

	Molecular weights of				
Virus	RNA-1 (MD)	Translation products of RNA-1 (kD)	RNA-2 (MD)	Translation products of RNA-2 (kD)	Ref.
Tomato black ring	2.8	220, 190	1.6	160^a	[30]
Grapevine fanleaf	2.4^b	220, 195	1.4^b	125, 68, 58^c	[75]
Tobacco ringspot	2.8	225, 207, 180, 37	1.3	116^a, 88^a, 77^a, 53^c, 40, 23	[22, 58]
Raspberry ringspot	2.8	200	1.4	114	[59]
Cherry rasp leaf	2.6	200	1.3	102	[59]
Cherry leafroll	2.8	220, 190, 33	2.3	165^a, 115^a, 90^a, 77^a, 54^c, 37, 24	d

[a] Reacting with antiserum to virus particles (other than coat protein).
[b] Estimates of non-denatured RNA.
[c] Coat protein.
[d] C. Hellen and J. I. Cooper, personal communication.

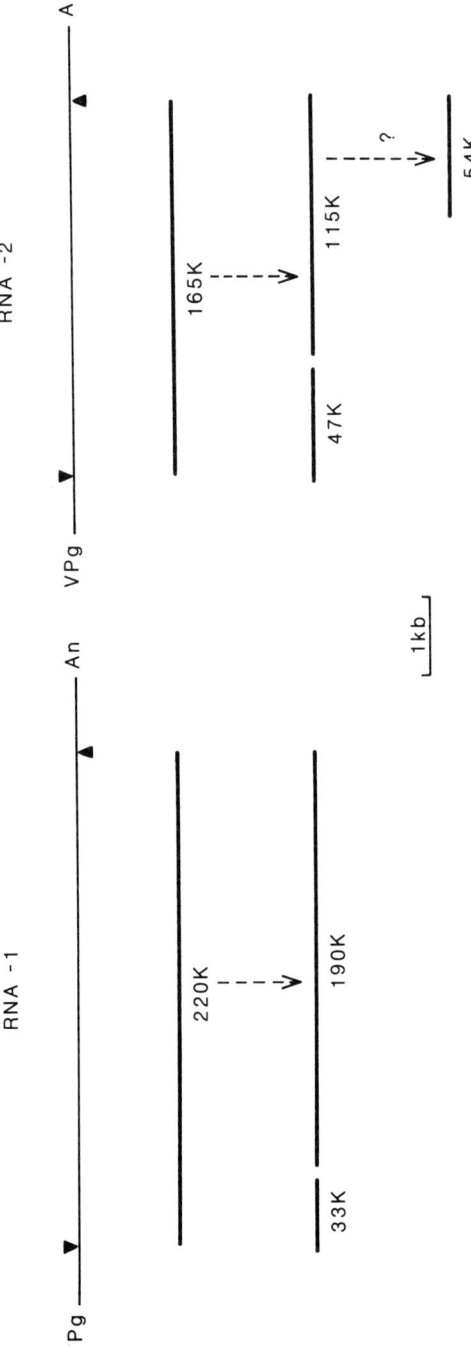

Figure 11.3. Diagram of RNA and translation products of CLRV. Details are as in Fig. 11.1. The placings of the translation products are all speculative as is the C-terminal location of the 54 kD coat protein in the 115 kD precursor. Smaller products are made (Table 11.3), but there is no information as to their location. (K = kilodalton.)

tentative interpretation of the arrangement of translation products of CLRV (C. Hellen and J. I. Cooper, personal communication).

There is little published evidence of the translation products formed *in vivo* other than the detection of coat protein and, because RNA is infective, VPg. Polypeptides of 190 kD and 170 kD have been detected in protoplasts infected with CLRV and TBRV respectively (C. Hellen and J. I. Cooper, personal communication; M. A. Mayo, unpublished data), suggesting that large polyprotein precursors are formed *in vivo*.

D. Results from Sequence Determinations

The only nepovirus genome RNA sequence known at present is that of TBRV RNA-2 (Meyer *et al.*, 1986). This sequence confirms the translation evidence in that it contains one large open reading frame potentially capable of coding for a 150 kD polypeptide. No evidence was found for a sub-genomic mRNA for coat protein, and the amino acid composition of the coat protein most resembled that of the *C*-terminal third of the 150 kD protein, suggesting that the coat protein arises from this part of the polyprotein (Meyer *et al.*, 1986).

Unlike the non-coding regions of CPMV RNA (see Section II), those of TBRV RNA-2 were not complementary. Rather some oligonucleotide sequences, often involving the sequence UUUUC, were found in both 5′ and 3′ regions.

E. Sequence Comparisons with Other Viruses

Several oligonucleotide sequences in the 3′ half of the 3′ non-coding region of TBRV RNA-2 resembled those in the 3′ non-coding region of CPMV M RNA; matches of 10/12, 13/14, 15/16 and 17/17 were present (Dodd, 1983; Meyer *et al.*, 1986). However there was much less similarity between these regions of TBRV RNA-2 and those of either its dependent satellite RNA (Meyer *et al.*, 1984) or TobRV (Dodd 1983). Thus it is not clear what significance to attribute to these very improbable matches.

The amino acid sequence of the 150 kD protein contains a region near the centre of the molecule that showed some similarity with the sequence of the 30 kD gene of some tobamoviruses and a more limited similarity was detected between this region and that of the *N*-terminal side of the coat protein genes in the 105 kD translation product from M-RNA of CPMV (Meyer *et al.*, 1986). The 30 kD protein of TMV has been thought to play a role in cell-to-cell spread of infection (Ohno *et al.*, 1983; Zimmern and Hunter, 1983) and it has been speculated that the non-coat protein region of the 105 kD protein of CPMV might also have a role in virus transport (Rezelman *et al.*, 1982). TBRV RNA-2 seems to contribute the ability of infections to spread from cell to cell and this ability might be coded for by this part of the 150 kD protein.

V. FUNGUS-TRANSMITTED ROD-SHAPED VIRUSES

These viruses are transmitted by the plasmodiophoromycete fungi *Polymyxa* spp. or *Spongospora subterranea* and comprise the proposed furovirus group (Shirako and Brakke, 1984). The most studied viruses of this type are soil-borne wheat mosaic (SBWMV), beet necrotic yellow vein (BNYVV) and peanut clump (PCV) viruses. Particles are about 20 nm in diameter but, whereas the largest of BNYVV are 390 nm long, those of PCV and SBWMV are about 250–300 nm in length.

Infectivity of SBWMV and PCV depend on both RNA-1 (mol. wt 2.3 or 1.9 MD) and RNA-2 (mol. wt 1.2 or 1.6 MD) being present in inocula (Shirako and Brakke, 1984; Reddy et al., 1985). The coat protein gene of BNYVV is in RNA-2 (mol. wt 1.8 MD), which has little sequence homology with RNA-1 (mol. wt 2.3 MD) (Richards et al., 1985), suggesting that it too has a bipartite genome. The coat proteins of SBWMV (Tsuchizaki et al., 1975; Hsu and Brakke, 1985a) and PCV (Mayo and Reddy, 1985) are also determined or coded for by RNA-2, and translation products of SBWMV RNA-2 are thought to determine the type of inclusion body formed in infected tissue (Tsuchizaki et al., 1975).

Although these viruses are similar in several properties, their RNA molecules differ in important properties. BNYVV RNA has a 5′ cap structure and is polyadenylated (Putz et al., 1983), whereas SBWMV RNA has neither 5′ cap nor 5′ genome-linked protein and will accept phosphate in a kinase-mediated reaction. Moreover, neither SBWMV RNA (Hsu and Brakke, 1985b) nor PCV RNA (Mayo and Reddy, 1985) is polyadenylated. An unusual point of similarity between SBWMV and BNYVV is that RNA-2 molecules are sometimes found with some sequence deleted (Hsu and Brakke, 1985b; Bouzoubaa et al., 1986).

The results of *in vitro* translation of RNA from SBWMV, BNYVV and PCV are summarised in Table 11.4. RNA-1 is translated to give a large polypeptide of about 200 kD or more and a prominent product of about 150 kD is also typical. However, Shirako and Ehara (1986) report that the 150 kD product of SBWMV RNA-1, formed in reticulocyte lysates is not a product of translation in wheat germ extracts, and the sequence of BNYVV RNA-1 does not show a termination codon appropriate for a 150 kD translation product. Products of less than about 200 kD may therefore be formed by proteolytic processing, or are artefacts of translation *in vitro*. RNA-2 codes for a protein that comigrates with coat protein and reacts with antiserum to virus particles. However, antiserum to SBWMV or BNYVV particles also reacts with larger translation products of the respective RNA-2 species (Hsu and Brakke, 1985a; Richards et al., 1985) and it was proposed that the coat protein gene is located at the 5′ end of these molecules and that read-through translation leads to the synthesis of larger products.

The sequence of BNYVV RNA-2 confirms that the coat protein cistron is at the 5′ end of RNA-2 and that read-through of a UAG codon gives rise

Table 11.4. Translation products and RNA species of rod-shaped, fungus-transmitted viruses

		Molecular weights of					
Virus	RNA-1 (MD)	Translation products of RNA-1 (kD)	RNA-2 (MD)	Translation products of RNA-2 (kD)	Other RNA species (MD)	Translation products of other RNA (kD)	Ref.
SBWMV	2.3	180, 152, 135, 80, 45, 220, 150	1.2	$90^{a,b}$, 28^a, 20^c 100^a, 46, 25^a, 19^c	—	—	[54] [101]
BNYVV	2.3	>220^f, 150, 50	1.8	85^a, 22^c	0.9^d 0.7^e 0.6^e	42 27 31	[11] [94, 113]
PCV	1.9	195, 143	1.6	25^c	0.5–0.9 0.2	50 20	[67]

[a] Reacting with antiserum to virus particles (except coat protein).
[b] Smaller from RNA from deletion mutants.
[c] Coat protein.
[d] Sub-genomic fragment of RNA-2.
[e] Possibly satellite molecules.
[f] Calculated from partial sequence data (K. E. Richards, personal communication).

to the 85 kD product (see Fig. 11.4). Indeed, the sequence around this codon resembles that around a similarly leaky UAG codon in TMV RNA (Bouzoubaa et al., 1986).

The 90 kD, 28 kD and coat proteins of SBWMV were detected by immune precipitation in extracts of infected wheat plants (Hsu and Brakke, 1985a), which suggests that the read-through products found *in vitro* are made in infected cells.

By contrast, translation of RNA-2 of PCV yielded only coat protein (Mayo and Reddy, 1985). However, Ziegler et al. (1985) have suggested that the *in vitro* system used may have lacked the appropriate suppressor tRNA.

Preparations of BNYVV and PCV contain particles shorter than those containing RNA-2 (Putz, 1977; I. M. Roberts, cited in Mayo and Reddy, 1985) and RNA from these particles act as mRNA *in vitro*. The 0.7 MD (RNA-3) and 0.6 MD (RNA-4) species of BNYVV have little sequence homology with RNA-1 and RNA-2 (Richards et al., 1985) and RNA-3 was eliminated from a culture of BNYVV without affecting the infectivity of the culture. One or both may be satellite RNA species.

The sequence of BNYVV RNA-2 shows that, in addition to the open reading frame for the 85 kD polypeptide that contains the coat protein, there are three open reading frames for polypeptides of about 14 kD and one for a 42 kD polypeptide that overlaps the coding sequence for the

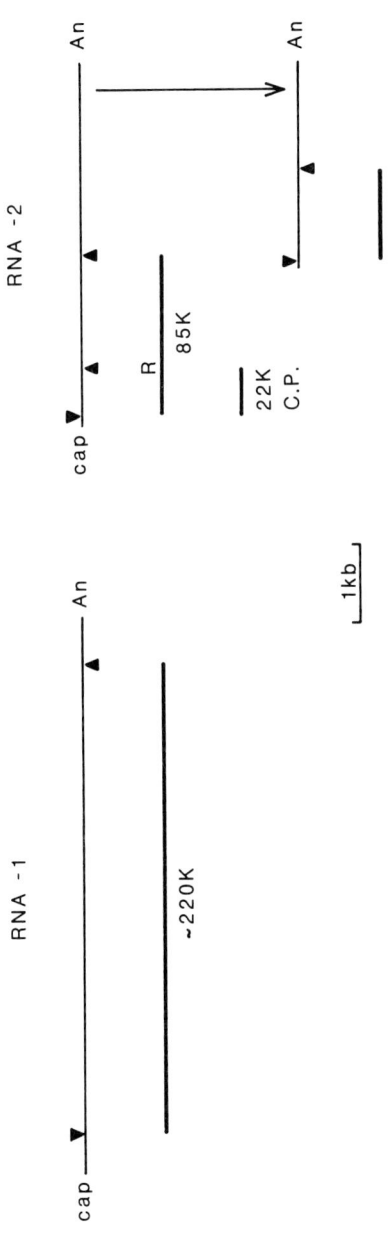

Figure 11.4. Diagram of RNA and translation products of BNYVV. Details are as in Figs 11.1 and 11.2. The positioning and size of the translation product of RNA-1 are speculative. (K = kilodalton.)

85 kD product (Bouzoubaa et al., 1986). Using cloned DNA complementary to this region, Bouzoubaa et al. (1986) isolated an mRNA from infected tissue that was translated to give a 42 kD product. This is evidence for the use of a sub-genomic mRNA corresponding to an overlapping out-of-phase coding region of the genome RNA, in a manner similar to that by which mRNA for TMV 30 kD protein is generated (Goelet et al., 1982). Figure 11.4 shows a diagram of the translation products of BNYVV RNA and their relation to the genome RNA species.

VI. DIANTHOVIRUSES

Viruses of this small group have isometric particles about 33 nm in diameter that sediment as one component at about 133S. However, particles contain two species of RNA, presumably either one molecule of RNA-1 (mol. wt 1.5 MD) or three of RNA-2 (mol. wt 0.5 MD) (Tremaine and Dodds, 1985). The coat protein has a molecular weight of about 38 kD and the RNA species are not polyadenylated (Tremaine and Dodds, 1985). There is no sequence homology between the RNA species and both are needed for infectivity (Dodds et al., 1977; Gould et al., 1981).

Pseudo-recombinant isolates have been made by mixing RNA species either from different strains of carnation ringspot virus (CRSV; Dodds et al., 1977; Okuno et al., 1983) or of red clover necrotic mosaic virus (RCNMV; Okuno et al., 1983; Osman et al., 1986) or from each virus (Lommel and Morris, 1982). These experiments showed that the coat protein gene is in RNA-1 and that, for RCNMV, RNA-2 determines the ability of the infection to spread systemically and the rate of spread of local lesions, both characters that may be related to cell-to-cell movement (Osman et al., 1986).

Translation of RCNMV RNA in rabbit reticulocyte lysates confirmed that RNA-1 codes for the coat protein (Morris-Krsinich et al., 1983a). RNA-1 is also translated to give a prominent 36 kD product and small amounts of 130 kD and 56 kD polypeptides.

Evidence from the translation of size-fractionated RNA suggested that the coat protein was translated from a sub-genomic RNA. Subsequent work (A. Marriott and K. W. Buck, personal communication) has shown that RNA-1 hybrid-selected with a particular cloned cDNA does not translate to give coat protein, confirming the idea that a sub-genomic RNA is mRNA for coat protein and therefore suggesting that this is located at the 3′ end of RNA-1. The 36 kD protein is the major product of RNA-1 and the larger polypeptides probably arise from read-through of the 36 kD termination codon.

RNA-2 is translated as a monocistronic mRNA to give a 34 kD protein. Osman et al. (1986) speculate, on the basis of the behaviour of pseudo-recombinant isolates, that this might be a "transport" protein analogous in function to similar-sized proteins thought to be involved in cell-to-cell movement (Ohno et al., 1983; Zimmern and Hunter, 1983).

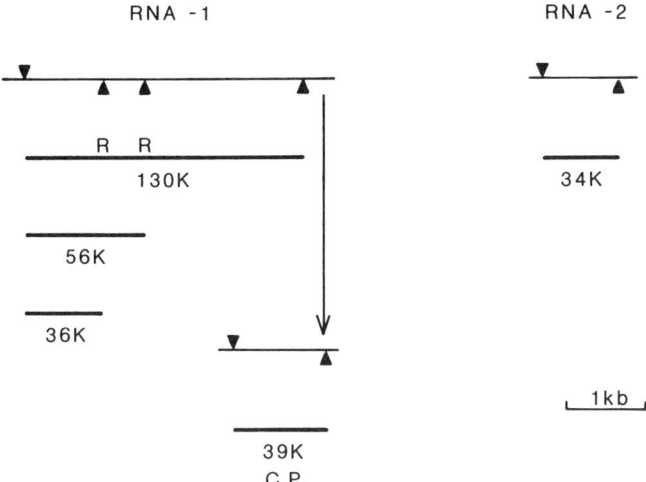

Figure 11.5. Diagram of RNA and translation products of RCNMV. Details are as in Figs. 11.1 and 11.2. The positioning of initiation and termination codons and the size of the subgenomic RNA are speculative. (K = kilodalton.)

Figure 11.5 shows the translation products and genome RNA species of RCNMV.

VII. PEA ENATION MOSAIC VIRUS

Pea enation mosaic virus (PEMV) is transmitted by aphids and has isometric particles about 26 nm in diameter. Virus particles sediment either at about 99S or 112S and contain RNA with molecular weights of about 1.7 MD (RNA-1) and 1.1–1.5 MD (RNA-2) respectively (Peters, 1982). The molecular weights of RNA species differ between different strains of PEMV. When pseudo-recombinants were made by inoculating plants with mixtures of RNA-1 and RNA-2 from different strains it was found that the coat protein gene is on RNA-1, but no phenotypic characters were associated with RNA-2 (Hull and Lane, 1973).

PEMV RNA is not polyadenylated and is not aminoacylatable *in vitro* (German *et al.*, 1978). Molecules of each species are attached at their 5′ ends to a 17.5 kD protein (VPg) (Reisman and de Zoeten, 1982). The coat protein comprises mainly a 21.8 kD species, but small amounts of protein of about 28 kD and 58 kD have been found in protein from particles of aphid-transmissible strains. These minor components are thought to confer transmissibility (Hull, 1977; Adam *et al.*, 1979). Adam *et al.* (1979) also found that RNA-1 of an aphid-transmissible isolate was about 0.4 kb larger than RNA-1 of the aphid-non-transmissible variant and suggested that a deletion mutation was responsible for the loss of transmissibility.

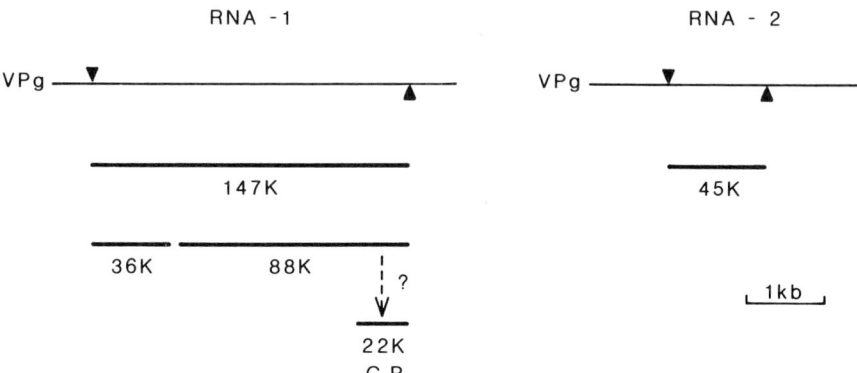

Figure 11.6. Diagram of RNA and translation products of PEMV. Details are as in Figs 11.1 and 11.2. The positioning of initiation and termination codons, and the location of the 22 kD coat protein in the 88 kD polypeptide are speculative. (K = kilodalton.)

Translation of PEMV RNA-1 in rabbit reticulocyte lysates and wheat germ extracts yields two main products of 36 kD and 88 kD and a minor high molecular weight product of 147 kD (Gabriel and de Zoeten, 1984). The 88 kD protein reacted with antiserum of PEMV particles, confirming that the coat protein gene is in RNA-1. No sub-genomic RNA was detectable by Northern blot hybridization and the 36 kD and 88 kD proteins did not share tryptic peptides. Radiolabelled amino acid was incorporated into the 36 kD polypeptide more rapidly than into the 88 kD one and, although there was no evidence for protease action, the most likely interpretation is that protease action cleaves a precursor polyprotein to release 88 kD protein, which is further cleaved to yield coat protein. RNA-2 is a monocistronic mRNA for a 45 kD protein, leaving about 2.5 kb of RNA unaccounted for.

Figure 11.6 illustrates the translation products and the genome RNA of PEBV to scale, although most of the positioning is purely speculative. The coat protein was put at the C-terminus of the 88 kD protein because this is the location in polyprotein precursors of potyviruses (see Chapter 10, this volume), comoviruses (see Section II) and probably nepoviruses (see Section IV).

VIII. DISCUSSION

The variety of translation strategies used by viruses with bipartite genomes is apparent from a comparison of Figs 11.1 to 11.6. Of the six patterns only those of comoviruses (Fig. 11.1) and nepoviruses (Fig. 11.3) could be described as similar. Both groups of viruses are characterized by having monocistronic mRNA which translates to give polyprotein precursors that

11. Strategies of Bipartite Genome, Plant Viruses

are then cleaved by viral proteases to give functional proteins. Also the sizes and relative positions of the proteins in the precursors are similar in the two groups. It is possible that translation of PEMV RNA also gives a polyprotein from which coat protein is derived proteolytically. In contrast, tobravirus RNA-1 and "furovirus" RNA-2 contain more than one initiation and termination site, both are transcribed to give subgenomic mRNA from which the 3' genes on the genome RNA are translated, and both, at least *in vitro*, contain termination sites that are read-through. Dianthovirus RNA-1 also seems to be in this category. However, whereas the coat protein of tobraviruses is produced from a monocistronic RNA, it is the coat protein termination codon that is read-through in BNYVV RNA and SBWMV RNA.

Comparison of Figs 11.2, 11.4, 11.5 and 11.6 also show that, in each, one RNA species is a monocistronic mRNA but the other has a more elaborate multigenic organization involving, in some cases at least, the use of more than one reading frame. The monocistronic species may be larger or smaller, thus RNA-2 of BNYVV (Fig. 11.4) and RNA-1 of RCNMV (Fig. 11.5) appear somewhat similar but are associated with quite different sizes of complementing RNA species. It is tempting to speculate that the more complex genetic arrangement is more ancient and that the monocistronic RNA species is a relatively recent acquisition by the complex species, such as by satellite capture or transduction of host material.

Although BNYVV and SBWMV have been discussed together under the name "furovirus" (Shirako and Brakke, 1984) the RNA species of these viruses have quite different terminal structures. If the termini of RNA molecules are considered symptomatic of a particular strategy then these otherwise genetically similar viruses should be distinguished. Whatever the resolution of this taxonomic question, it is apparent that bipartite genome viruses use a wide range of translation strategies, a range that seems to be as wide as that exhibited by monopartite genome viruses (see Chapter 10). However, not all possibilities are represented. The strategy of translation of the bipartite RNA genome of black beetle virus (see Chapter 12) differs from all those described in this chapter.

It is striking that knowledge of the nucleotide sequences of the RNA species concerned has led to a refinement and sometimes an enhancement of knowledge of translation behaviour as for example with TRV (Cornelissen *et al.*, 1986; Boccara *et al.*, 1986) and BNYVV (Bouzoubaa *et al.*, 1986). It seems certain that the many provisional details in Figs. 11.1 to 11.6 will be modified to a greater or lesser extent by such studies, and indeed this is one reason to embark on them.

Assuming that the bipartite genome arrangement has evolved from the monopartite, the different types of bipartite genome suggest strongly that this evolutionary step has occurred several times. Thus the sizes and arrangement of the gene products of RNA-1 of TRV and of TMV are remarkably similar (Boccara *et al.*, 1986), whereas the production of a polyprotein translation product from polyadenylated RNA with a 5'-

terminal VPg is common to comoviruses, nepoviruses and to potyviruses such as tobacco etch virus (Hari, 1981; Chapter 10). One advantage of having a bipartite rather than a monopartite genome is that potentially more genetic information can be transmitted. Table 11.2 shows that the total amount of genetic material of bipartite genome viruses, except for dianthoviruses and pelargonium zonate spot virus, is as much as or greater than that in monopartite genome viruses except carlaviruses (7.5 kb), potyviruses (8.3–9.7 kb) and closteroviruses (6.1–13 kb) (see Chapter 10). Viruses in these groups have filamentous particles, whereas no bipartite genome virus does (Table 11.1). Perhaps the amount of RNA that can be packaged in a simple isometric particle, or that could survive as a rod-shaped particle, is by necessity only exceeded by viruses with either multipartite genomes or filamentous particles.

Table 11.5. Assignment of characters to the RNA species of bipartite genomes

	Characters determined by		
	RNA-1	*RNA-2*	*Reference*
Comoviruses (CPMV)	Lesion type in cowpea	Systemic infection of Beka bean	[12, 26, 103]
	Systemic symptoms in cowpea	Symptom type in various legumes	
	Replication	Coat protein	
	VPg	?transport	
	Proteases for protein processing		
Tobraviruses	Lesion type in *Nicotiana* spp.	Symptoms in solanaceous hosts	[48]
	Lesion formation in *Petunia hybrida*	Coat protein	
	Systemic invasion of *Nicotiana glutinosa* and *P. hybrida*		
	Replication		
Nepoviruses (RRV, TBRV)	Systemic symptoms in *Chenopodium quinoa*	Yellowing and chloroplast changes in *P. hybrida*	[46, 75, 98]
	Ability to infect Lloyd George raspberry		
	Systemic invasion of *Phaseolus vulgaris*	Nematode transmissibility	
	Competitive ability in *C. quinoa*	Coat protein	
	Replication	?transport	
	VPg		
	Proteases for protein processing		
SBWMV	Symptoms in rye and *Tetragonia expansa*	Coat protein	[54, 106]
		Inclusion body type	
RCNMV	Systemic invasion of sweet clover at 26°C	Rate of spread of infection	[43, 80, 81]
	Coat protein		
PEMV	Coat protein	?	[56, 57]
	Aphid transmissibility		

A disadvantage of a bipartite genome is that, at least for those viruses discussed here, it is necessary that two different particles are present in an inoculated cell to initiate an infection. It is notable in Table 11.1 that most groups of soil-borne viruses, and both groups of nematode-borne viruses, have bipartite genomes. Possibly the amount of inoculum put into roots by the vectors of these viruses is greater than that normally delivered by aerial vectors such as aphids, and this minimizes the chance of infection failing for want of a complementary particle.

Table 11.5 lists some characters of viruses and their infections which have been assigned to the individual RNA species. For the more extensively studied viruses it is clear that the symptoms induced by infection can be determined by either species, and sometimes both. Thus a natural reassortment of RNA species between strains of a virus could well lead to the immediate appearance of a novel virus phenotype. In the laboratory this was shown in work with strains of RRV (Harrison et al., 1974), and RCNMV (Okuno et al., 1983). For example, with RRV two pseudo-recombinants were prepared from heterologous mixtures of RNA that were either more or less virulent than either parent isolate and with RCNMV a number of novel symptom reactions were induced by pseudo-recombinants between strains. That pseudo-recombination occurs in nature is suggested from the properties of an isolate of TRV that has RNA-1 molecules typical of TRV but RNA-2 molecules typical of PEBV (Harrison and Robinson, 1986). These studies suggest that enhanced variability resulting from such pseudo-recombination could well be one genetic advantage to a virus of having a bipartite genome.

From the standpoint of the virologist, the experiments illustrated in Table 11.5 demonstrate the advantage of working with a divided genome when trying to understand virus behaviour. The impact of knowledge of the nucleotide sequences of these viruses is apparent in the sections on comoviruses, tobraviruses and "furoviruses". The coupling of this knowledge with the ability to study parts of genomes offers the prospect of rapid advance in the understanding of RNA plant virus genetics.

ACKNOWLEDGEMENTS

I should like to thank Drs D. Baulcombe, H. Beier, K. W. Buck, J. W. Davies, R. W. Goldbach, W. Hamilton, K. E. Richards and P. Vos for supplying information prior to publication. I also thank Miss Maureen McMaster for preparing the diagrams.

REFERENCES

1. Abou Haidar, M. and L. Hirth (1977). *Virology* **76**, 173–185.
2. Adam, G., E. Sander and R. J. Shepherd (1979). *Virology* **92**, 1–14.
3. Ahlquist, P. and P. Kaesberg (1979). *Nucl. Acids Res.* **7**, 1195–1204.
4. Argos, P., G. Kamer, M. J. H. Nicklin and E. Wimmer (1984). *Nucl. Acids Res.* **12**, 7251–7267.

5. Ball, L. A., A. C. Minson and D. S. Shih (1973). *Nature New Biol.* **246**, 206–208.
6. Barker, H. and M. A. Mayo (1982). *Ann. Rep. Scottish Crop Res. Inst. for 1981*, p. 109.
7. Beier, H., O. G. Issinger, M. Deuschle and K. W. Mundry (1981). *J. Gen. Virol.* **54**, 379–390.
8. Bergh, S. T., M. G. Koziel, S. Huang, R. A. Thomas, D. P. Gilley and A. Siegel (1985). *Nucl. Acids Res.* **13**, 8507–8518.
9. Bisaro, D. M. and A. Siegel (1980). *Virology* **107**, 194–201.
10. Boccara, M., W. D. O. Hamilton and D. C. Baulcombe (1986). *EMBO J.* **5**, 223–229.
11. Bouzoubaa, S., V. Zeigler, D. Beck, H. Guilley, K. Richards and G. Jonard (1986). *J. Gen. Virol.* **67**, 1689–1700.
12. Bruening, G. (1977). *In* "Comprehensive Virology" (Eds H. Fraenkel-Conrat and R. R. Wagner), vol. 11, pp. 55–141. Plenum Press, New York and London.
13. Bruening, G. (1978). CMI/AAB Descriptions of Plant Viruses No. 199.
14. Chu, P. W. G., G. Boccardo and R. I. B. Francki (1981). *Virology* **109**, 428–430.
15. Cornelissen, B. J. C., H. J. M. Linthorst, F. T. Brederode and J. F. Bol (1986). *Nucl. Acids Res.* **14**, 2157–2169.
16. Daubert, S. D. and G. Bruening (1979). *Virology* **98**, 246–250.
17. Davies, J. W., J. Stanley and A. van Kammen (1979). *Nucl. Acids Res.* **7**, 493–500.
18. Dodd, S. M. (1983). Ph.D. Thesis, University of Dundee.
19. Dodd, S. M. and D. J. Robinson (1984). *J. Gen. Virol.* **65**, 1731–1740.
20. Dodds, J. A., J. H. Tremaine and W. P. Ronald (1977). *Virology* **83**, 322–328.
21. Doel T. (1975). *J. Gen. Virol.* **26**, 95–108.
22. Forster, R. L. S. and B. A. M. Morris-Krsinich (1985). *Virology* **144**, 516–519.
23. Francki, R. I. B. (1985). *Annu. Rev. Microbiol.* **39**, 151–174.
24. Franssen, H. (1984). Ph.D. Thesis, Agricultural University, Wageningen.
25. Franssen, H., R. Goldbach, M. Broekhuijsen, M. Moerman and A. van Kammen (1982). *J. Virol.* **41**, 8–17.
26. Franssen, H., R. Goldbach and A. van Kammen (1984a). *Virus Res.* **1**, 39–49.
27. Franssen, H., J. Leunissen, R. Goldbach, G. Lomonosoff and D. Zimmern (1984b). *EMBO J.* **3**, 855–861.
28. Franssen, H., M. Moerman, G. Rezelman and R. Goldbach (1984c). *J. Virol.* **50**, 183–190.
29. Fritsch, C., M. A. Mayo and L. Hirth (1977). *Virology* **77**, 722–732.
30. Fritsch, C., M. A. Mayo and A. F. Murant (1980). *J. Gen. Virol.* **46**, 381–389.
31. Gabriel, C. J. and G. A. De Zoeten (1984). *Virology* **139**, 223–230.
32. Gabriel, C. J., K. S. Derrick and D. S. Shih (1982). *Virology* **122**, 476–480.
33. Gallitelli, D., A. Quacquarelli and G. P. Martelli (1983). CMI/AAB Descriptions of Plant Viruses No. 272.
34. German, T. L., G. A. De Zoeten and T. C. Hall (1978). *Intervirology* **9**, 226–230.
35. Ghabrial, S. A. and R. M. Lister (1973). *Virology* **52**, 1–12.
36. Goelet, P., G. P. Lomonosoff, P. J. G. Butler, M. E. Akam, M. J. Gait and J. Karn (1982). *Proc. Natl. Acad. Sci. USA* **79**, 5818–5822.

37. Goldbach, R. W. (1986). *Annu. Rev. Phytopathol.* **24**, 289–310.
38. Goldbach, R. and J. Krijt (1982). *J. Virol.* **43**, 1151–1154.
39. Goldbach, R. and G. Rezelman (1983). *J. Virol.* **46**, 614–619.
40. Goldbach, R., G. Rezelman and A. van Kammen (1980). *Nature (London)* **286**, 297–300.
41. Goldbach, R. W., J. G. Schilthuis and G. Rezelman (1981). *Biochem. Biophys. Res. Commun.* **99**, 89–94.
42. Goldbach, R., G. Rezelman, P. Zabel and A. van Kammen (1982). *J. Virol.* **42**, 630–635.
43. Gould, A. R., R. I. B. Francki, T. Hatta and M. Hollings (1981). *Virology* **108**, 499–506.
44. Hanada, K. and B. D. Harrison (1977). *Ann. Appl. Biol.* **85**, 79–92.
45. Hari, V. (1981). *Virology* **112**, 391–399.
46. Harrison, B. D. and A. F. Murant (1977a). CMI/AAB Descriptions of Plant Viruses No. 185.
47. Harrison, B. D. and A. F. Murant (1977b). *Ann. Appl. Biol.* **86**, 209–212.
48. Harrison, B. D. and D. J. Robinson (1978). *Adv. Virus Res.* **23**, 25–77.
49. Harrison, B. D. and D. J. Robinson (1981). In "Handbook of Plant Virus Infections and Comparative Diagnosis" (Ed. E. Kurstak), pp. 515–540. Elsevier/North Holland, Amsterdam.
50. Harrison, B. D. and D. J. Robinson (1986). In "The Plant Viruses, Vol. 2: The Rod-shaped Plant Viruses" (Ed. M. van Regenmortel), pp. 339–369. Plenum Press, New York.
51. Harrison, B. D., A. F. Murant and M. A. Mayo (1972). *J. Gen. Virol.* **16**, 339–348.
52. Harrison, B. D., A. F. Murant, M. A. Mayo and I. M. Roberts (1974). *J. Gen. Virol.* **22**, 233–247.
53. Hiebert, E. and D. E. Purcifull (1981). *Virology*, **113**, 630–636.
54. Hsu, Y. H. and M. K. Brakke (1985a). *Virology* **143**, 272–279.
55. Hsu, Y. H. and M. K. Brakke (1985b). *J. Gen. Virol.* **66**, 915–919.
56. Hull, R. (1977). *J. Gen. Virol.* **34**, 183–187.
57. Hull, R. and L. C. Lane (1973). *Virology*, **55**, 1–13.
58. Jobling, S. A. and K. R. Wood (1985). *J. Gen. Virol.* **66**, 2589–2596.
59. Jones, A. T., M. A. Mayo and S. J. Henderson (1985). *Ann. Appl. Biol.* **106**, 101–110.
60. Kamer, G. and P. Argos (1984). *Nucl. Acids Res.* **12**, 7269–7282.
61. Koenig, I. and C. Fritsch (1982). *J. Gen. Virol.* **60**, 343–353.
62. Lommel, S. A. and T. J. Morris (1982). *Phytopathology* **72**, 955 (abstr.).
63. Lomonossoff, G. P. and M. Shanks (1983). *EMBO J.* **2**, 2253–2258.
64. Lomonossoff, G. P., M. Shanks, H. D. Matthes, M. Singh and M. J. Gait (1982). *Nucl. Acids Res.* **10**, 4861–4872.
65. Matthews, R. E. F. (1982). *Intervirology* **17**, 1–199.
66. Mayo, M. A. (1982). *Intervirology* **17**, 240–246.
67. Mayo, M. A. and D. V. R. Reddy (1985). *J. Gen. Virol.* **66**, 1347–1351.
68. Mayo, M. A. and D. J. Robinson (1975). *Intervirology* **5**, 313–318.
69. Mayo, M. A., C. Fritsch and L. Hirth (1976). *Virology* **69**, 408–415.
70. Mayo, M. A., H. Barker and B. D. Harrison (1979). *J. Gen. Virol.* **43**, 603–610.
71. Mayo, M. A., H. Barker and B. D. Harrison (1982). *J. Gen. Virol.* **59**, 149–162.

72. Meyer, M., O. Hemmer and C. Fritsch (1984). *J. Gen. Virol.* **65**, 1575–1583.
73. Meyer, M., O. Hemmer, M. A. Mayo and C. Fritsch (1986). *J. Gen. Virol.* **67**, 1257–1271.
74. Morris-Krsinich, B. A. M., R. L. S. Forster and D. W. Mossop (1983a). *Virology* **124**, 349–356.
75. Morris-Krsinich, B. A. M., R. L. S. Forster and D. W. Mossop (1983b). *Virology* **130**, 523–526.
76. Murant, A. F. and M. A. Mayo (1982). *Annu. Rev. Phytopathol.* **20**, 49–70.
77. Murant, A. F., M. Taylor, G. H. Duncan and J. H. Raschke (1981). *J. Gen. Virol.* **53**, 321–332.
78. Najarian, R. C. and G. Bruening (1980). *Virology* **106**, 301–309.
79. Ohno, T., N. Takamatsu, T. Meshi, Y. Okada, M. Nishiguchi and Y. Kiho (1983). *Virology* **131**, 255–258.
80. Okuno, T., C. Hiruki, D. V. Rao and G. C. Figueiredo (1983). *J. Gen. Virol.* **64**, 1907–1914.
81. Osman, T. A. M., S. M. Dodd and K. W. Buck (1986). *J. Gen. Virol.* **67**, 203–207.
82. Pelham, H. R. B. (1979a). *Virology* **96**, 463–477.
83. Pelham, H. R. B. (1979b). *Virology* **97**, 256–265.
84. Peters, D. (1982). CMI/AAB Descriptions of Plant Viruses No. 257.
85. Putz C. (1977). *J. Gen. Virol.* **35**, 397–401.
86. Putz C., L. Pinck, M. Pinck and C. Fritsch (1983). *FEBS Lett.* **156**, 41–56.
87. Ramirez-Baudrit, M. P. (1981). Thèse de 3ème cycle, Université Louis Pasteur de Strasbourg.
88. Randles, J. W., B. D. Harrison, A. F. Murant and M. A. Mayo (1977). *J. Gen. Virol.* **36**, 187–193.
89. Reddy, D. V. R., D. J. Robinson, I. M. Roberts and B. D. Harrison (1985). *J. Gen. Virol.* **66**, 2011–2016.
90. Reisman, D. and G. A. De Zoeten (1982). *J. Gen. Virol.* **62**, 187–190.
91. Rezaian, M. A. and R. I. B. Francki (1974). *Virology* **59**, 275–280.
92. Rezelman, G., R. Goldbach and A. van Kammen (1980). *J. Virol.* **36**, 366–373.
93. Rezelman, G., H. J. Franssen, R. W. Goldbach, T. S. Ie and A. van Kammen (1982). *J. Gen. Virol.* **60**, 335–342.
94. Richards, K., G. Jonard, H. Guilley, V. Ziegler and C. Putz (1985). *J. Gen. Virol.* **66**, 345–350.
95. Robinson, D. J. (1983). *J. Gen. Virol.* **64**, 657–665.
96. Robinson, D. J. and B. D. Harrison (1985a). *J. Gen. Virol.* **66**, 171–176.
97. Robinson, D. J. and B. D. Harrison (1985b). *J. Gen. Virol.* **66**, 2003–2009.
98. Robinson, D. J., H. Barker, B. D. Harrison and M. A. Mayo (1980). *J. Gen. Virol.* **51**, 317–326.
99. Robinson, D. J., M. A. Mayo, C. Fritsch, A. T. Jones and J. H. Raschke (1983). *J. Gen. Virol.* **64**, 1591–1599.
100. Shirako, Y. and M. K. Brakke (1984). *J. Gen. Virol.* **65**, 119–127.
101. Shirako, Y. and Y. Ehara (1986). *J. Gen. Virol.* **67**, 1237–1245.
102. Stanley, J., P. Rottier, J. W. Davies, P. Zabel and A. van Kammen (1978). *Nucl. Acids Res.* **5**, 4505–4522.
103. Stanley, J., R. Goldbach and A. van Kammen (1980). *Virology* **106**, 180–182.
104. Taylor, R. H. and L. L. Stubbs (1972). CMI/AAB Descriptions of Plant Viruses No. 81.

11. Strategies of Bipartite Genome, Plant Viruses

105. Tremaine, J. H. and J. A. Dodds (1985). AAB Descriptions of Plant Viruses, No. 308.
106. Tsuchizaki, T., H. Hibino and Y. Saito (1975). *Phytopathology* **65**, 523–532.
107. Van Kammen, A. (1968). *Virology* **34**, 312–318.
108. Van Wezenbeek, P., J. Verver, J. Harmsen, P. Vos and A. van Kammen (1983). *EMBO J.* **2**, 941–946.
109. Vos P., J. Verver, P. van Wezenbeek, A. van Kammen and R. Goldbach (1984). *EMBO J.* **3**, 3049–3053.
110. Wellink, J., M. Jaegle and R. Goldbach (1987). *J. Virol.* **61**, 236–238.
111. Wellink, J., G. Rezelman, R. Goldbach and K. Beyreuther (1986). *J. Virol.* **59**, 50–58.
112. Zabel, P., M. Moerman, F. van Straaten, R. Goldbach and A. van Kammen (1982). *J. Virol.* **41**, 1083–1088.
113. Ziegler, V., K. Richards, H. Guilley, G. Jonard and C. Putz (1985). *J. Gen. Virol.* **66**, 2079–2087.
114. Zimmern, D. and T. Hunter (1983). *EMBO J.* **2**, 1893–1900.

12. Organization of Bipartite Insect Virus Genomes: The Genome of Black Beetle Virus

Paul Kaesberg

Biophysics Laboratory and Biochemistry Department, University of Wisconsin, Madison, WI 53706, U.S.A.

I. Introduction	207
II. The Nodaviridae	207
III. The emergence of BBV	209
IV. The molecular biology of BBV	209
V. Plaque assay for BBV	211
VI. Sequencing of BBV RNA1, RNA2 and RNA3, and the genetic map of BBV	211
VII. Infectious RNA derived by transcription from cloned BBV cDNA	215
References	217

I. INTRODUCTION

This chapter concerns the structure and functions of the genome of black beetle virus (BBV), a member of the virus family Nodaviridae. BBV, a bipartite RNA insect virus, is a relative newcomer to molecular virology. However, its properties suggest that it will prove to be immensely useful both as a model virus and as a vector for the expression of foreign genes.

II. THE NODAVIRIDAE

The type member of the Nodaviridae is Nodamura virus. Until 1985 BBV was the only other family member (Matthews, 1982). Recently, Arkansas bee virus, Boolara virus, Endogenous Drosophila line virus and Flock house virus have been accepted to membership. The Nodaviridae are the only known messenger-sense bipartite viruses of higher or lower animals. Plant bipartite viruses are known, but their genome parts are encapsidated in separate virions.

Nodamura virus is named after the village in Japan where it was first detected in swine by Scherer and Hurlbut (1967). Although its primary natural host is the mosquito, Nodamura virus has also been found in ticks and other insects. The virus is lethal to honey bees (Bailey and Scott, 1973) and mice, but evidently does not infect humans (Scherer et al., 1968). The natural host of BBV is the black beetle (*Heteronychus arator*). It has no known vertebrate hosts. Thus, because of its known host range, Nodamura virus is to be regarded as a virus of higher animals with an insect vector, while BBV is to be regarded as an insect virus. "Aliens" visiting Earth might have a less chauvinistic classification scheme.

The molecular biology of Nodamura virus is known largely from the careful work of John Newman and Fred Brown (Newman and Brown, 1973, 1976). The virus resembles picornaviruses in size and appearance in the electron microscope, in acid stability and in buoyant density, and its pathology in mice is somewhat similar to that of the group A coxsackie viruses (Murphy et al., 1970). However, Nodamura virus differs markedly from the picornaviridae and caliciviruses (Matthews, 1982) both in genomic structure and in expression. The genome consists of two RNA species: RNA1, a 22S species with a molecular weight of 1.1 MD; and RNA2, a 15S species with a molecular weight of 0.46 MD, both encapsidated in the same virion. Both RNAs are required for infectivity and replicative forms of both exist in infected, baby hamster cells, indicating that the two RNAs have independent genetic functions.

The interesting similarities and differences between Nodamura virus and picornaviruses led to John Newman's coming to the Biophysics Laboratory at Wisconsin to join in a collaboration with Roland Rueckert's and my laboratories to see just how different Nodamura and picornaviruses were. There, he and colleagues showed that Nodamura virus RNA1 and RNA2 were active messengers in cell-free extracts of wheat embryo and HeLa cells lysates (Newman et al., 1978); RNA1 gives a protein, p105 (mol. wt 105 kD) and RNA2 gives a protein, p43 (mol. wt 43 kD) that is a precursor of mature coat protein. In these early studies, processing to give mature coat protein was not evident. Amit Ghosh and Ranjit Dasgupta in my laboratory undertook physical and sequence studies of the RNAs and showed that, unlike picornaviral RNAs, Nodamura RNAs 1 and 2 were capped at their 5′ termini and lacked 3′ terminal poly(A). It was thus clear that Nodamura virus was a quite novel RNA virus and warranted further study. However, common lines of cultured animal cells did not support the growth of Nodamura virus and the consequent necessity of growing the virus in suckling mice or later in waxmoth larvae suggested that related, more tractable, viruses be sought.

III. THE EMERGENCE OF BBV

The successful candidate was BBV, which Longworth and Archibald (1975) had isolated from larvae of the black beetle. Longworth and Carey (1976) carefully delineated the physical and chemical properties of BBV and showed that it closely resembled Nodamura virus. BBV and Nodamura virus are indistinguishable in the electron microscope. They have similar sizes, densities, and sedimentation rates. Their major capsid proteins migrate similarly upon electrophoresis in SDS-polyacrylamide gels. Each has two RNAs of similar molecular weight, both RNAs are encapsidated in the same virion. However they are not serologically related. BBV has a wide host range among insects, but it does not replicate in mice.

Learning of the fine work of the Longworth laboratory, Roland Rueckert obtained an isolate of BBV, and his student Paul Friesen soon found that BBV could be grown readily in wax moth larvae, in which it reaches a very high titre and which are therefore suitable for growing quantities of virus appropriate both for molecular biology and physical characterization. Although this was sufficient incentive to begin a programme of study with BBV, progress in its molecular biology accelerated following the important discovery by Friesen et al. (1980) that the virus grows in cultured cells of Drosophila melanogaster. BBV multiplies exceedingly well in Schneider's Drosophila line 1. Wild-type infection is not very cytolytic; the infected cells remain intact for periods of four or more days after infection. The yield of virus is very high, ranging from 1 to 2 mg per 10^8 cells within 3 days of infection. By the third day of infection the virus accounts for about 20% of the total cell-associated protein, a 100-fold greater yield than that typically obtained from picornavirus-infected cells.

IV. THE MOLECULAR BIOLOGY OF BBV

The ability of BBV to grow in cultured Drosophila cells prompted several studies in vitro and in vivo. Linda Guarino (Guarino et al., 1981), and later also Saunders and Kaesberg (1985), isolated an RNA polymerase from infected cells that was specific for BBV RNAs, indicating the existence of a genome-coded replicase function. Guarino and colleagues also developed a cell-free protein synthesizing system from Drosphila (Guarino et al., 1981). Translation of BBV RNA in that system resulted in the synthesis of protein A of molecular weight about 120 kD from RNA1 and protein α of molecular weight about 46 kD from RNA2. Protein α is found in relatively small amounts in BBV virions and, as we shall see below, is the precursor of BBV major coat protein β.

However, the existence and the roles of the BBV-encoded proteins was delineated most clearly in three definitive papers from the Rueckert laboratory (Friesen and Rueckert, 1981, 1982; Gallagher et al., 1983). The unusually high yields of BBV produced in *Drosophila* cells made feasible the studies *in vivo*. Virus-specific proteins are present in infected cells in quantities sufficient to permit detection simply by staining polyacrylamide electropherograms. Friesen and Rueckert (1981) exposed cells to serial dilutions of stock virus, lysed them 24 h later with hot SDS, and subjected the contents to electrophoresis on SDS-polyacrylamide gels. Virus-specific proteins were readily distinguished from the background of host *Drosophila* proteins in Coomassie Brilliant Blue-stained electropherograms. Three prominent protein bands (α, β and B) and a fainter band (A) were evident within 24 h after infection. Proteins α and β were identified as BBV coat proteins. Protein B had not been described previously. B and α were detected at input multiplicities as low as 10 virus particles per cell. An input multiplicity of 300 particles per cell was sufficient for their maximum synthesis.

Protein α was the major viral product synthesized during the entire course of infection. Virion protein β appeared 18 h after infection but had accumulated to the same concentration as protein α by 48 h, leading to the suggestion that protein α is the precursor of coat protein β. Proteins A and B appeared as early as 6 h and were present for at least 48 h after infection. Infection was accompanied by a significant decline in the rate of host protein synthesis occurring as early as 12 h after infection. The constant staining intensity of host proteins throughout infection suggested that cell lysis, which occurs only late in infection, was not the major cause of the decline.

Pulse-chase experiments conducted during the period of maximum coat protein synthesis indicated that protein α is indeed cleaved to generate protein β. Precursor α has an unusually long half-life (about 1.5 days) compared, for example, with picornaviral coat precursor, with a half-life of only 7–10 min (Rueckert, 1976). A previously undescribed protein, γ, appeared late in the chase period. The small size of this protein and its simultaneous appearance with protein β suggested that it also is generated by cleavage of protein α. High-specific-activity radiolabelling showed that protein γ is in fact also a BBV virion protein. Additional pulse-chase experiments indicated that maximum synthesis of non-capsid proteins A and B occurred during the first 12 h of infection. Each protein was stable for at least 36 h after its synthesis. Thus, the observed increase and subsequent decrease in the rates of radiolabel incorporation by these proteins were due to differences in the rates of synthesis and not to rapid cleavage or breakdown of newly synthesized protein.

The total complement of BBV encoded proteins thus includes proteins A, B, α, β and γ.

The products of cell-free protein synthesis programmed with the genomic RNAs as messengers do not include protein B. However, when RNA extracted from infected cells directed the cell-free synthesis, protein B was detected (Friesen and Rueckert, 1982). When polysomal RNA was fractionated in a sucrose density gradient, the messengers for proteins A

12. The Genome of Black Beetle Virus

and α cosedimented with viral RNAs 1 and 2, respectively. However, the messenger for protein B was a 9S RNA (RNA3) not found in purified virions. Like the synthesis of viral RNAs 1 and 2, intracellular synthesis of RNA3 was not affected by the drug actinomycin D at concentrations that blocked synthesis of host-cell RNA. Thus RNA3 is a virus-specific subgenomic RNA and protein B is a virus-encoded protein.

Gallagher *et al.* (1983) next showed that RNA1 can be expressed in infected cells in the absence of RNA2. Cells infected with only RNA1 produced RNA3 and in proportions 10- to 20-fold greater than cells infected with virions. The overproduction of RNA3 decreased with increasing proportions of RNA2 in the infecting RNA1. Thus RNA1 is the previously unidentified progenitor of subgenomic RNA3, and RNA2 regulates the amount of RNA3 produced in infected cells.

The Rueckert laboratory used protein synthesis *in vitro* to allow analysis of the mechanism by which translation of RNA1 is modified by the existence of RNA2 (Friesen and Rueckert, 1984). The results revealed no evidence for control by virus-encoded proteins or by virus-induced modification of mRNAs by the cell-free system. Rather, with increasing RNA2 concentration, translation of RNA1 was inhibited. This suggests that the early shut-off in intracellular synthesis of protein A is due to decreasing ability of RNA1 to compete for a rate-controlling translational factor as the concentration of viral RNAs accumulates within the infected cell.

V. PLAQUE ASSAY FOR BBV

Despite its ability to grow vigorously in culture, wild-type BBV is not very cytolytic for *Drosophila* cells, possibly because *Drosophila* is not the natural host for BBV. Unfortunately, the only available strain of black beetle cells grows so slowly in culture as to be impractical for routine work. Therefore, Selling and Rueckert (1984) developed a more cytolytic strain of BBV by selecting for a laboratory strain of virus that grew more rapidly in cultured *Drosophila* cells. They passaged wild-type virus repeatedly, infecting at low multiplicity to minimize the formation of defective interfering particles. Their efforts yielded BBV strain W17, alluded to further below. Selling and Rueckert also discovered conditions for a reproducible plaque assay for infectivity—an assay all-important for the analyses of transcript infectivity described below. The plaque assay revealed a linear relationship between PFU and the number of particles supporting the contention that infection can be initiated by a single virion, i.e. that both RNA1 and RNA2 are contained in the same particle.

VI. SEQUENCING OF BBV RNA1, RNA2 AND RNA3, AND THE GENETIC MAP OF BBV

Genomic RNAs 1 and 2 and subgenomic RNA3 were sequenced by Dasmahapatra *et al.* (1985), Dasgupta *et al.* (1984) and Guarino *et al.* (1984)

respectively. The bulk of the sequences were readily determined, largely by DNA methods. However the structure at the 3' terminus of all three RNAs presented some difficulty. The 3' terminal base does not have a free 3' hydroxyl and will not accept pCp in a ligase reaction nor A in a polyadenylation reaction, nor can a DNA primer be annealed to the 3' proximal bases for reverse transcription. However, priming succeeds if the RNAs are first treated with a protease, thus suggesting that the blocking moiety is a protein. To date no other such 3' terminal blocking groups are known.

The sequence data, the RNA translations *in vitro* and the analysis *in vivo* of induced new proteins and nucleic acids all yield a consistent picture of the organization and expression of the BBV genome.

I summarize the sequence data and recapitulate the above results *in vitro* and *in vivo* leading to a genetic map. RNA1 is 3106 bases long. The published sequence of RNA1, given as 3105 bases, requires an additional base (namely A) following base 2688. The longest open reading frame follows the first AUG codon at bases 39–41 and terminates with a UGA codon at bases 3033–3035. The 999-amino-acid sequence coded by this frame corresponds to a protein of molecular weight 112,290 daltons, in good agreement with previous estimates of the molecular weight of protein A by gel electrophoresis (Guarino *et al.*, 1981). As described above, protein A is made in cell-free extracts programmed with RNA1 and is made *in vivo* upon transfection of *Drosophila* cells. The second in-frame, 5' proximal, initiating codon occurs at bases 564–566 and would correspond to a protein too small to be protein A. The other two reading frames are tightly closed in the region 1–2700, thus precluding their coding for other proteins of substantial size. The RNA3 sequence (389 bases long) starts at position 2718. A second open reading frame exists in that RNA3 region following AUG codons at bases 2737–2739. The third frame is tightly closed. The second frame is 318 bases in length, encoding a protein of molecular weight 11,633 daltons, corresponding to protein B found in BBV-infected cells. Recall that cell-free extracts programmed with RNA3 make protein B and that infected or transfected cells synthesize RNA3 and its encoded protein B. Thus RNA3 is the subgenomic messenger RNA for protein B. Both RNA1 and RNA3 have a 5' terminal capping group.

RNA2 is 1399 bases long. It has a large open reading frame corresponding to protein α which is made *in vitro* upon programming with RNA2. Protein α is proteolytically processed into virion proteins β and γ. The BBV virion has 180 copies of each of protein β and protein γ. The amino termini of α and β are very basic suggesting that they interact with viral RNA in virions. The 5' terminus of RNA2 is capped. A second open reading frame for a putative protein 72 amino acid residues long begins at base 1110; however that open frame is absent in BBV W17. Thus RNA2 is the mRNA for the virion structural proteins and probably no other proteins. No other large open reading frames exist.

The genetic map resulting from the data *in vitro* and *in vivo* and from the sequence data is shown in Fig. 12.1.

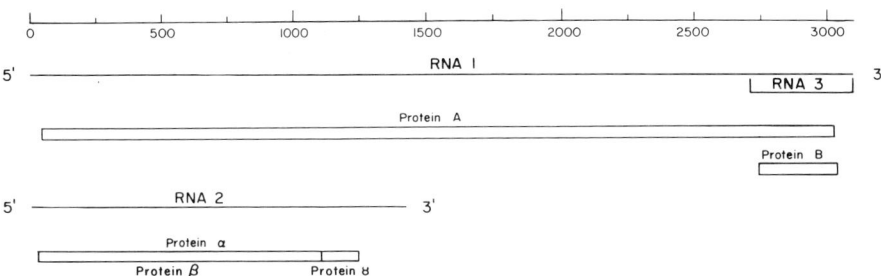

Figure 12.1. Genetic map of BBV showing the coding region of protein A and protein B on RNA1 and proteins α, β, and γ on RNA2.

Application of a secondary structure programme (Zuker and Stiegler, 1981) to RNA2 gave a complicated collection of stems and loops that suggest that BBV RNA2 has a remarkably stable and distinctive secondary structure in solution (Dasgupta et al., 1984). As shown in Fig. 12.2, the 3' half of the RNA can fold into a single, long, imperfect hairpin that has several short stem and loop regions and a large (200 bases long) multiply bifurcated set of stems and loops projecting from its side, the base of the hairpin being stabilized by an exceptionally long (38 bases) helix. The 5' half can fold into an intricate, multiply bifurcated structure also of high predicted stability.

No striking homology was observed among the 5'-terminal or among the 3'-terminal sequences of RNA1 and RNA2. Neither were there strong homologies in the coding regions.

RNAs 1 and 2 are encapsidated into the same virions and might be expected to have similar sequence regions serving as sites for initiation of encapsidation. Our inability to find such common sequences suggests the possibility that RNA1 and RNA2 may interact with each other so that only a single recognition site for coat protein would be needed. We thus looked for sequences on the two RNAs that would have the potential for interaction. The region of strongest base-pairing between RNA1 and RNA2 involves a set of 16 out of 17 base-pairs providing a helical interaction of greater than one turn (Dasmahapatra et al., 1985).

It has been mentioned above that cells transfected with RNA1 alone produce not only RNA1 but also large quantities of RNA3 and protein B. Synthesis of RNA3 and protein B becomes progressively less with increasing amounts of RNA2. The inhibitory effect of RNA2 on the production of RNA3 is insensitive to cycloheximide, a potent inhibitor of translation. Thus the possibility exists that the inhibition of RNA3 production occurs by a specific base-pairing with RNA2 itself, rather than an interaction with its translation product. We have analysed base-pairing possibilities between RNA2 and negative strand RNA1 (Dasmahapatra et al., 1985). The longest region of Watson–Crick base-pairing is 13 base pairs long and, on RNA1, occurs just prior to the start of the RNA3

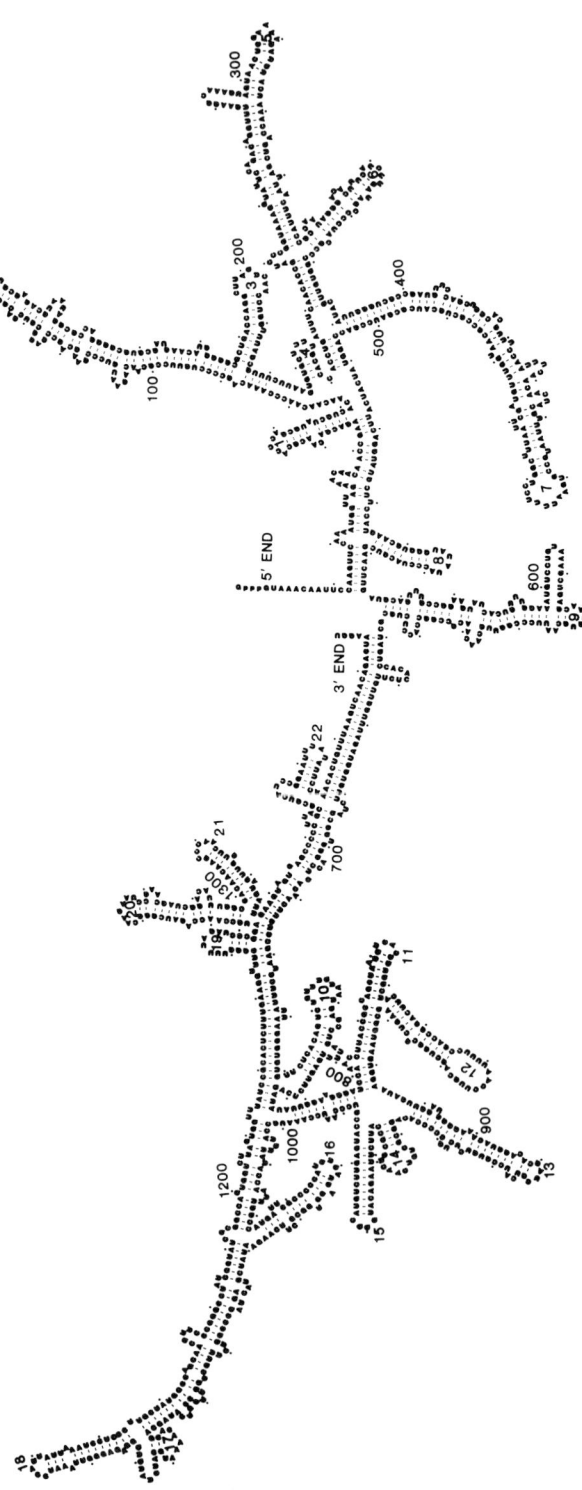

Figure 12.2. Computer generated secondary structure of BBV RNA2. Stem-loop structures are numbered consecutively from the 5' end. (Taken from Dasgupta *et al.*, 1984.)

sequence; two other long regions of possible interaction occur just to either side. These regions also contain a direct repeat and a sequence complementary to the direct repeat that can form a stable stem and loop structure. Possibly these sequences act as a recognition site for replicase to initiate RNA3 synthesis, and the base-pairing with the RNA2 sequence impairs the recognition process so as to inhibit RNA3 production.

VII. INFECTIOUS RNA DERIVED BY TRANSCRIPTION FROM CLONED BBV cDNA

Detailed analysis of the functions of the encoded proteins and of the RNA non-coding regions requires the ability to specifically modify the BBV genomes. Thus we turned our attention to the construction and modification of cDNA forms of the BBV genomic RNAs with the objective of producing specifically modified viral genomes. Such methods are intrinsically available for retroviruses, which have DNA as an intermediate in their synthesis. The methods have become applicable to RNA phage $Q\beta$ (Taniguchi et al., 1978) and to poliovirus (Racaniello and Baltimore, 1981), whose cDNAs have been shown to be infectious, and now to brome mosaic virus (Ahlquist et al., 1984), for which infectious RNA has been made by transcription from cloned viral cDNA. Infectious cDNAs (Cress et al., 1983) and infectious transcripts (Ohno et al., 1983) are also known for viroids.

Armed with the plaque assay of Selling and Rueckert (1984) and the infectious RNA methodology of Gallagher et al. (1983), Bimal Dasmahapatra and Ranjit Dasgupta set out to obtain infectious transcripts from the viral cDNAs they had made previously. Keith Saunders, who had been successful in getting high yields of full-length transcripts for his replicase studies, joined in the effort. They, together with Bernie Selling and Tom Gallagher, found that RNA transcripts derived *in vitro* from DNA copies of the genomic RNAs of BBV are highly infectious to cultured cells of *Drosophila melanogaster*, and thus BBV also is modifiable by recombinant DNA methods (Dasmahapatra et al., 1986).

Double-stranded DNA forms of BBV W17 RNAs 1 and 2 were cloned in the transcription vector pSP64. DNA from such clones were transcribed to produce complete copies of BBV genomes. All SP6 transcripts have the same four additional non-viral nucleotides at their 3' ends and the same 20 additional non-viral nucleotides at their 5' ends headed by a capping group, or in some experiments by a methylated capping group or a triphosphate.

Since we were dealing with two independent genome parts, we first tested with one transcript and one authentic RNA, generally under conditions optimal for transfection with authentic RNAs. We thus identified several RNA1 and RNA2 transcripts that were infectious at about 10–20% the level of authentic RNA isolated from virions. An uncapped RNA1 transcript had approximately 0.1% the infectivity of virion RNA1.

Several of the capped RNA1 transcripts and several of the capped RNA2 transcripts checked by the above procedures had similar infectivity. We selected one combination for more detailed study. When cells were transfected with this transcript combination, plaque assays indicated infectivity about 2% that of authentic RNAs.

The virus arising from the combination, designated BBV K1, was phenotypically indistinguishable from BBV W17, the progenitor virus from which it was derived, as judged by infectivity analyses and various physical and biochemical analyses of the virions and their RNAs and proteins.

Sequence analysis showed that the additional 20 non-viral bases present at the 5' ends of the transcript RNAs were not reproduced in the K1 virion RNAs. Except for base 15 in K1 RNA1, both RNA1 and RNA2 of the K1 virion RNAs have 5' terminal sequences identical to their W17 counterparts. Position 15 in K1 RNA1 and the corresponding position in transcript RNA1 (position 35) have U while W17 RNA1 has C, indicating a transitional mutation, presumably introduced during cloning. Thus 5' non-viral sequences in transcripts are not reproduced in progeny viral RNA, but internal viral sequence changes introduced into transcripts were reproduced in the progeny virus they generate. Thus by our present criteria, the new virus (designated K1) and its progenitor virus W17 are phenotypically indistinguishable but have at least one genotypic difference.

The transcript RNAs are less infectious than their W17 virion RNA counterparts, most likely because of the structural differences at their 5' and 3' termini. Nevertheless, they should provide an effective vehicle for introduction of defined sequence changes into the viral genome by the methods of recombinant DNA technology. Moreover, the fact that the RNA1 transcripts are functional allows study of their replication in the complete absence of RNA2. Previously, studies of transfection with RNA1 alone were limited by the difficulty of completely eliminating RNA2, since even a minute admixture of virion RNA2 led to completion of the infectious process.

Although in our assays the biological activity of BBV RNA was not observably affected by the presence of extra non-viral bases at the 5' terminus, we should emphasize that this result probably obtains only under a limited set of conditions. It may be that the infectivity of our RNA transcripts is relatively insensitive to additional non-viral 5' bases. Alternatively, other structural features of our transcripts, such as their 3' termini, may be limiting factors in infectivity and may prevent discrimination of activity differences due to the 5' end. In our experiments only about 0.1% of treated cells are productively infected. This level of infectivity is certainly sufficient to provide a genetically engineered virus and also to investigate some features of the course of virus synthesis. However, meaningful study of events early in virus synthesis will require a higher level of infection, possibly obtainable with correctly terminated RNA1.

BBV and its host, *Drosophila*, comprise a system of great potential for elucidating the molecular biology of RNA viruses. BBV grows to exceptionally high titre in cultured cells. *Drosophila* is an organism whose genetic characteristics are well understood. A reliable plaque assay is available. The crystallography of its virion is well underway (Hosur *et al.*, 1984). Since RNA1 can replicate independently of RNA2, replicative functions can be studied in the absence of RNA2 or with extensively modified RNA2. Similarly, modification of RNA2 can lead to an understanding of the functions of coat protein and the process of assembly.

Thus, although BBV must be regarded as a brash newcomer to virology, its promise is considerable.

ACKNOWLEDGEMENTS

I am grateful for the support of the U.S. National Institutes of Health under Grants AI-1466 and AI-15342 and Career Award AI-21942. I am indebted to Professor Roland R. Rueckert and especially to my laboratory colleagues Ranjit Dasgupta, Amit Ghosh, Linda Guarino, Bimal Dasmahapatra, and Keith Saunders for expert advice and unfailing support.

REFERENCES

1. Ahlquist, P., R. French, M. Janda and S. Loesch-Fries (1984). *Proc. Natl Acad. Sci. USA* **81**, 7066–7070.
2. Bailey, L. and H. A. Scott (1973). *Nature (London)* **241**, 545.
3. Cress, D., M. Kiefer and R. Owens (1983). *Nucl. Acids Res.* **11**, 6821–6835.
4. Dasgupta, R., A. Ghosh, B. Dasmahapatra and P. Kaesberg (1984). *Nucl. Acids Res.* **12**, 7215–7223.
5. Dasmahapatra, B., R. Dasgupta, A. Ghosh and P. Kaesberg (1985). *J. Mol. Biol.* **182**, 183–189.
6. Dasmahapatra, B., R. Dasgupta, K. Saunders, B. Selling, T. Gallagher and P. Kaesberg (1986). *Proc. Natl Acad. Sci. USA* **83**, 63–66.
7. Friesen, P. and R. R. Rueckert (1981). *J. Virol.* **37**, 876–886.
8. Friesen, P. and R. R. Rueckert (1982). *J. Virol.* **42**, 986–995.
9. Friesen, P. and R. R. Rueckert (1984). *J. Virol.* **49**, 116–124.
10. Friesen, P., P. Scotti, J. Longworth and R. R. Rueckert (1980). *J. Virol.* **35**, 741–747.
11. Gallagher, T. M., P. D. Friesen and R. R. Rueckert (1983). *J. Virol.* **46**, 481–489.
12. Guarino, L. and P. Kaesberg (1981). *J. Virol.* **40**, 379–386.
13. Guarino, L., D. Hruby, L. Ball and P. Kaesberg (1981). *J. Virol.* **37**, 500–505.
14. Guarino, L., A. Ghosh, B. Dasmahapatra, R. Dasgupta and P. Kaesberg (1984). *Virology* **139**, 199–203.
15. Hosur, M., T. Schmidt, R. Tucker, J. Johnson, B. Selling and R. Rueckert (1984). *Virology* **133**, 119–127.
16. Longworth, J. F. and R. D. Archibald (1975). *N.Z. J. Zool.* **2**, 233–236.
17. Longworth, J. F. and G. P. Carey (1976). *J. Gen. Virol.* **33**, 31–40.
18. Matthews, R. E. F. (1982). *Intervirology* **17**, 1–199.
19. Murphy, F. A., W. F. Scherer, A. K. Harrison, H. W. Dunne and G. W. Gary (1970). *Virology* **40**, 1008–1021.

20. Newman, J. F. E. and F. Brown (1973). *J. Gen. Virol.* **21**, 371–384.
21. Newman, J. F. E. and F. Brown (1976). *J. Gen. Virol.* **30**, 137–140.
22. Newman, J. F. E., T. Matthews, D. R. Omilianowski, T. Salerno, P. Kaesberg and R. Rueckert (1978). *J. Virol.* **25**, 78–85.
23. Ohno, T., M. Ishikawa, N. Takamatsu, T. Meshi, Y. Okada, T. Sano and E. Shikata (1983). *Proc. Japan Acad.* **59** Ser.B, 251–254.
24. Racaniello, V. and D. Baltimore (1981). *Science* **214**, 916–919.
25. Rueckert, R. R. (1976). *In* "Comprehensive Virology" (Eds H. Fraenkel-Conrat and R. R. Wagner), vol. 6, pp. 131–213. Plenum Press, New York.
26. Saunders, K. and P. Kaesberg (1985). *Virology* **147**, 373–381.
27. Scherer, W. F. and H. S. Hurlbut (1967). *Am. J. Epidemiol.* **86**, 271–285.
28. Scherer, W. F., J. E. Verna and G. W. Richter (1968). *Am. J. Trop. Med. Hyg.* **17**, 120–128.
29. Selling, B. H. and R. R. Rueckert (1984). *J. Virol.* **51**, 251–253.
30. Taniguchi, T., M. Palmieri and C. Weissmann (1978). *Nature (London)* **274**, 223–228.
31. Zuker, M. and P. Stiegler (1981). *Nucl. Acids Res.* **9**, 133–148.

13. Organization of Tripartite Plant Virus Genomes: The Genome of Brome Mosaic Virus

Paul Kaesberg

Biophysics Laboratory and Biochemistry Department, University of Wisconsin, Madison WI 53706, U.S.A.

I.	Introduction	220
II.	Early history of BMV and other divided-genome viruses	220
III.	Virion structure	221
IV.	Translation of BMV RNAs	223
V.	Silent cistrons and subgenomic RNAs: The eukaryotic rule	224
VI.	BMV sequences	224
	A. Near-identity of 3′ RNA secondary structure in bromoviruses and cucumber mosaic virus	225
	B. Striking similarities in amino acid sequence among non-structural proteins encoded by BMV, AlMV, TMV and Sindbis virus	227
VII.	BMV RNA replication	231
VIII.	BMV infection derived from cloned viral cDNA	232
IX.	RNA recombination	234
	References	234

Abbreviations: AlMV, alfalfa mosaic virus; BBMV, broad bean mottle virus; BMV, brome mosaic virus; BPMV, bean pod mottle virus; CCMV, cowpea chlorotic mottle virus; CMV, cucumber mosaic virus; CPMV, cowpea mosaic virus; SBMV, southern bean mosaic virus; SqMV, squash mosaic virus; STNV, satellite tobacco necrosis virus; TBSV, tomato bushy stunt virus; TMV, tobacco mosaic virus; TNV, tobacco necrosis virus; TYMV, turnip yellow mosaic virus; WCMV, wild cucumber mosaic virus.

THE MOLECULAR BIOLOGY OF THE
POSITIVE STRAND RNA VIRUSES   ISBN 0-12-599930-5

Copyright © 1987 Academic Press Inc. (London) Ltd.
All rights of reproduction in any form reserved

I. INTRODUCTION

This chapter concerns the structure and functions of the genome of brome mosaic virus (BMV). I will present a historical perspective based on the studies in my laboratory and then discuss more recent discoveries that I believe will substantially alter the course of RNA virus research.

BMV has been of some moment in the history of molecular virology. The icosahedral structure of viruses was first discovered in BMV (Kaesberg, 1956). With it, subgenomic RNAs and silent cistrons were discovered (Shih and Kaesberg, 1973). The sequence of its coat protein mRNA (BMV RNA4) delineated the first eukaryotic ribosome binding site (Dasgupta et al., 1975). Its coat protein was the first complete protein made in the wheat embryo protein synthesizing system (Shih and Kaesberg, 1973). Its RNAs were early examples of terminal capping (Dasgupta et al., 1976) and aminoacylatability (Hall et al., 1972). Its genome was the first to be reproduced by transcription of cloned cDNA to yield infectious RNA (Ahlquist et al., 1984b). Moreover, its sequence studies have shown a wide-ranging relationship with other tripartite viruses, with monopartite tobacco mosaic virus (TMV), and with the alpha-viruses of animals and insects (Haseloff et al., 1984).

II. EARLY HISTORY OF BMV AND OTHER DIVIDED-GENOME VIRUSES

Our work with BMV began more than three decades ago. W. W. Beeman, J. W. Anderegg and I had been attempting to adapt the theory and practice of the new method of small-angle x-ray scattering to the study of proteins and other biological macromolecules. At that time the theory had been worked out most explicitly for spherically symmetric particles (Beeman et al., 1956) and so we sought appropriate biological examples of such structures. We were greatly impressed with Robley Williams' electron microscopic studies of tomato bushy stunt virus (TBSV), the physicochemical studies of southern bean mosaic virus (SBMV) by Max Lauffer, and the chemical and biological work on turnip yellow mosaic virus (TYMV) by Roy Markham, all of which suggested that plant viruses were small spheres. We began x-ray scattering studies of these viruses and tobacco necrosis virus (TNV), both because they seemed to be ideal test objects and because the x-ray method held promise of providing their precise and detailed structural characterization. We showed that these were exquisitely uniform and nearly spherically symmetric and that all of them had a hollow centre containing only the buffer in which they were dispersed (Leonard et al., 1953). With TYMV, studied together with its associated nucleic-acid-free particle (called top component), we verified Roy Markham's hypothesis that the protein portion of viruses is in the form of a shell enclosing a centrally located nucleic acid (Schmidt et al., 1954). Later x-ray scattering studies showed that in some symmetric viruses, notably BMV, protein and RNA interdigitate (Anderegg et al., 1963).

Because of the considerable interest generated by my laboratory's TYMV studies we had numerous requests to supply the virus. TYMV was not indigenous to America and thus its delivery required that any recipient first receive formal permission from the United States Department of Agriculture. This procedure was sometimes slow, sometimes unproductive, always embarrassing. The delivery problem was later circumvented by the discovery of an American isolate just outside a famous virology laboratory, but meanwhile it prompted me to search for "American" viruses that we could study and also distribute freely. We looked for viruses that were exceptionally small (so as to be structurally and genetically simple) or had a TYMV-like top component. We obtained isolates of wild cucumber mosaic virus (WCMV) which has a top component and other similarities to TYMV, squash mosaic virus (SqMV), the first example of a divided-genome virus, a bipartite virus related to cowpea mosaic virus (CPMV), later to become noteworthy as a result of the work of van Kammen and his colleagues, and more pertinently for this review, we examined BMV and alfalfa mosaic virus (AlMV) as examples of viruses that were exceptionally small. AlMV derives its fame from the pioneering studies of the Leiden group (Bosch, Jaspars, van Vloten-Doting, Bol and their associates). Our original inference that the small size of BMV and AlMV particles implied a small genome turned out to be simplistic, since their genomes are divided among several virions. Indeed, TNV and SBMV, with larger virions but a single-component genome, have the smallest known plant virus RNA genome.

III. VIRION STRUCTURE

BMV was the first example of an icosahedral virus. We observed its icosahedral structure in the electron microscope (Kaesberg, 1956) after freeze-drying by the method developed by Robley Williams (1953) and this was followed two years later by visualization by Professor Williams of the icosahedral insect virus, tipula iridescent virus, a virus some 200 times the mass of BMV (Williams and Smith, 1958). Thus these two original examples span almost the entire range of sizes of "spherical viruses". By now, numerous intermediate sizes have been reported and, indeed, it is noteworthy when a small symmetric virus does not appear polyhedral.

The AlMV virion also has an interesting history of structural studies. John Bancroft and I examined it in the belief it was even smaller than BMV. Its sedimentation coefficient was reputed to be in the range 50–70S, i.e. much smaller than 86S, the value for BMV. It turned out, however, that there were three (later, at least five) major kinds of sedimenting virions differing as a consequence of their different masses and shapes (Kelley and Kaesberg, 1962). The original sedimentation values had been obtained in one of the first Svedberg ultracentrifuges with absorption optics rather than the more recent Schlieren or interference optics and the existence of the AlMV slowest-sedimenting component α (68S) had completely obscured the presence of the larger AlMV virions. Indeed, we were a bit disappointed that our AlMV preparations were heterogeneous and sedimented rapidly.

Realization that the genomes of some of these viruses were multicomponent took some time to develop. Initially, the existence of the several RNAs of SqMV (Mazzone et al., 1962), BMV (Bockstahler and Kaesberg, 1965) and AlMV was taken to indicate partial degradation rather than the presence of authentic genome parts. The reconstruction experiments of Bockstahler showing that addition of BMV RNA3 to preparations of BMV RNA1 plus 2 increased infectivity, was taken as evidence of the protective effect of RNA3 (Bockstahler, 1964). Our AlMV infectivity experiments produced variable results, which we attributed to virus fragility rather than to the variable admixtures of the virions containing AlMV RNA2 and RNA3 which were in fact essential to make the bottom component preparations, containing RNA1, infectious.

Beginning with the work of Wood and Bancroft (1965) on bean pod mottle virus (BPMV), another relative of SqMV and with AlMV, there was a gradual realization by virologists that in some instances more than one virion might be required for infectivity. However, it took the ingenious work of the Leiden virologists, van Vloten-Doting, Dingjan-Versteegh and Jaspars, a half dozen years later to show conclusively that all three AlMV components were required for infectivity and thus to reveal the true nature of divided-genome viruses (Van Vloten-Doting et al., 1970).

Subsequently, Leslie Lane showed that BMV consists of three virions containing RNA1, RNA2, and RNA3 plus RNA4 (Lane and Kaesberg, 1971) respectively. The three virions, or alternatively their isolated RNAs 1, 2, and 3, were required for infectivity (Fig. 13.1). RNA4 was not needed.

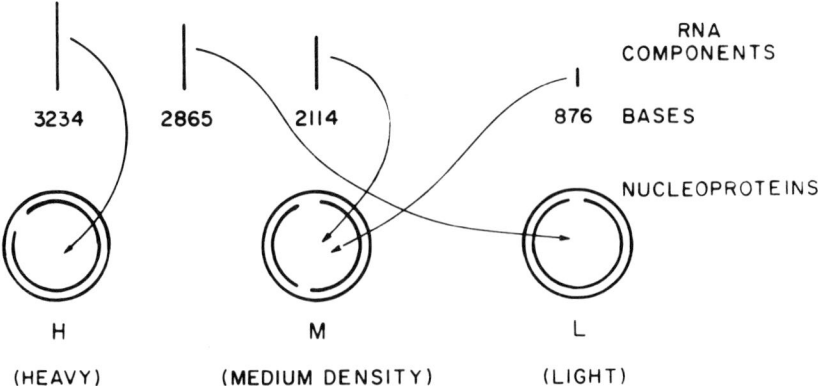

Figure 13.1. Relationships between the RNA and nucleoprotein components of BMV. The RNA components are schematically represented with size decreasing from left to right, and the nucleoproteins are represented with density decreasing from left to right.

Complementation experiments showed that RNA3 contained the gene for BMV coat protein and sequence analyses showed that RNA3 carried the sequence of RNA4 (Shih et al., 1972).

IV. TRANSLATION OF BMV RNAs

With the development of the *Escherichia coli* cell-free protein synthesizing system by Nirenberg and Matthaei (1961), and with the demonstration by Nathans et al. (1962) that the system could be programmed with phage RNA as a messenger, it was evident that cell-free protein synthesis would become a powerful method for determining the gene content of RNA viruses. However, the system apparently failed in the crucial test of the synthesis of TMV coat protein (Aach et al., 1964) (at that time the only really well-characterized viral protein) and it was widely concluded that such systems required homologous messengers.

In 1965, Clark et al. showed that satellite tobacco necrosis virus (STNV) coat protein sequences were synthesized in the *E. coli* system. In 1967, Stubbs and I programmed the *E. coli* system with the BMV RNAs and showed that the smallest BMV RNA induced synthesis of proteins some of whose tryptic peptides resembled those of BMV coat protein. However, the available analytical procedures were cumbersome and, moreover, no identifying characteristics of other BMV proteins were known. Several years later, when the importance of the Leiden AlMV work and our work with BMV had sunk in, and with the advent of polyacrylamide gel electrophoresis, we reopened the question of what the BMV RNAs encoded.

In the interim, Marcus (1970) had shown that wheat embryo extracts programmed with TMV RNA readily incorporated amino acids into protein although TMV coat protein was not among the products. Klein et al. (1972) had shown that wheat embryo extracts could be programmed with STNV RNA to synthesize a product resembling STNV coat protein in composition. Wheat is a host of BMV, but not of TMV or STNV, and so, after tutelage by Professor John Clark, we tested BMV in the wheat embryo system in the belief that what was really needed was an homologous system.

From our complementation and structural studies, we had expected that programming of the wheat system with RNA3, and possibly with RNA4, would lead to BMV coat protein synthesis. Indeed, we soon found that RNA4 catalysed a remarkably high synthesis of a single product, completely indistinguishable from authentic coat protein (Shih and Kaesberg, 1973). However, RNA3 induced synthesis of a single protein (designated protein 3A) that was not related to coat protein. It was shown subsequently that RNA1 and RNA2 each catalyse synthesis of single proteins (designated protein 1 and protein 2, respectively) (Shih and Kaesberg, 1976). Thus RNA1, RNA2, and RNA4 are monocistronic messenger RNAs and RNA3 is a dicistronic mRNA, containing cistrons 3A and 3B, the latter of which (i.e. the coat protein cistron) is silent. So RNA3 is the *genomic RNA* for coat protein while RNA4 is the *messenger RNA* for coat protein synthesis. We expended considerable effort and no little verbiage in unsuccessful attempts to translate 3B (J. Davies, unpublished).

V. SILENT CISTRONS AND SUBGENOMIC RNAS: THE EUKARYOTIC RULE

It is now clear for viruses indigenous to eukaryotes that a major feature in expression of their genomes is the existence of silent cistrons (Hunter et al., 1976; Pleij et al., 1976) analogous to the coat protein cistron in BMV RNA3 and the creation of subgenomic RNAs analogous to BMV RNA4. This has been generalized into the eukaryotic rule: messenger RNAs of eukaryotes are functionally monocistronic. The variety in expression of viral genomes results from their conformity to the eukaryotic rule. Obviously, in eukaryotes there is a positive feature of expression that makes it advantageous to prohibit translation of internal cistrons in general. Possibly internal initiation exists in instances in which it is important for an organism to have a temporal or a quantitative relationship between the translation products of a 5′ proximal cistron and an adjacent internal cistron.

An early example of conformity to the eukaryotic rule is the case of wild-type TMV whose coat protein cistron is silent in the genomic RNA and whose coat protein subgenomic mRNA is not encapsidated. The failure to synthesize TMV coat protein (see above) was not any inadequacy of the *E. coli* cell-free protein synthesizing system but resulted simply from the absence of the coat protein messenger (i.e. the subgenomic RNA) in the virions from which TMV RNA had been isolated. With BMV, a fortunate feature had been that its coat protein mRNA (RNA4) is encapsidated into virions and thus can be isolated from them. What would the course of molecular biology have been if RNA from the cowpea strain of TMV had been used (in which the encapsidation origin exists on the subgenomic coat mRNA as well as the genomic RNA)? Surely in those early years plant and other RNAs heterologous to *E. coli* would have been studied extensively in the Nirenberg and Matthaei system and other (perhaps more nuclease-free) heterologous systems.

VI. BMV SEQUENCES

The genomic RNAs of several of the viruses mentioned above have been sequenced in the last several years, among them monopartite TMV, bipartite CPMV and tripartite BMV, AlMV and cucumber mosaic virus (CMV) (see references, below). With the advances of recombinant DNA technology and the ease of generation of viral cDNAs, such sequence determinations are now to be regarded as routine. Nowadays the expenditure of time for the determination of such sequences is mostly occupied in generating DNA forms and in defining the RNA termini. This was not so earlier. Much of the RNA methodology evolved together with the emerging data. Four years ago, determination of the sequence of RNA4 of cowpea chlorotic mottle virus (CCMV), a close relative of BMV took us a matter of weeks. Our sequencing of BMV RNA4 (876 bases), completed a few months earlier, had occupied us for almost five years. With some sadness we decided to publish the two sequences in the same paper (Dasgupta and Kaesberg, 1982).

A. Near-Identity of 3' RNA Secondary Structure in Bromoviruses and Cucumber Mosaic Virus

Divided-genome viruses should be especially useful in defining the signals for RNA encapsidation, RNA replication and other functions crucial for viral infectivity, because these signals should exist on each of the genomic RNAs with the degree of similarity of primary and secondary structure required for effective interaction with the relevant proteins. The 3' noncoding regions of the bromovirus RNAs have characteristics one might expect of such signal sites: all the 3' termini are enzymatically tyrosylatable, indicating the existence of a precise spatial configuration capable of specific interaction with proteins (Kohl and Hall, 1974); a nearly identical fragment especially resistant to nucleases can be quantitatively cleaved from the 3' termini of each of the RNAs, indicating an evolutionarily well-conserved structure (Dasgupta and Kaesberg, 1977).

Thus we examined the 3' terminal sequences of RNAs 1, 2, 3 and 4 from each of the three bromoviruses (BMV, CCMV, and BBMV (broad bean mottle virus)) and also from CMV, and found that they display considerable inter-viral sequence similarity in addition to strong intra-viral homology (Ahlquist *et al.*, 1981). The inter-viral similarity is much more dramatic when RNA secondary structures, rather than primary structures, are compared. As shown in Fig. 13.2 the last 190 or so bases in each of the BMV, CCMV and CMV RNAs can be folded into highly base-paired structures strikingly similar in their arrangement of helical stems and unpaired loops (Ahlquist *et al.*, 1981). The 3' ends of the BBMV RNAs can be folded into related structures that differ principally from the other viral RNAs in lacking hairpin *d* (see Fig. 13.2). This difference corresponds exactly to the 20 base gap in the homologous alignment of primary sequences.

The structures of Fig. 13.2 conserve not only a general shape but also the position of some well-conserved sequences within that shape. Examples include the extreme 10–15 bases of each RNA, the UACA sequence which appears in loop *f* in each RNA except BBMV RNA1 (where it is replaced by UCCA), and the single-stranded sequence UUC(Pu)GAA which appears between hairpins *d* and *e* in the BMV, CCMV and CMV RNAs and is truncated to UUC(Pu)A in the BBMV RNAs.

An additional common feature of the structures shown in Fig. 13.2 is

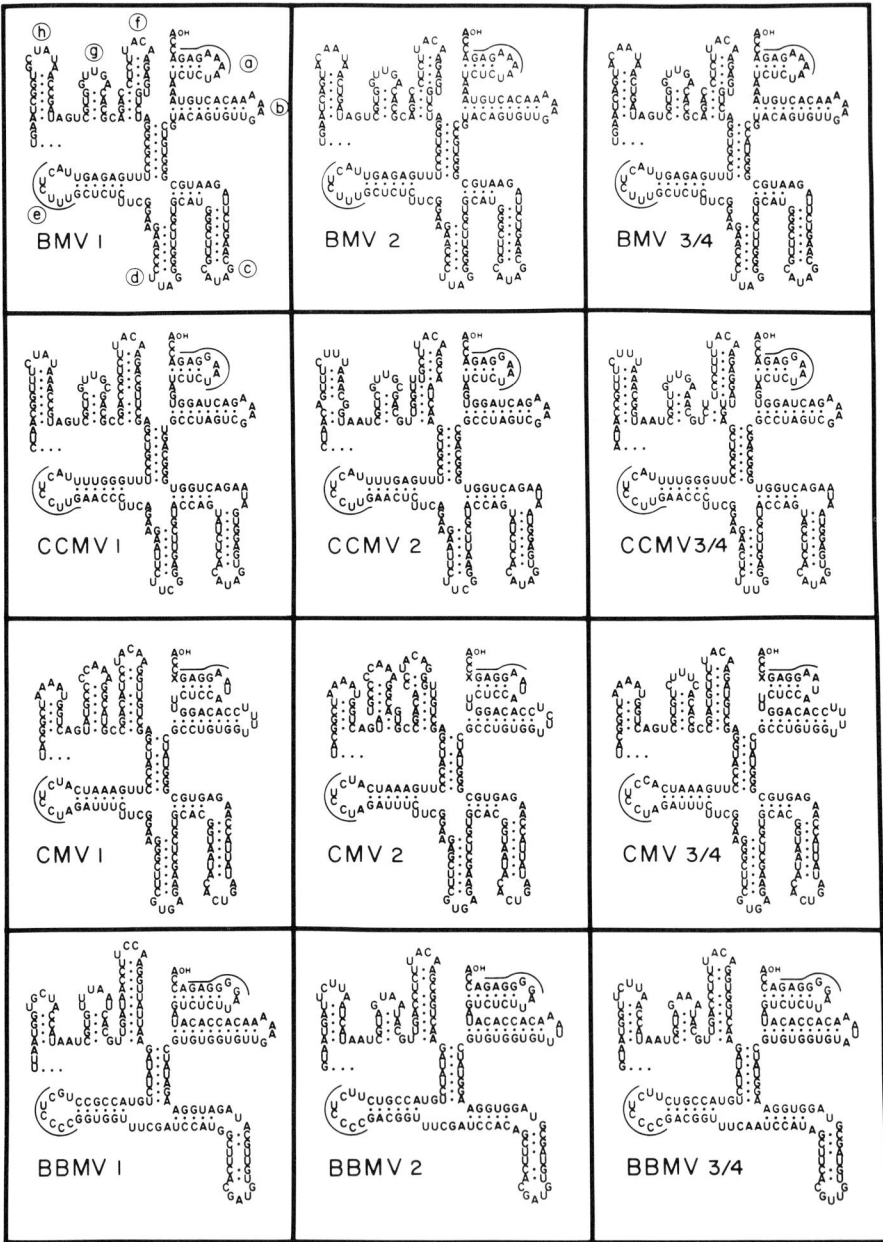

Figure 13.2. Proposed secondary structure for the RNA sequences discussed in the text. The overlined sequences in each structure are complementary. (Taken from Ahlquist *et al.*, 1981.)

that the extreme 3' end of each can be folded into two configurations, one locally base-paired to form hairpin *a*, the other with the over-lined 3' proximal sequence base-paired to its over-lined complement in the loop of hairpin *e*. The structure resulting from the second pairing is analogous to that originally proposed by Symons (1979) for the CMV RNAs and it leaves the 3' terminal-ACCA completely unpaired as it is in tRNAs, facilitating aminoacylation of the BMV 3' termini (Joshi et al., 1983). Rietveld et al. (1983, 1984) have studied chemical modification and selective enzyme digestion of TYMV RNA and BMV RNA2 and have presented ingenious models of tertiary structure that preserve the suggested base pairings and show a striking similarity to the known tertiary configuration of tRNA (see Fig. 13.3).

Serological, host-range and symptomatic variations, as well as the pattern of gaps and non-homologous bases in the aligned sequences, make it clear that substantial evolutionary divergence has occurred among the RNAs of BMV, CCMV, BBMV and CMV. Nevertheless, features of the 3' terminal structure in these RNAs have been strongly conserved, indicating that the bromoviruses and cucumoviruses are related and that their RNA 3' ends are subject to strong functional constraints.

B. Striking similarities in amino acid sequence among non-structural proteins encoded by BMV, AlMV, TMV and Sindbis virus

With the availability of the complete sequences of TMV, AlMV and BMV, our colleagues at the MRC Laboratory in Cambridge, the Biochemistry Department in Leiden and Paul Ahlquist, Ranjit Dasgupta and I in Madison independently undertook study of the question of the similarities of the sequences of the encoded proteins (Haseloff et al., 1984; Cornelissen and Bol, 1984). We learned of the Cambridge work at a relatively early stage and published jointly with them. Nevertheless, we all reached essentially the same conclusions at the same time. Probably none of us expected at the outset that our analyses would lead to such general conclusions.

I will summarize some of the pertinent information derived from the sequences. The complete sequences of BMV and AlMV have verified that their genome consists of three RNA segments. The two larger RNAs of each virus are monocistronic. The smallest is dicistronic, with the 3' proximal gene in both cases encoding coat protein that is translated from a subgenomic mRNA. Although both viruses require all three RNAs for infection, AlMV, unlike BMV, also requires either coat protein or the subgenomic mRNA for coat protein. The genome of TMV is a single RNA molecule. It encodes four known proteins in three open reading frames. That nearest the 5' end contains an in-phase amber termination codon that is partly suppressed during translation *in vitro* or *in vivo* to give two products, the larger (p183) being a read-through extension of the smaller (p126). The template for translation of both of these proteins is the

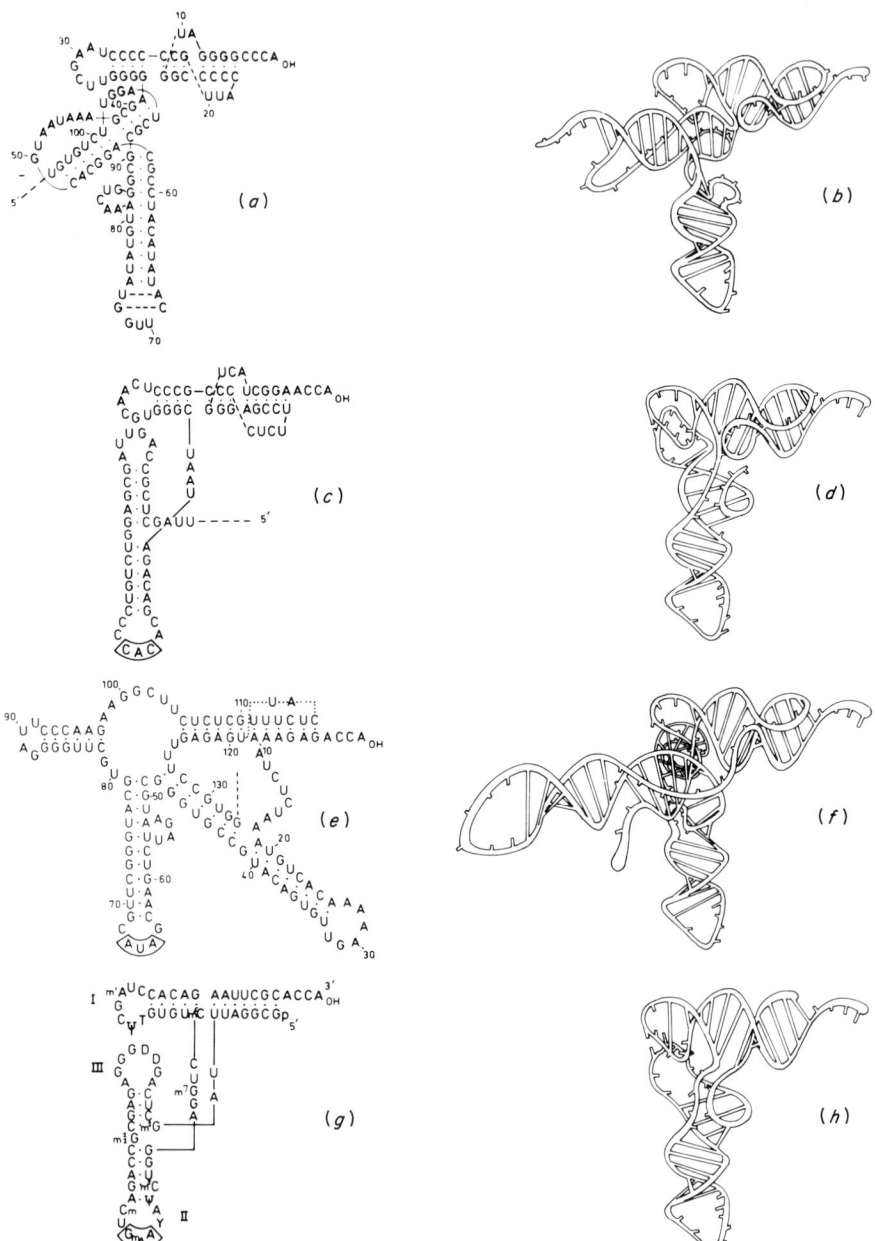

Figure 13.3. L-arrangements and three-dimensional foldings of the 3' ends of TMV RNA (*a, b*), TYMV RNA (*c, d*), BMV RNA (*e, f*) and tRNA^{Phe} from yeast (*g, h*). (Taken from Rietveld *et al.*, 1984.)

genomic RNA, the two remaining genes being expressed via subgenomic RNAs. All three BMV RNAs are aminoacylatable with tyrosine. TMV RNA accepts either histidine or valine according to the strain. The AlMV RNAs are not acylatable enzymatically. Each virus has a different morphology, TMV being rod-shaped, AlMV bacilliform, and BMV icosahedral.

Because of some common features of genome expression Haseloff et al. (1984) also made comparisons with the recently sequenced alphavirus Sindbis (Straus et al., 1984). Sindbis virus is a positive-strand RNA animal virus with enveloped particles of considerably more complexity than the capsids of the above plant viruses (Schlesinger, 1980). Partly because of the greater number of structural polypeptides encoded, the single genomic RNA of Sindbis virus is much larger (11.7 kilobases) than the individual genomes of the three plant viruses (6.4 to 8.2 kb). Sindbis genomic RNA encodes two polyproteins, a non-structural p270 polyprotein translated directly from genomic RNA and a structural p130 polyprotein translated from a subgenomic mRNA. Both polyproteins are post-translationally cleaved, the non-structural polyprotein p270 being cleaved into four polypeptides from which the C-terminal nsP4 is generated by translational readthrough (see Chapter 6, this volume).

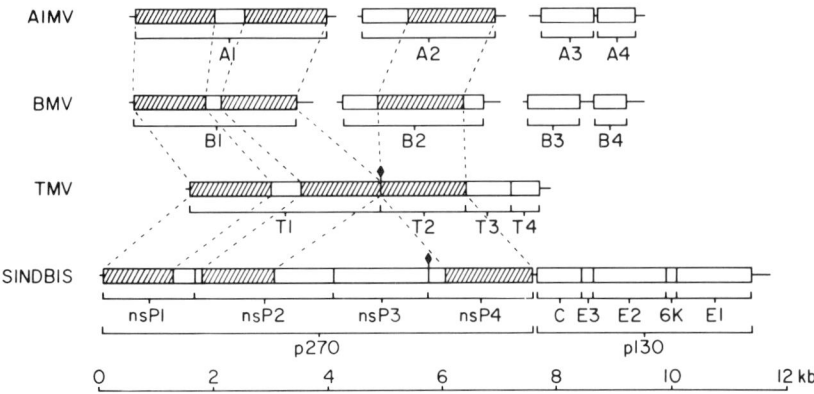

Figure 13.4. Schematic, drawn to scale, of the genomic RNAs of AlMV, BMV, TMV, and Sindbis virus. Subgenomic mRNAs are shown below their parent genomic RNAs. (The TMV T3 subgenomic mRNA is assumed to be capped by analogy with the TMV coat protein mRNA.) Boxed regions denote protein coding sequences. Inter-viral amino acid homologies are represented by regions of like shading.

The products of AlMV RNAs 1 and 2 and of BMV RNAs 1 and 2 are designated as A1 and A2 and as B1 and B2, respectively, and the products of each dicistronic RNA3 as A3 and B3 (products of the 5' proximal genes) and A4 and B4 (the coat proteins), respectively. The TMV open reading frames in their 5' to 3' order along the RNA are designated as T1 (p126), T2 (the remainder of the first open reading frame), T3, and T4 (the coat protein).

Although all four of these viruses are thus distinguishable by conventional criteria, the amino acid sequences of some of their encoded proteins are strikingly similar (see Fig. 13.4). The proteins encoded by AlMV RNA1 and BMV RNA1 and TMV p126 are similar. There are two distinct domains of similarity, one near the amino terminus, the other near the carboxy terminus. The proteins encoded by AlMV RNA2 and BMV RNA2 are similar to each other and to the COOH-terminal read-through portion of TMV protein p183. This homology has now also been shown to extend to CMV (Rezaian et al., 1984). Even more surprisingly, the Sindbis non-structural proteins nsP1 and nsP2 are homologous to the two conserved domains, of the AlMV 1a, BMV 1a, and TMV p126 proteins. Also the domain, conserved among the AlMV and BMV 2a proteins and the read-through portion of TMV p183, is homologous to the Sindbis non-structural read-through protein nsP4.

No homology has been detected among the structural proteins of the four viruses even though they obviously have a common function. Nor has any statistically significant homology been demonstrated in the non-structural proteins A3, B3, T3, and nsP3 in which a commonality of function is to be anticipated.

The existence of these conserved domains strongly supports the notion that all of these viruses have fundamental similarities in their mechanisms of replication and may be related evolutionarily. It seems clear that the observed conservation must result from selection of functions central to virus survival. Because two of the conserved domains are linked on a single protein in the plant viruses but separated on distinct mature proteins in Sindbis, it appears that either the two domains function independently on multifunctional plant virus proteins or the two Sindbis proteins, nsP1 and nsP2, function together as subunits in an enzyme complex.

Several studies indicate that the products encoded by AlMV and BMV RNAs 1 and 2 are required for viral RNA replication. Although RNAs 1, 2, and 3 are required for a productive infection by BMV, transfection of barley protoplasts with BMV RNAs 1 and 2, alone, resulted in synthesis of B1 and B2 translation products consistent with amplification of their templates (Kiberstis et al., 1981). Similarly, transfection of cowpea protoplasts with AlMV particles containing only RNAs 1 and 2 results in symmetric synthesis of plus and minus viral RNA strands (Nassuth and Bol, 1983). Mutations in AlMV RNAs 1 and 2 interfere with RNA synthesis (Roosien and Van Vlaten-Doting, 1982; Nassuth et al., 1981). Mutants in RNA1 of the bromovirus, CCMV, are deficient in RNA replication (Dawson, 1981). All four mature Sindbis non-structural polypeptides are involved in RNA replication, with roles in RNA elongation as well as initiation of minus-strand, plus-strand genomic, and plus-strand subgenomic RNA synthesis (Strauss and Strauss, 1983). It is thus probable that the homologous domains in these four viruses correspond to homologous RNA replication functions.

The evolutionary implications of these homologies have been discussed in some detail by Haseloff et al. (1984) and by Ahlquist et al. (1985). The structural similarities among these proteins must reflect some combination of convergent and divergent evolution. The requirements of RNA replication may account for some resemblance among the responsible proteins. However, these proteins bear no obvious homology to two other virus groups that have related proteins implicated in viral nucleic acid replication, i.e. cauliflower mosaic virus and the reverse transcriptases of retroviruses and hepadnaviruses (Toh et al., 1983), and the replicases of cowpea mosaic virus and picornaviruses (Franssen et al., 1985). Thus, since each group uses a distinctive set of proteins to achieve the common end of replicating an RNA template, it seems more attractive to conclude that each group of proteins owes its common features to descent from a common ancestor.

Possibly, a TMV-like virus was generated by the fusion of the three RNA segments of a tripartite virus to form a single RNA, with the generation of control sequences appropriate to the expression of T2 and T3 with the virion structure becoming adapted to carry a larger RNA. Conversely, a segmented virus could have been derived from a TMV-like progenitor by fission, in which the fragments became able to replicate and be encapsidated. The three plant viruses and the alphaviruses might also have descended from a common viral ancestor able to replicate in both plant and animal cells, like the reo- and rhabdoviruses of plants that also multiply in their insect vectors (Matthews, 1981).

However, an intriguing scenario that seems most plausible is that similar genes were incorporated independently into different viral genomes by recombination from a separate common source, presumably cellular genes. Thus Haseloff et al. (1984) contend: "On a formal basis, the differences in genomic organization of these four viruses can be regarded as permutations of modules of related genetic information and of controlling elements appropriate to their distribution along one or more viral RNAs. Ultimately, all of the modules of information whose reassortment we observe as viruses with different structures may be cellular in origin. In that case, the sequence conservation displayed by the proteins considered here may reflect strong cellular evolutionary conservation, maintained after transduction by residual functional constraints and by the need to interact with the products of other genes that remain in the more slowly evolving host cell."

VII. BMV RNA REPLICATION

Studies of BMV RNA replicase and other plant viral replicases have been reviewed recently (Hall et al., 1982) and I will not do so here. I will mention several important studies that bear on the work discussed above, and on the recombinant studies discussed below.

Much of our knowledge of the replication of BMV RNAs comes from the elegant work of Timothy Hall and his associates. His laboratory has prepared BMV RNA replicase in a template-dependent form (Bujarski et al., 1982) and has used it to study its mechanism of action. It seems now well established that the signals directing template selection *in vitro* lie in the tRNA-like structure formed by, at most, the 153 3' proximal bases of the BMV RNAs. Initiation of synthesis begins at the 3' end of RNAs that serve as a template and involves initiation *de novo* rather than template self-priming. Indeed, BMV replicase is incapable of elongating RNA or DNA primers, suggesting that specific template sequences must be recognized (Ahlquist et al., 1984a).

To delineate further the functions of the tRNA-like regions, sequence modifications were introduced into their cDNA and RNA transcripts from such cloned DNA were then analysed for replicase template activity and also for their ability to undergo aminoacylation (Dreher et al., 1984; Bujarski et al., 1985). Specific regions could be identified that selectively altered one or the other activity.

BMV replicase has also been used by Miller, Dreher, and Hall to study generation of RNA4 *in vitro* (Miller et al., 1985). They prepared negative-strand RNA3 and several modified negative strands by transcription from cloned BMV RNA3 viral cDNA and used these as templates for RNA synthesis with replicase. Full-length negative strand served as a template for RNA4 synthesis, although synthesis of full length plus-strand RNA3 was not apparent. Thus the replicase has the ability to recognize an internal site on negative-strand RNA3 so as to correctly initiate synthesis of RNA4. Similarly, the enzyme induces synthesis of RNA4 with a truncated negative-strand RNA3 as a template, provided at least 20 bases prior to the RNA4 sequence are present, showing that the recognition site for RNA4 synthesis lies in that 20 base region or in the RNA4 sequence region itself. Thus the Hall replicase can use both positive-strand BMV RNA (to make negative-strand RNAs) and negative-strand RNA3 (for RNA4 synthesis) as templates, but the appropriate conditions for synthesizing full-length positive strands *in vitro* are not yet known.

VIII. BMV INFECTION DERIVED FROM CLONED VIRAL cDNA

The application of recombinant DNA technology to the study and manipulation of organisms with DNA genomes has had spectacular success. Such success has only recently begun to extend to studies of RNA viruses. An important initial block to this work was the lack of general methods of reconverting cDNA to RNA in a way that preserves features of the genomic RNAs important to their biological activity, particularly their end structures.

In some instances this has not been necessary because the DNA forms, themselves, have been infectious. In addition to the intrinsically infectious DNA forms of retroviruses, it has been shown that the cDNAs of the RNA pathogens phage (Taniguchi et al., 1978), poliovirus (Racaniello and

Baltimore, 1981), and potato spindle tuber viroid (Cress et al., 1983) are infectious. However, the mechanism by which cDNA expresses its infectivity is not known, so that the generality of DNA infection is difficult to assess.

Cloned cDNA forms of the BMV genome are not infectious. However, Ahlquist and his colleagues have shown that BMV cDNA can be ligated to a general transcription vector that can be transcribed to yield infectious BMV RNA (Ahlquist et al., 1984b). Since such transcripts can be compared precisely with authentic RNA, or if necessary can be made identical to authentic RNA, the procedure for generating infectivity from cDNA is devoid of the mysteries of cDNA infection. Their transcription vector (Ahlquist and Janda, 1984), designated pPM1, allows transcription initiation to be directed to the first nucleotide of any attached 5′-terminated DNA fragment. Moreover, inclusion of m7GpppG in the reaction mixture produces RNA capped in the manner of authentic BMV RNA. Under their assay conditions, infectivity requires the existence of cap but tolerates a small number of additional non-viral 3′-terminal bases generated by transcription runoff. Transcripts from mixtures of appropriate BMV cDNA clones were infectious both to barley plants and to barley protoplasts. Infectivity depended on the presence of transcripts of all three BMV genetic components.

Direct RNA sequencing showed that a deletion in the non-coding region of one infectious BMV clone was preserved in viral RNA from plants systemically infected with transcripted mixtures representing that clone, showing unequivocally that mutations can be introduced by systematic alteration of the viral cDNA.

Because of its generality, this method of producing virus infection from cloned cDNA represents a fundamental advance in RNA virology. Moreover, unlike classical production of local lesions, which may not resolve strain mixtures (Garcia-Arenal et al., 1984), cDNA expression allows the production of genetically well-defined RNA plant virus strains. Studies have already begun that utilize designed modifications of the BMV genomes to examine BMV replication and gene expression and the use of BMV as a vector for the directed expression of new proteins (French et al., 1986).

Beginning with the constructs originally used for studies of RNA replication in vitro (see above and Dreher et al. (1984) and Bujarski et al. (1985)), several of the BMV laboratories have collaborated in the construction of a modified BMV RNA3 which has a 20-base deletion (stem and loop d) in its 3′-terminal, tRNA-like, non-coding region (J. J. Bujarski, P. Ahlquist, T. C. Hall, T. W. Dreher and P. Kaesberg, unpublished). This region includes the promoter for replication and is common to all three RNAs. Modified RNA3 showed increased template activity in vitro, and together with wild-type RNA1 and wild-type RNA2 was infectious to protoplasts or whole plants. Sequence analysis confirmed that progeny virus had retained the modified RNA3 sequence.

IX. RNA RECOMBINATION

In view of the fine work on recombination reported by Andrew King (see Chapter 9, this volume), I will mention an additional recent study begun in my laboratory. Jozef Bujarski, considered the possibility that, although the deletion in RNA3 (see above) does not abolish infectivity, it may nevertheless inflict a selective disadvantage and that, because similar sequences exist in wild-type RNA1 and wild-type RNA2 such disadvantage could be reversed by recombination *in vivo*. We followed the time-course of infection for the modified virus: initially, the progeny viral RNA3 retained this deletion. However, virus isolated from the same plants two weeks later had RNA3 whose deleted region had been restored. Sequence analysis of virus isolated from a single plant showed that its RNA3 had an insertion in the originally modified region whose sequence matched that of wild-type RNA2, suggesting that this insertion was introduced by recombination between modified RNA3 and wild-type RNA2. Use of independently derived transcripts as inoculum precluded the possibility of recombination prior to transfection. These observations suggest that recombinants between distinct viral RNAs can be selected readily and imply a high level of viral RNA recombination. It supports the view, expressed by Haseloff *et al.* (1984) and Ahlquist *et al.* (1985), that RNA recombination is an important mechanism in virus evolution.

Clearly, BMV has been a distinguished member of the virological community and will continue to contribute to our understanding of biology.

ACKNOWLEDGEMENTS

I gratefully acknowledge research support by the U.S. National Institutes of Health extending over a period of more than 30 years and the receipt from them of a lifetime Research Career Award in 1962. I am also grateful for the research support of the U.S. Department of Agriculture beginning with its inception of a competitive grants programme. I dedicate this writing to Professor William W. Beeman, my major professor and life-long friend, a physicist of brilliant insight and an extraordinary teacher.

REFERENCES

1. Aach, H., G. Funatsu, M. Nirenberg and H. Fraenkel-Conrat (1964). *Biochemistry* **3**, 1362.
2. Ahlquist, P. and M. Janda (1984). *Mol. Cell. Biol.* **4**, 2876–2882.
3. Ahlquist, P., R. Dasgupta and P. Kaesberg (1981). *Cell* **23**, 183–189.
4. Ahlquist, P., J. J. Bujarski, P. Kaesberg and T. C. Hall (1984a). *Plant Mol. Biol.* **3**, 37–44.
5. Ahlquist, P., R. French, M. Janda and S. Loesch-Fries (1984b). *Proc. Natl Acad. Sci. USA* **81**, 7066–7070.
6. Ahlquist, P., E. G. Strauss, C. M. Rice, J. H. Strauss, J. Haseloff and D. Zimmern (1985). *J. Virol.* **53**, 536–542.

7. Anderegg, J. W., M. Wright and P. Kaesberg (1963). *Biophys. J.* **3**, 175–182.
8. Beeman, W. W., P. Kaesberg, J. W. Anderegg and M. B. Webb (1956). *Handbuch der Physik* **32**, 321–442.
9. Bockstahler, L. (1964). Thesis, University of Wisconsin.
10. Bockstahler, L. E. and P. Kaesberg (1965). *J. Mol. Biol.* **13**, 127–137.
11. Bujarski, J. J., S. T. Hardy, W. A. Miller and T. C. Hall (1982). *Virology* **119**, 465–473.
12. Bujarski, J. J., T. W. Dreher and T. C. Hall (1985). *Proc. Natl Acad. Sci. USA* **82**, 5636–5640.
13. Clark J. M., A. Y. Chang, S. Spiegelman and M. E. Reichmann (1965). *Proc. Natl Acad. Sci. USA* **54**, 1193–1197.
14. Cornelissen B. and J. Bol (1984). *Plant Mol. Biol.* **3**, 379–384.
15. Cress, D. E., M. C. Kiefer and R. A. Owens (1983). *Nucl. Acids Res.* **11**, 6821–6835.
16. Dasgupta, R. and P. Kaesberg (1977). *Proc. Natl Acad. Sci. USA* **74**, 4900–4904.
17. Dasgupta, R. and P. Kaesberg (1982). *Nucl. Acids Res.* **10**, 703–713.
18. Dasgupta, R., D. S. Shih, C. Saris and P. Kaesberg (1975). *Nature (London)* **256**, 624–628.
19. Dasgupta, R., F. Harada and P. Kaesberg (1976). *J. Virol.* **18**, 260–267.
20. Davies, J. Unpublished soliloquy: 3B or not 3B, that is the question.
21. Dawson, W. O. (1981). *Virology* **115**, 130–136.
22. Dreher, T. W., J. J. Bujarski and T. C. Hall (1984). *Nature (London)* **311**, 171–174.
23. Franssen, H., J. Leunissen, R. Goldbach, G. Lomonossoff and D. Zimmern (1985). *EMBO J.* **3**, 855–861.
24. French, R., M. Janda and P. Ahlquist (1986). *Science*, **231**, 1294–1297.
25. Garcia-Arenal, F., P. Palukaitis and M. Zaitlin (1984). *Virology* **132**, 131–137.
26. Hall, T. C., D. S. Shih and P. Kaesberg (1972). *Biochem. J.* **129**, 969–976.
27. Hall T. C., W. A. Miller and J. J. Bujarski (1982). *Adv. Plant Pathol.* **I**, 179–211.
28. Haseloff, J., P. Goelet, D. Zimmern, P. Ahlquist, R. Dasgupta and P. Kaesberg (1984). *Proc. Natl Acad. Sci. USA* **81**, 4358–4362.
29. Hunter T., T. Hunt, J. Knowland and D. Zimmern (1976). *Nature (London)* **260**, 759–764.
30. Joshi, R., S. Joshi, F. Chapeville and A. Haenni (1983). *EMBO J.* **2**, 1123–1127.
31. Kaesberg, P. (1956). *Science* **124**, 626–628.
32. Kelley, J. J. and P. Kaesberg (1962). *Biochim. Biophys. Acta* **61**, 865–871.
33. Kiberstis, P. A., L. S. Loesch-Fries and T. C. Hall (1981). *Virology* **112**, 804–808.
34. Klein, W., C. Nolan, J. Lazar and J. Clark (1972). *Biochemistry* **11**, 2009–2014.
35. Kohl, R. J. and T. C. Hall (1974). *J. Gen. Virol.* **25**, 257–261.
36. Lane, L. C. and P. Kaesberg (1971). *Nature New Biol.* **232**, 40–43.
37. Leonard, B. R. Jr., J. W. Anderegg, S. Shulman, P. Kaesberg and W. W. Beeman (1953). *Biochim. Biophys. Acta* **12**, 449–507.
38. Marcus, A. (1970). *J. Biol. Chem.* **245**, 962–966.
39. Matthews, R. E. F. (1981). *Plant Virology*, 2nd ed., pp. 591–598. Academic Press, London and Orlando.
40. Mazzone, H. M., N. L. Incardona and P. Kaesberg (1962). *Biochim. Biophys. Acta* **55**, 164–175.

41. Miller, W. A., T. W. Dreher and T. C. Hall (1985). *Nature (London)* **313**, 68–70.
42. Nassuth, A. and J. F. Bol (1983). *Virology* **124**, 75–85.
43. Nassuth, A., F. Alblas and J. F. Bol (1981). *J. Gen. Virol.* **53**, 207–214.
44. Nathans, D., G. Notani, J. Schwartz and N. Zinder (1962). *Proc. Natl Acad. Sci. USA* **48**, 1424–1431.
45. Nirenberg, M. and J. H. Matthaei (1961). *Proc. Natl Acad. Sci. USA* **47**, 1588–1602.
46. Pleij, C. W. A., A. Neeleman, L. Van Vloten-Doting and L. Bosch (1976). *Proc. Natl Acad. Sci. USA* **73**, 4437–4441.
47. Racaniello, V. and D. Baltimore (1981). *Science* **214**, 916–919.
48. Rezaian, M., R. Williams, K. Gordon, A. Gould and R. Symons (1984). *Eur. J. Biochem.* **143**, 277–284.
49. Rietveld, K., Pleij W. and L. Bosch (1983). *EMBO J.* **2**, 1079–1085.
50. Rietveld, K., K. Linschooten, C. Pleij and L. Bosch (1984). *EMBO J.* **3**, 2613–2619.
51. Roosien J. and L. van Vloten-Doting (1982). *J. Gen. Virol.* **63**, 189–198.
52. Schlesinger, R. (Ed.) (1980). "The Togaviruses". Academic Press, London and Orlando.
53. Schmidt, P., P. Kaesberg and W. W. Beeman (1954). *Biochim. Biophys. Acta* **14**, 1–11.
54. Shih, D. S. and P. Kaesberg (1973). *Proc. Natl Acad. Sci. USA* **70**, 1799–1803.
55. Shih, D. S. and P. Kaesberg (1976). *J. Mol. Biol.* **103**, 77–88.
56. Shih, D. S., L. C. Lane and P. Kaesberg (1972). *J. Mol. Biol.* **64**, 353–362.
57. Strauss, E. G. and J. H. Strauss (1983). *Current Topics Microbiol. Immunol.* **105**, 1–98.
58. Strauss, E. G., C. M. Rice and J. H. Strauss (1984). *Virology* **133**, 92–110.
59. Stubbs J. D. and P. Kaesberg (1967). *Virology* **33**, 385–397.
60. Symons, R. (1979). *Nucl. Acids Res.* **7**, 825–837.
61. Taniguchi, T., M. Palmieri and C. Weissmann (1978). *Nature (London)* **274**, 223–228.
62. Toh, H., H. Hayashida and T. Miyata (1983). *Nature (London)* **305**, 827–829.
63. Van Vloten-Doting, L., A. Dingjan-Versteegh and E. Jaspars (1970). *Virology* **40**, 419–430.
64. Williams, R. (1953). *Exp. Cell Res.* **4**, 188–201.
65. Williams, R. C. and K. M. Smith (1958). *Biochim. Biophys. Acta* **28**, 464–469.
66. Wood, H. A. and J. B. Bancroft (1965). *Virology* **27**, 94–102.

14. Molecular Architecture and Assembly of Tobacco Mosaic Virus Particles

P. J. G. Butler and M. A. Mayo*

*Laboratory of Molecular Biology, Hills Road, Cambridge CB2 2QH, U.K. and *Scottish Crop Research Institute, Invergowrie, Dundee DD2 5DA, U.K.*

I.	Introduction	237
II.	The coat protein	239
	A. Polymorphism of coat protein aggregates	239
	B. Bonding between protein molecules in the disk aggregate	240
III.	Structure model of the virus particle	242
IV.	Assembly	243
	A. Initiation: the role of the disk	245
	B. Initiation at a specific site on virus RNA	247
	C. Elongation in two directions	248
	D. Elongation: experiments with cross-linked RNA	253
	E. Binding of oligonucleotides to disks	255
V.	Concluding remarks	256
	References	256

I. INTRODUCTION

The study of tobacco mosaic virus (TMV) predates that of molecular biology by a very long way. TMV was one of the first viruses to be studied and it has figured in many of the studies that have directed plant virologists to think in molecular terms (Fraenkel-Conrat, 1981). Particles of TMV were the first to be shown to be nucleoprotein (Bawden and Pirie, 1937) and TMV was the first virus shown to depend solely on its RNA for its infectivity (Fraenkel-Conrat, 1956; Gierer and Schramm, 1956). More pertinently to this review, TMV particles were the first virus particles to be reassembled from protein and RNA (Fraenkel-Conrat and Williams, 1955).

In this chapter the extensive literature on the structure of TMV has been greatly condensed into a brief introductory section. The process of reassembly is then discussed in some detail to illustrate the functional role of the polymorphic structures of the protein aggregate and, finally, results of some recent experiments are described. These results demonstrate that the study of TMV can still generate controversy and, more importantly, that TMV can still offer an experimental challenge to molecular biologists. Other recent reviews of this area are to be found in Hirth and Richards (1981), Butler (1984), Holmes (1984) and Stubbs (1984).

TMV has rod-shaped particles 300 nm long and 20 nm in diameter, each with a central hole about 4 nm in diameter (Caspar, 1963). The ends of the particles are not identical; one end is convex, or pointed, and the other is either concave or flat (Fig. 14.1(a)). The particle structure is a single helix (Watson, 1954) comprising about 2140 protein molecules of molecular weight 17500 daltons around an RNA molecule of about 6396 nucleotides (Goelet et al., 1982) (Fig. 14.1(b)). There are three nucleotides associated

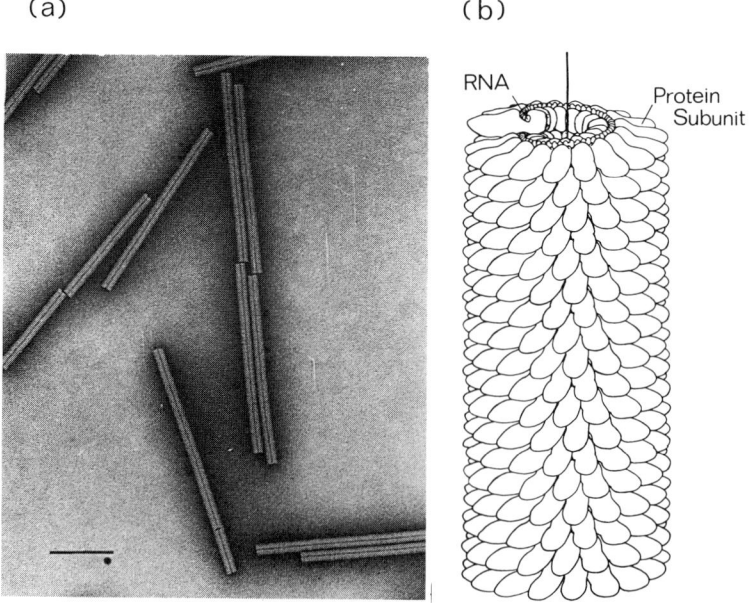

Figure 14.1. (a) Tobacco mosaic virus (TMV) particles in the electron microscope after negative staining with uranyl acetate (courtesy of Dr. J. T. Finch) (scale bar is 100 nm). (b) Schematic drawing showing overall organisation of RNA and coat protein. (Adapted from Caspar, 1963.)

14. Architecture and Assembly of TMV

with each protein subunit and the length of the particle is determined by the length of the RNA molecule encapsidated. Lengthwise growth of the virus rod ceases when all the RNA has been encapsidated, but nothing is known about the detailed structure of the end turns of the helix. TMV RNA packaged in the virus particle is exceptionally resistant to potentially inactivating conditions such as exposure to ribonuclease solutions or air-drying, and it is therefore probable that the ends of the RNA molecule are fully protected. Efforts to find a special capping protein have not succeeded and it seems likely that the virus coat protein itself forms protective ends.

II. THE COAT PROTEIN

A. Polymorphism of Coat Protein Aggregates

The coat protein is very polymorphic; it forms a large number of different types of aggregate (Fig. 14.2) (Durham et al., 1971). The main controlling

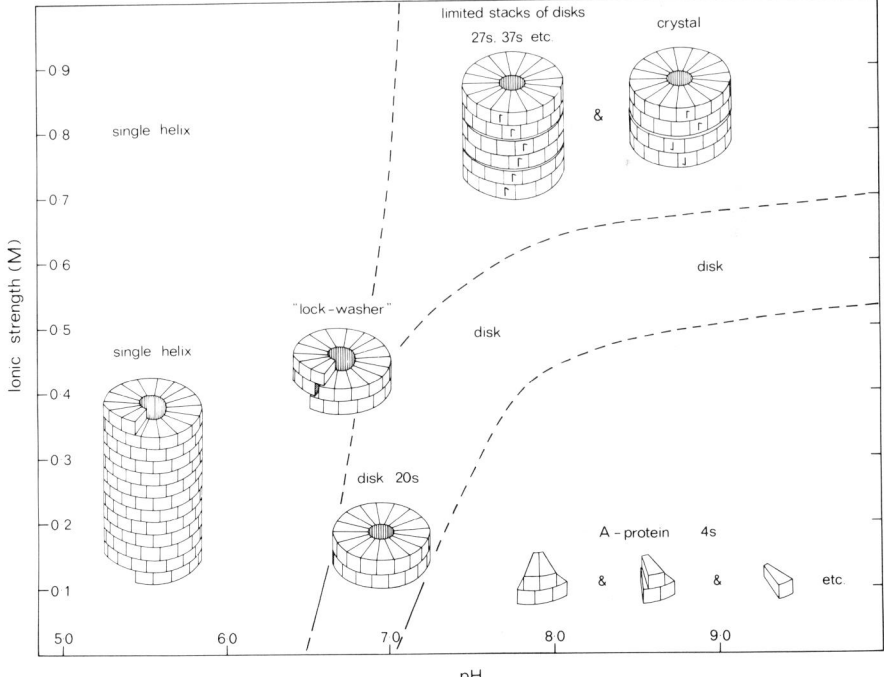

Figure 14.2. Dependence of aggregation of TMV protein upon pH and ionic strength. The "map" shows the range of conditions over which a given type of aggregate predominates at 20°C and a protein concentration of 5 mg ml^{-1}. (Adapted from Durham et al., 1971.)

variable in this "phase diagram" is pH. At low pH, protein forms a helix that is like the virus particle helix in having $16\frac{1}{3}$ protein subunits per turn, although occasionally helices are formed which have $17\frac{1}{3}$ subunits per turn (Mandelkow et al., 1976, 1981). At higher pH the protein forms smaller aggregates often referred to as A-protein ("alkalischer Protein"; Schramm and Zillig, 1955). However, as the ionic strength increases A-protein forms larger aggregates, the first being a disk of protein subunits with a sedimentation coefficient of about 20S. This disk structure comprises two rings of 17 subunits, each in the same orientation. There is no axis, or series of axes, of diassymmetry within the disk (Bloomer et al., 1978).

B. Bonding between Protein Molecules in the Disk Aggregate

As the ionic strength of a solution of TMV protein is raised, disks aggregate into limited stacks and also in a head-to-head arrangement which, when suitably treated, can crystallize to give a true three-dimensional crystal that is suitable for structural analysis. Some years ago such crystals were analysed by x-ray crystallography and the map obtained gave detail at a resolution of about 0.28 nm (Bloomer et al., 1978). Figure 14.3 shows a section in which the fit can be seen for the atomic model built into the map. The structure is of two α-helices joined together by a rather tight loop at the inner end. The basic folding (Fig. 14.4) is of four packed α-helices tied together at the outer ends by a β-sheet; there is a less regular outer region. Both N- and C-termini are at high radius on the outside of the virus particle.

The overall bonding is dominated by the non-polar region at high radius. Figure 14.4 includes a view from above a ring of subunits illustrating the extension of this hydrophobic region to form a "girdle" round the outside of the particle linking the subunits firmly together. Lauffer and colleagues (Lauffer, 1975) have shown that the driving force in TMV protein aggregation is hydrophobic and it is likely that this region is the site of the major hydrophobic bonding. There are, however, other more central contacts comprising both hydrophobic and hydrophilic interactions. Comparisons between these regions in proteins of different strains have shown that the boundaries between these hydrophilic and hydrophobic "patches" are differently disposed in different strains. Thus the inability of heterologous proteins to coaggregate into helices (Sarkar, 1960; Novikoff et al., 1974) can probably be explained by incompatible interactions in these regions.

The vertical contacts between proteins in the disk are much weaker and consist mainly of some hydrogen bonds and a rather extended salt bridge, as shown in Fig. 14.5. The carboxyl oxygen of the glutamate residue forms two hydrogen bonds, and arginine and lysine residues also contribute to hydrogen bonding. The salt bridge is unusually stable; disks are crystallized at an ionic strength of 1 M and do not disintegrate even in 3 M ammonium sulphate.

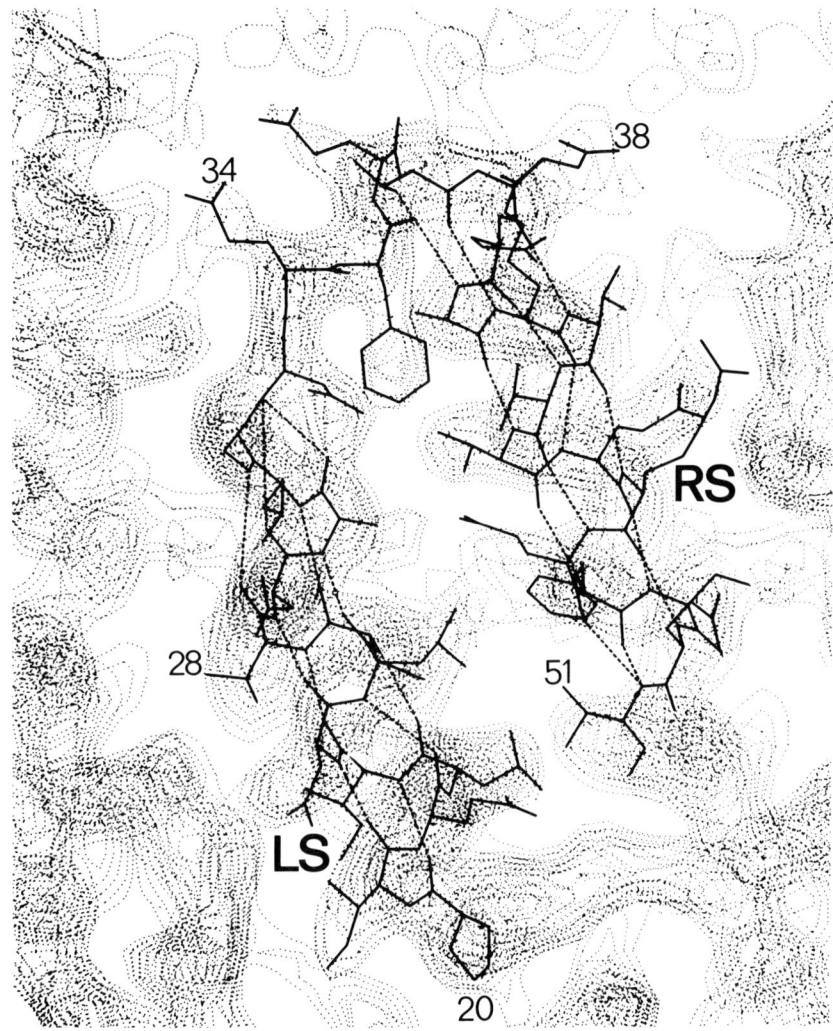

Figure 14.3. Fit of the skeletal atomic model to the electron density map for TMV protein. Some sections of the density map viewed along the disk axis are shown, containing the left-slewed (LS) and right-slewed (RS) α-helices and the hairpin loop between them, together with the skeletal model for this section of the subunit with some side chains numbered. The disk axis is off the top of the picture and some density from the neighbouring subunits in the ring is visible on each side of the central subunit.

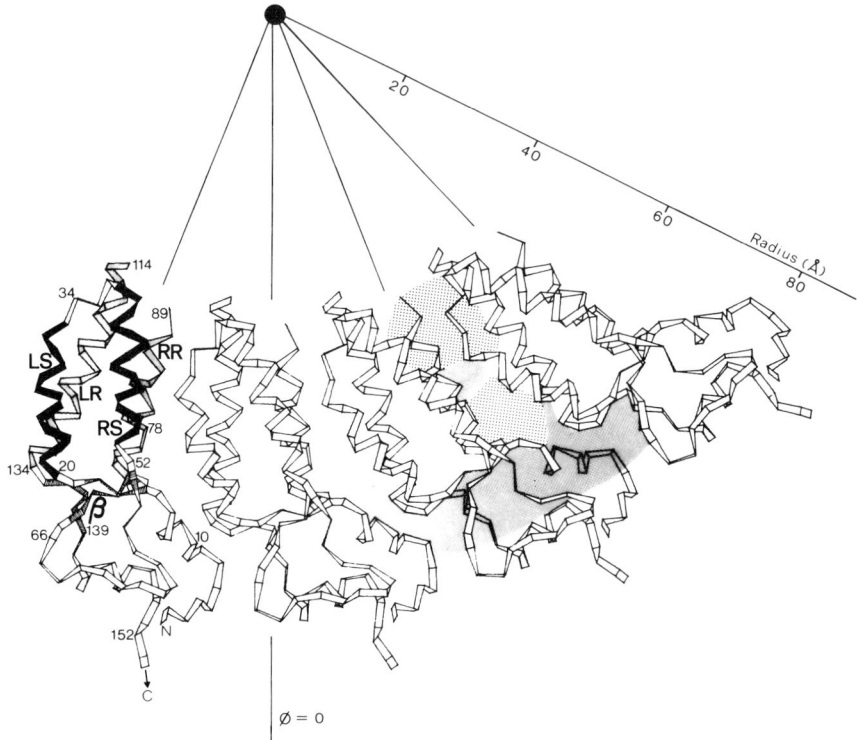

Figure 14.4. Ribbon drawing of α-carbon positions in the protein subunits in the TMV protein disk. Four subunits of one ring are shown, with the disk axis (the heavy circle) at the top. Locations along the polypeptide chain are marked on the left-hand subunit, together with the main secondary structure of α-helices (left-slewed, LS; right-slewed, RS; left-radial, LR; and right-radial, RR) and a small β-sheet (marked β). The nature of the contact between neighbouring subunits is shown with the right-hand subunits, stippling indicating polar contacts and shading non-polar or hydrophobic contacts. The "hydrophobic girdle" around the ring at a radius of 8 nm (80Å) both within and between subunits is clearly visible. (Adapted from Bloomer *et al.*, 1978.)

It is the vertical contacts between proteins that are greatly altered when the structure of the aggregate changes from disk to helix.

III. STRUCTURE MODEL OF THE VIRUS PARTICLE

The structure of the virus helix has been determined by workers in Heidelberg (Stubbs *et al.*, 1977) and more recently Brandeis (Namba *et al.*, 1985), although not yet at sufficiently high resolution to permit an independent model to be built. However, by taking the subunit structure from the disk and fitting this to the electron density map of the virus it has

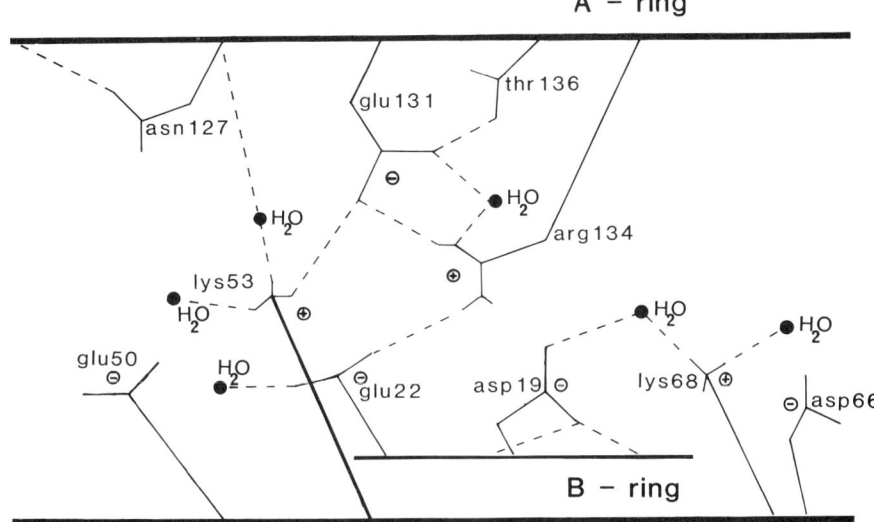

Figure 14.5. Diagram of the extended salt-bridging between subunits in the two rings of the protein disk. This shows the extended hydrogen bonding and charge neutralization between the side-chains of amino acids coming from each ring and leading to a structure that is stable even at high ionic strengths.

been possible to obtain a model. The major difference between the helix and the disk (Fig. 14.6) is in the vertical direction — whereas in disks one ring of subunits is flat and the other slopes down towards the central hole and the axis, in the virus particle all protein subunits tilt upwards towards the axis. Also the inner region of the protein, which is flexible and disordered in the disk, becomes ordered in the virus particle. Presumably the ordering of this region prevents access to the RNA and the RNA binding sites.

The other major feature in virus particles is the RNA molecule. The three nucleotides that bind to each subunit lie on three surfaces of one of the α-helices (Holmes, 1979; Mandelkow et al., 1981). The RNA sugar-phosphate backbone is in a partially extended configuration and is shown running through two protein subunits in Fig. 14.7.

IV. ASSEMBLY

The simple structure of TMV particles led to the expectation that the particles would be assembled by an equally simple mechanism, and that the process might be autonomous. Indeed, the earliest work on virus particle assembly was that of Fraenkel-Conrat and Williams (1955) with TMV. However, the reactions observed were rather slow; several hours

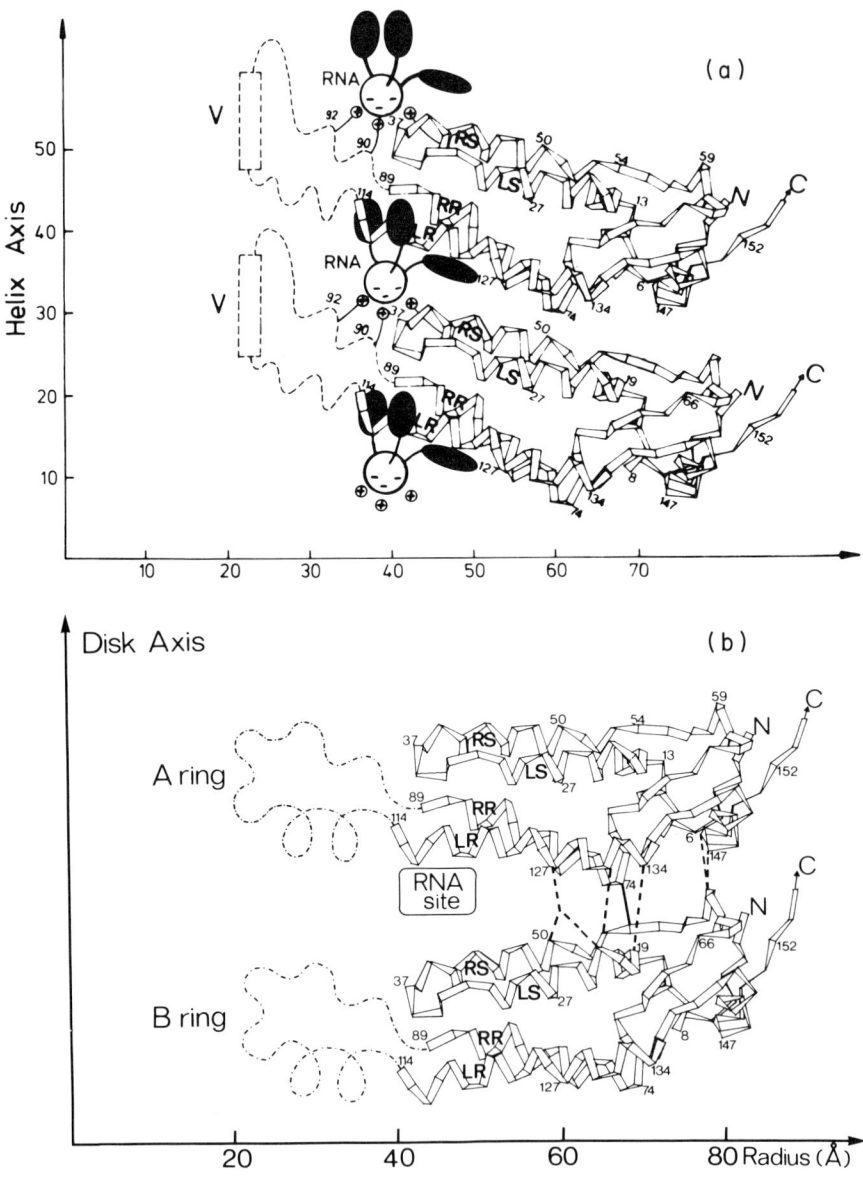

Figure 14.6. Comparison of the vertical packing of subunits in the virus helix and disk. Two axially adjacent subunits are shown for (*a*) the virus and (*b*) the disk, as ribbon drawings joining the α-carbon positions. The axial orientation is the same in both cases. (Adapted from Holmes, 1982, and Bloomer *et al.*, 1978.)

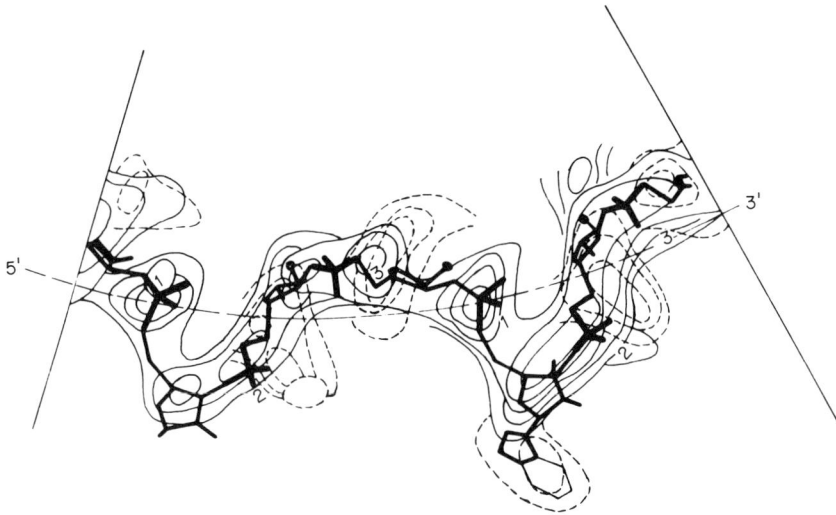

Figure 14.7. Path of the RNA backbone in the virus. The electron density map and a skeletal model for the sugar-phosphate backbone of the nucleotides corresponding to two protein subunits are shown, with the phosphate positions for each subunit numbered from the 5'-end (1, 2 and 3). Most of the electron density for the nucleotide bases is out of the plane of these sections, but in one case such density is visible and a generalized base has been fitted since the varying sequence for each subunit will lead to an "average base" in the structure at each site. (From Holmes, 1979.)

were necessary for completion and this prompted further research into the process.

A. Initiation: The Role of the Disk

The assembly of a helix can be thought of as initiation followed by elongation. It is initiation that is the problematical phase. One theoretical means of obtaining rapid initiation would be for a jig to be formed, and disks were proposed as such a structure (Butler and Klug, 1971). When a neutral solution of disks is acidified to pH5, the disk structure changes rapidly, apparently dislocating and forming a protohelix. These then seem to stack and form a structure with discontinuities, a "nicked helix" in which the ends of the protohelices are not in register (Durham et al., 1971). The discontinuities seem then gradually to disappear and regular helices are formed that are of indefinite length. Micrographs, together with a schematic interpretation, of these structures, are shown in Fig. 14.8.

The disk or protohelix appeared to be a suitable candidate for the role of a "jig" or "former", so experiments were made in which TMV RNA was added to protein in different aggregation states and the formation of rods

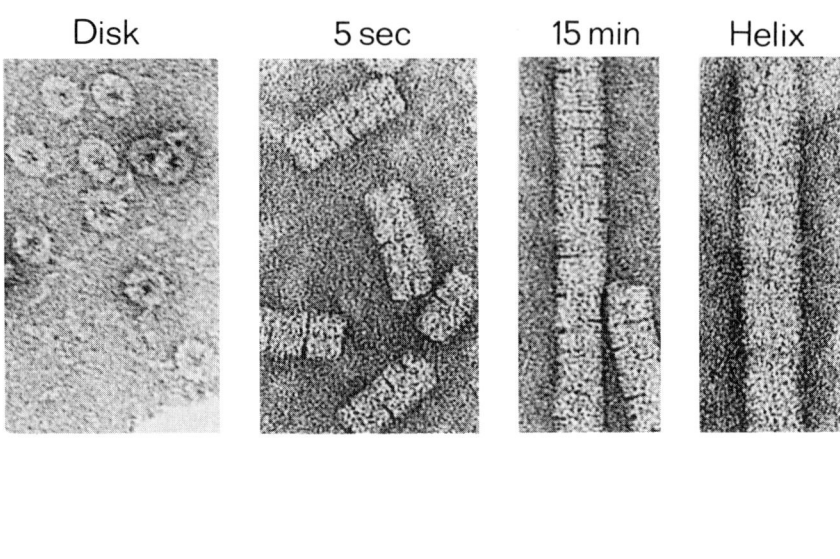
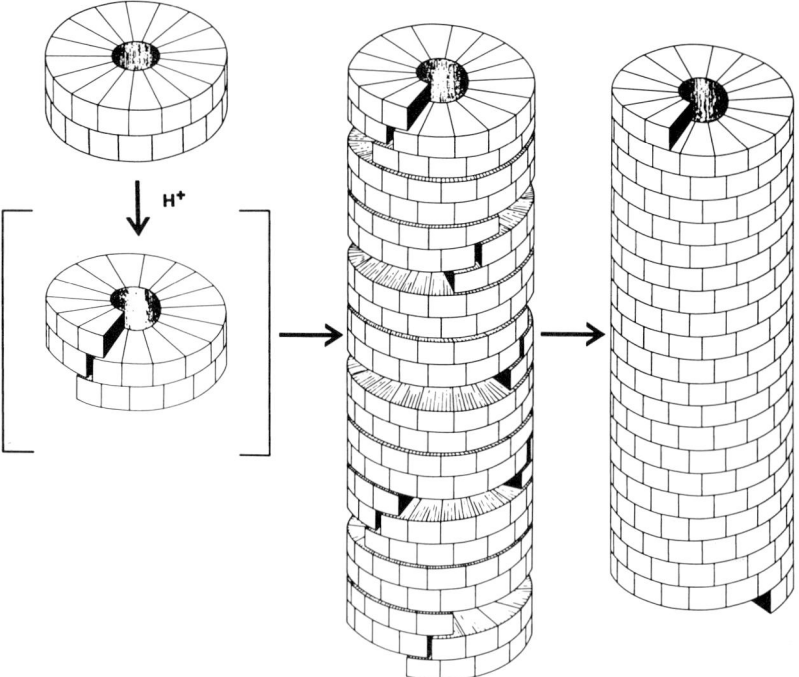

Figure 14.8. Effect of a rapid drop in pH upon the structure of the protein disk. The upper panels show electron micrographs of a disk preparation at pH 7 and then at times of 5 s, 15 min and several hours (by which time perfect helices had formed) after the pH was dropped to pH 5. The lower drawings show a schematic interpretation of the structures deduced from the micrographs. (Electron micrographs by courtesy of Dr. J. T. Finch, adapted from Durham et al., 1971.)

was measured as an increase in turbidity. When the protein was in the form of A-protein the reaction was barely detectable, but the addition of a disk preparation caused rods to form rapidly. Furthermore, omission of additional A-protein (although disk preparations are an equilibrium mixture of disks and a relatively small amount of A-protein) did not greatly affect the reaction rate. This rapidity suggested that the disks were being added directly to the growing rods, although other workers have contested this idea (reviewed in Butler, 1984).

B. Initiation at a Specific Site on Virus RNA

Evidence that nucleation was specific came from experiments in which radioactively labelled RNA was mixed with small amounts of protein and then treated with RNase so that only the encapsidated RNA remained intact. The oligonucleotides present in the shortest piece of protected RNA were present in all the other sizes of protected RNA, and were thus characteristic of a specific initiation site (Zimmern and Butler, 1977). The RNA and the protein helix are polar structures and the simplest model would be of unidirectional elongation following initiation at one end. However, it is now known from enzyme protection studies and encapsida-

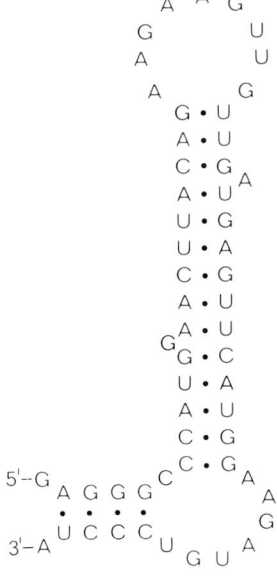

Figure 14.9. The nucleotide sequence of the minimum core of the origin of assembly on TMV RNA. The diagram shows how it could be folded to give a weakly base-paired stem and a single-stranded loop with a G residue in each third position. (Adapted from Zimmern, 1977.)

tion of oligonucleotides that the site of initiation is about 1000 nucleotides from the 3'-end of the RNA (Zimmern and Wilson, 1976; Guilley et al., 1979). Thus elongation occurs in both directions along the RNA, although more rapidly towards the 5'-end.

The sequence of the nucleation region is shown in Fig. 14.9, folded into a weakly base-paired stem and a single-stranded loop. The sequence of the loop has a G in every third position and, perhaps, a preference for A in all other positions. And it is known that TMV coat protein has an affinity for poly(A) and purine-rich synthetic oligonucleotides (Fraenkel-Conrat and Singer, 1964; Butler and Klug, 1971).

The structure of the disk suggested a possible model for nucleation by the insertion of this loop of RNA (Butler et al., 1976) between the two rings of subunits which open, like jaws, onto the central hole of the disk (Fig. 14.10). This might involve the flexible loop moving out of the way for the RNA loop to insert into the protein. The RNA–protein interaction would then give sufficient energy to "melt" the rather weak base-pairing allowing a single turn to bind. Provided this interaction also causes the dislocation to form a protohelix, the process has become one of elongation. The prediction of this model (Fig. 14.11), that partially assembled rods should have exposed RNA tails emerging from one end of the particle, was confirmed by electron microscopy (Butler et al., 1977; Lebeurier et al., 1977). Close examination revealed that the RNA was emerging from the end of the rod destined to contain the 3'-end of the RNA and thus the 5'-tail was doubled back along the rod, presumably through the central hole.

C. Elongation in Two Directions

Elongation is more rapid in the 5'-direction than the 3'-direction. Analysis of RNA fragments protected from nuclease digestion after partial elongation suggested that elongation is quantized because the RNA fragments form bands in gels (Fig. 14.12) (Butler and Lomonossoff, 1978). Fourier analysis of densitometer traces of such gels shows repeats of around 50 or 100 nucleotides. Furthermore, in similar experiments with elongation of TMV protein on poly(A) molecules, bands were detected with a repeat of 100 or 200 nucleotides (Schoen and Mundry, 1984). The simplest explanation of these results is that protein is added as disks of 34 subunits protecting 51 or 102 nucleotides depending on whether one or both rings are rapidly incorporated. This model is illustrated in Fig. 14.13.

Elongation in the 3'-direction is slower and has been less studied than that in the 5'-direction. To study this process, RNA molecules were used that had been cut to comprise almost only the initiation site and the sequence to the 3'-end (Lomonossoff and Butler, 1980). Thus assembly in the 5'-direction was completed very quickly and caused little turbidity. With these nucleated rodlets as the RNA substrate, elongation, measured as an increase in turbidity, was more rapid when A-protein was used than when disks were used.

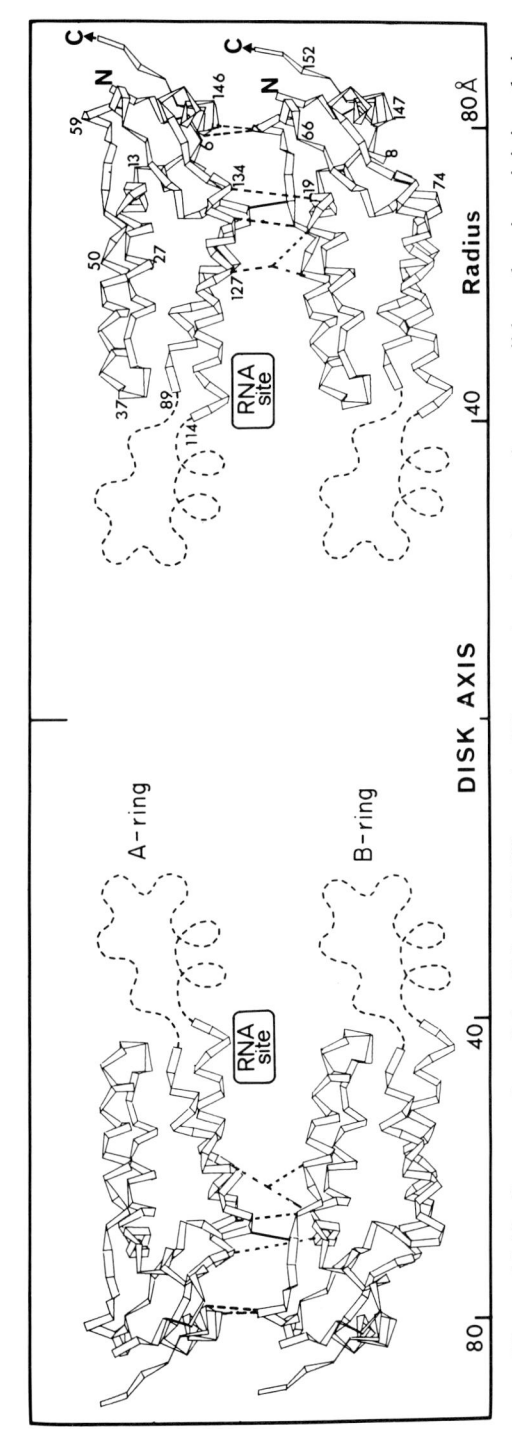

Figure 14.10. Cross-section of the disk of TMV protein. The protein subunits are shown as ribbon drawings joining their α-carbon positions and the "flexible loop" between the two radial α-helices is shown dotted. The position of the RNA binding site is marked, showing the ready access from the central hole of the disk. (Adapted from Bloomer et al., 1978.)

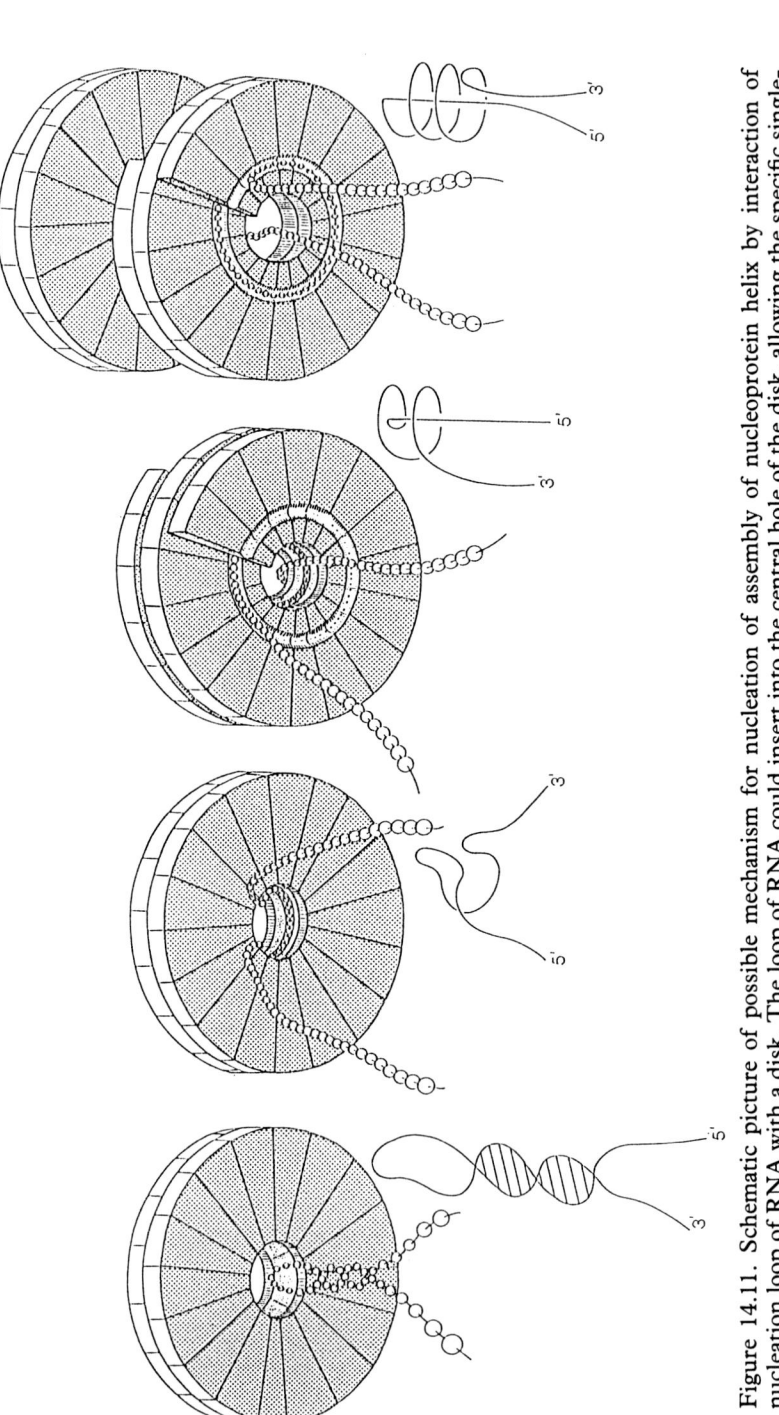

Figure 14.11. Schematic picture of possible mechanism for nucleation of assembly of nucleoprotein helix by interaction of nucleation loop of RNA with a disk. The loop of RNA could insert into the central hole of the disk, allowing the specific single-stranded sequence to interact with the RNA binding site and this interaction to propagate, melting the stem of the loop. Some feature of this interaction could cause the disk to "dislocate", entrapping the first turn of RNA in a protohelix. Subsequently the RNA could bind onto one or both surfaces of this protohelix and elongation could ensue. (From Butler, 1984.)

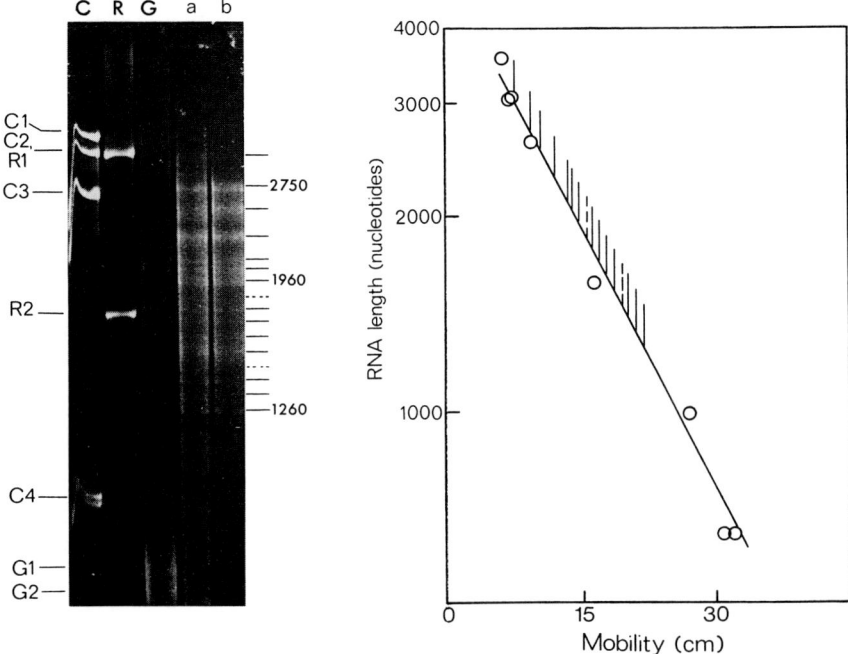

Figure 14.12. Lengths of RNA protected from nuclease digestion during reassembly with a disk preparation. The left-hand panel shows a picture of a gel stained with ethidium bromide; the right-hand panel shows the spacing of the bands seen for the protected TMV RNA. The markers are cowpea chlorotic mottle virus RNA (C1–C4), *E. coli* ribosomal RNA (R1–R2) and human globin mRNA (G1–G2). Reassembly of the TMV was for 1 min, followed by a digestion with micrococcal nuclease for (*a*) 5 min or (*b*) 60 min. (From Butler and Lomonossoff, 1978.)

Because elongation seems to use different protein forms and occurs in different directions at different rates it was of interest to determine whether elongation could occur in the two directions simultaneously. This point has been controversial for several years. Although early work (Butler and Finch, 1973) showed that complete 300 nm rods formed rapidly, other groups have proposed that no 3'-direction elongation occurs until that in the 5'-direction is complete (Lebeurier *et al.*, 1977; Otsuki *et al.*, 1977; Fukuda *et al.*, 1978). However, the virus strains and the conditions used in the reassembly reactions have been different in different laboratories. The observation by Lomonossoff and Butler (1979) that nucleotides characteristic of the 3'-side of the nucleation region were protected by coating before the 5'-directional coating was complete mean that at least in some circumstances bidirectional elongation occurs.

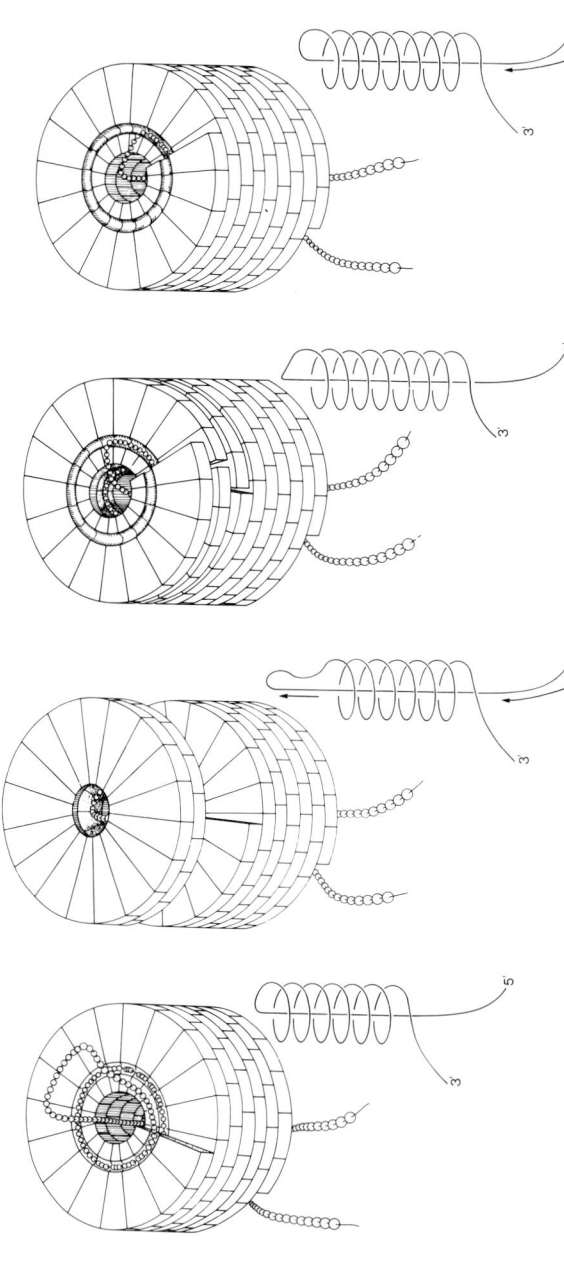

Figure 14.13. Schematic picture of possible mechanism for elongation in the major 5′-direction using a disk aggregate of protein. The "travelling loop" of RNA, formed where the 5′-tail is doubled back down the central hole of the growing rod, could insert between the two rings of a disk, through the central hole, and interact with the RNA binding site. This could lead to dislocation and binding of the disk, and two turns of RNA becoming packaged, one between the incoming disk and the helix surface and the other between the turns of the disk. Subsequent movement of the RNA tail up through the hole of the helix would regenerate the loop, allowing a continuation of the elongation. (From Butler, 1984.)

14. Architecture and Assembly of TMV

D. Elongation: Experiments with Cross-linked RNA

New evidence that elongation is bidirectional comes from experiments in which cloned cDNA of 100–300 bp has been covalently cross-linked photochemically to specific parts of the RNA (Fig. 14.14(a)) (Fairall et al., 1986). These hybrid molecules were then nucleated and allowed to elongate and the lengths of unencapsidated RNA in each hybrid were measured. Cross-linking to the initiation site largely eliminated nucleation, cross-linking to a site on the 3'-tail gave a strong major peak, and cross-

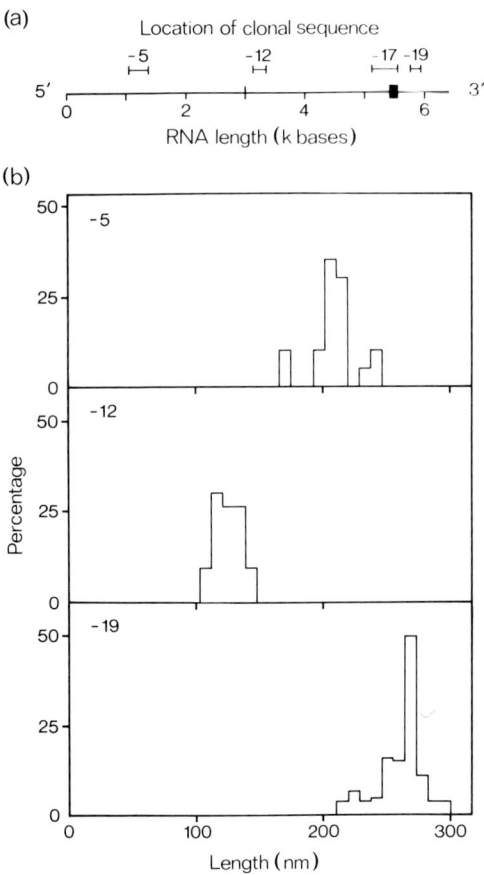

Figure 14.14. Reassembly employing RNA cross-linked to hybridized cDNA at various sites. (a) Map of the TMV genome showing the locations of the cloned cDNA used in each hybridization and cross-linking experiment (clones -5, -12, -17 and -19) and also the origin of assembly (solid block). (b) Histograms of the lengths of rods reassembled upon RNA hybridized and cross-linked to cDNA from clones -5, -12 and -19. (From Fairall et al., 1986.)

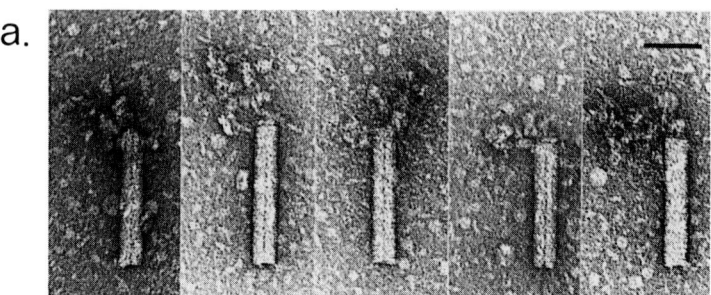

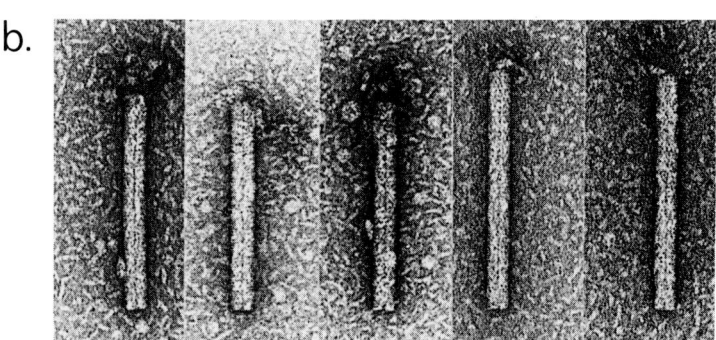

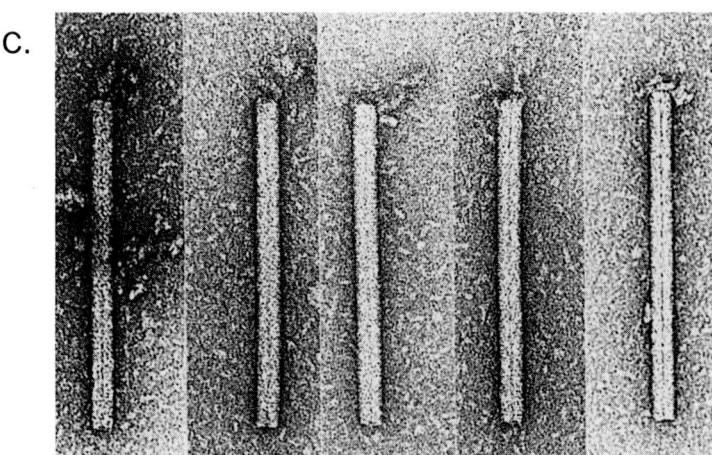

Figure 14.15. Electron micrographs of rods reassembled upon RNA hybridized and cross-linked to specific cloned cDNA. The cDNA used was from (*a*) clone −12, (*b*) clone −5, and (*c*) clone −19. The particles selected are ones where the two ends of the helix are distinguishable and they are orientated with their convex ends (i.e. containing the 3′-end of the RNA in a complete rod) upwards. Scale bar is 50 nm. (From Fairall *et al.*, 1986.)

linking to two sites on the 5'-tail gave broader peaks (Fig. 14.14(b)). Electron micrographs (Fig. 14.15) show RNA tails always at the end of the particle destined to be the 3'-end. Thus even when the cross-linking is on the 5'-tail, this still doubles back down the rod that is no longer growing. The length of rods blocked near the 3'-end shows that elongation occurs right up to the 3'-block, and the rods containing RNA blocked at the 5'-end show that although the blocked 5'-end does not allow completion of this elongation reaction, that in the 3'-direction is completed.

E. Binding of Oligonucleotides to Disks

One approach to simplifying the study of the elongation process has been to investigate the binding of oligonucleotides to preparations of disk aggregates (reviewed in Butler, 1984). Crystals of disks will bind the trinucleotide AAG or the hexamer AAGAAG (Graham and Butler, 1979; Butler and Lomonossoff, 1980), but crystallography disappointingly revealed no difference in three dimensions between crystals that had and those that had not bound oligonucleotides. Reactions between a solution of disks and AAGAAG gave a binding isotherm with AAGAAG that saturated at about 8 mol hexamers per disk and with an apparent dissociation constant of about 1.5 μM (Fig. 14.16) (Turner et al., 1986). In contrast, the binding to crystals was much weaker; even at 1 mM only one

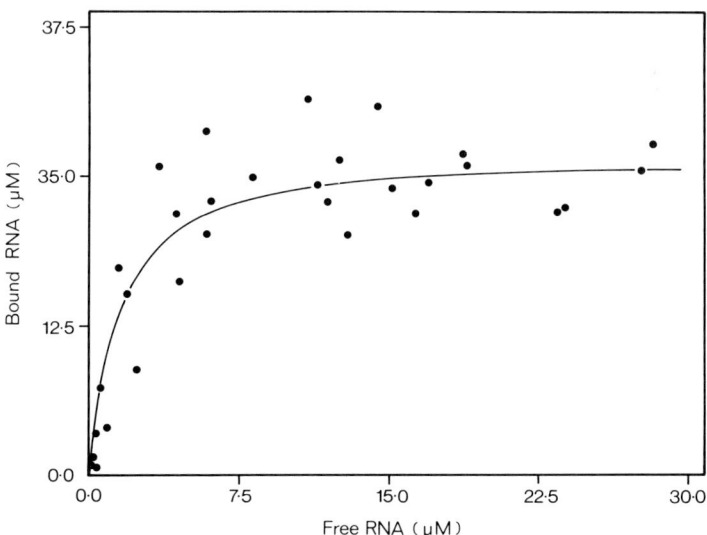

Figure 14.16. Binding curve for the hexanucleotide AAGAAG to disks of TMV protein. The curve shows the fit of a hyperbola to the data, corresponding to the binding of a maximum of about 8 hexamers per disk with an apparent dissociation constant (K_d) of 1.5 μM. (From Turner et al., 1986.)

or two hexamer molecules bound per disk. Moreover, the binding isotherm for the reaction in solution may be slightly sigmoidal, and when examined in an electron microscope such disks appear unlike untreated disks, and dislocated into protohelices. Possibly the initial binding of the oligonucleotide causes a dislocation and this cannot extend to allow further and more favoured binding when the disks are locked in a crystal.

Studies on the binding of artificial oligonucleotides were extended to include other molecules. As reported earlier (Schuster et al., 1980), AAG binds poorly to disks at pH7, whereas AAGAAGUUG bound more tightly than the hexamer, but not as tightly as might be expected, suggesting that UUG binds less well than AAG. Also d(AAGAAG) did not bind to disks at concentrations of up to 800 µM. Thus the 2'-hydroxyl in ribose is important for the interaction. Further work with hybrid oligonucleotides lacking some 2'-OH groups should resolve which of these groups is vital for the interaction.

V. CONCLUDING REMARKS

The aim of this chapter has been to introduce the topic of virus particle structure and assembly by reviewing the development of our ideas of the structures of TMV particles and their functional significance. These ideas have kept pace with the developments in technique in molecular biology and have sometimes led these developments. Molecular techniques have continued to advance. Thus we can expect the ability to make mutations at defined sites in cloned DNA and to synthesize experimentally useful quantities of predictably altered protein to open a new chapter in the now long-running story of TMV.

REFERENCES

1. Bawden, F. C. and N. W. Pirie (1937). *Proc. Roy. Soc.* **B123**, 274–320.
2. Bloomer, A. C., J. N. Champness, G. Bricogne, R. Staden and A. Klug (1978). *Nature (London)* **276**, 362–368. Macmillan Journals Limited.
3. Butler, P. J. G. (1984). *J. Gen. Virol.* **65**, 253–279.
4. Butler, P. J. G. and J. T. Finch (1973). *J. Mol. Biol.* **78**, 637–649.
5. Butler, P. J. G. and A. Klug (1971). *Nature New Biol.* **229**, 47–50.
6. Butler, P. J. G. and G. P. Lomonossoff (1978). *J. Mol. Biol.* **126**, 877–882.
7. Butler, P. J. G. and G. P. Lomonossoff (1980). *Biophys. J.* **32**, 295–312.
8. Butler, P. J. G., A. C. Bloomer, G. Bricogne, J. N. Champness, J. Graham, H. Guilley, A. Klug and D. Zimmern (1976). *Proc. 3rd John Innes Symp.* pp. 101–110.
9. Butler, P. J. G., J. T. Finch and D. Zimmern (1977). *Nature (London)* **265**, 217–219.
10. Caspar, D. L. D. (1963). *Adv. Protein Chem.* **18**, 37–121.
11. Durham, A. C. H., J. T. Finch and A. Klug (1971). *Nature New Biol.* **229**, 37–42. Macmillan Journals Limited.
12. Fairall, L., J. T. Finch, C.-F. Hui, C. R. Cantor and P. J. G. Butler (1986). *Eur. J. Biochem.* **156**, 459–465.

13. Fraenkel-Conrat, H. (1956). *J. Am. Chem. Soc.* **78**, 882–883.
14. Fraenkel-Conrat, H. (1981). *Intervirology* **15**, 177–189.
15. Fraenkel-Conrat, H. and B. Singer (1964). *Virology* **23**, 354–362.
16. Fraenkel-Conrat, H. and R. C. Williams (1955). *Proc. Natl Acad. Sci. USA* **41**, 690–698.
17. Fukuda, M., T. Ohno, Y. Okada, Y. Otsuki and I. Takebe (1978). *Proc. Natl Acad. Sci. USA* **75**, 1727–1730.
18. Gierer, A. and G. Schramm (1956). *Nature (London)* **177**, 702–703.
19. Goelet, P., G. P. Lomonossoff, P. J. G. Butler, M. E. Akam, M. J. Gait and J. Karn (1982). *Proc. Natl Acad. Sci. USA* **79**, 5818–5822.
20. Graham, J. and P. J. G. Butler (1979). *Eur. J. Biochem.* **93**, 333–337.
21. Guilley, H., G. Jonard, B. Kukla and K. E. Richards (1979). *Nucl. Acids Res.* **6**, 1287–1308.
22. Hirth, L. and K. E. Richards (1981). *Adv. Virus Res.* **26**, 145–199.
23. Holmes, K. C. (1979). *J. Supramolec. Struct.* **12**, 305–320.
24. Holmes, K. C. (1982). *Ciba Foundation Symposium* No. 93, pp. 116–138.
25. Holmes, K. C. (1984). *In* "Biological Macromolecules and Assemblies, Vol. 1: Virus Structures" (Eds F. A. Jurnak and A. McPherson), pp. 121–148. Wiley, New York.
26. Lauffer, M. A. (1975). "Entropy Driven Processes in Biology". Springer-Verlag, Berlin and New York.
27. Lebeurier, G., A. Nicolaieff and K. E. Richards (1977). *Proc. Natl Acad. Sci. USA* **74**, 149–153.
28. Lomonossoff, G. P. and P. J. G. Butler (1979). *Eur. J. Biochem.* **93**, 157–164.
29. Lomonossoff, G. P. and P. J. G. Butler (1980). *FEBS Lett.* **113**, 271–274.
30. Mandelkow, E., K. C. Holmes and U. Gallwitz (1976). *J. Mol. Biol.* **102**, 265–285.
31. Mandelkow, E., G. Stubbs and S. Warren (1981). *J. Mol. Biol.* **152**, 375–386.
32. Namba, K., D. L. D. Caspar and G. J. Stubbs (1985). *Science* **227**, 773–776.
33. Novikov, K. K., K. K. Sarakhan-Bek and J. G. Atabekov (1974). *Virology* **62**, 134–144.
34. Otsuki, Y., I. Takebe, T. Ohno, M. Fukuda and Y. Okada (1977). *Proc. Natl Acad. Sci. USA* **74**, 1913–1917.
35. Sarkar, S. (1960). *Z. Naturforsch.* **15b**, 778–786.
36. Schoen, A. and K. W. Mundry (1984). *Eur. J. Biochem.* **140**, 119–127.
37. Schramm, G. and W. Zillig (1955). *Z. Naturforsch.* **B10**, 493–499.
38. Schuster, T. M., R. B. Scheele, M. L. Adams, S. J. Shire, J. J. Steckert and M. Potschka (1980). *Biophys. J.* **32**, 313–329.
39. Stubbs, G. (1984). *In* "Biological Macromolecules and Assemblies, Vol. 1: Virus Structures" (Eds F. A. Jurnak and A. McPherson), pp. 149–202. Wiley, New York.
40. Stubbs, G., S. Warren and K. Holmes (1977). *Nature (London)* **267**, 216–221.
41. Turner, D. R., A. Mondragon, L. Fairall, A. C. Bloomer, J. T. Finch, J. H. van Boom and P. J. G. Butler (1986). *Eur. J. Biochem.* **157**, 269–274.
42. Watson, J. D. (1954). *Biochim. Biophys. Acta.* **13**, 10–19.
43. Zimmern, D. (1977). *Cell* **11**, 463–482.
44. Zimmern, D. and P. J. G. Butler (1977). *Cell* **11**, 455–462.
45. Zimmern, D. and T. M. A. Wilson (1976). *FEBS Lett.* **71**, 294–298.

15. Structure of Picornavirus Coat Proteins and their Antigenicity

P. D. Minor

National Institute for Biological Standards and Control, Holly Hill, Hampstead, London NW3 6RB, U.K.

I. Introduction	259
II. Antigenicity of viral and subviral particles and isolated coat proteins	261
III. Analysis of sequence data in the identification of antigenic sites in picorna viruses	262
IV. The use of immunogenic protein fragments and synthetic peptides in studies of the antigenic structure of FMDV	263
V. The use of synthetic peptide immunogens in the study of the antigenic structure of poliovirus	267
VI. The use of antigenic mutation in the study of the antigenic structures of polio virus and rhinovirus	267
VII. Identification of sites to which antibodies bind by cross-linking and peptide binding studies	269
VIII. Conclusions	276
References	277

I. INTRODUCTION

The family picornaviridae consist of small RNA-containing viruses with many structural features in common. They are classified into four groups on the basis of their genomic structure, stability at low pH and buoyant density in the presence of caesium ions. The four groups are the enteroviruses, typified by polioviruses and Coxsackie viruses, the rhinoviruses, the cardioviruses such as encephalomyocarditis virus (EMC) and mengovirus, and the aphthoviruses, or foot-and-mouth disease viruses (see Chapter 1, this volume). Antigenically distinct serotypes exist within a group with little cross-neutralization between types; for example there are three serotypes of poliovirus (types 1, 2 and 3), over 115 serotypes of human rhinovirus and seven serotypes of foot-and-mouth disease virus (A, O, C, SAT1, SAT2, SAT3 and Asia 1). There is also evidence for serologically different strains within serotypes, although this is more marked and clinically significant in the aphthoviruses than the polioviruses.

Picornavirus virions consist of a genome of single-stranded and messenger-sense RNA enclosed in a lipid-free icosahedral shell, 25–30 nm in diameter, composed of 60 copies each of the four virion proteins VP1, VP2, VP3 and VP4 (or α, β, γ and δ for cardioviruses). The genomic RNAs of representatives of each of the groups have now been sequenced (Skern et al., 1985; Stanway et al., 1984; Palmenberg et al., 1984; Forss et al., 1984; Carroll et al., 1984; Toyoda et al., 1984). These studies demonstrate the similarities in the organization of the genomes and that, except for foot-and-mouth disease virus, the sizes of the homologous structural proteins are very similar (Table 15.1). The architecture of the virions and the strategy of replication are consequently thought likely to be basically the same, although differing in detail.

This review is restricted to a summary of currently available information on the antigenic sites of picornaviruses that are involved in virus neutralization, although it is recognized that certain antibodies may bind to virion proteins without affecting viral infectivity. The mechanisms by which the attachment of an antibody to a specific site neutralizes the virus are as yet almost wholly unknown.

It has been established that neutralizing antibodies specific for certain viruses are predominantly directed against a single virion protein. Thus the haemagglutinin (HA) molecule is the principal target of antibodies able to neutralize influenza virus as neutralization by specific sera covaries with the haemagglutinin subtype and the HA molecule alone is able to induce neutralizing antibody.

A similar approach has met with mixed success with the picornaviruses, for which independent segregation of the genes coding for coat proteins is as yet unknown, and for which there is reason to believe that all four

Table 15.1. Number of amino acids in the structural proteins of picornaviruses

Virus	VP1 (α)	VP2 (β)	VP3 (γ)	VP4 (δ)
Poliovirus 1	302	271	239	70
Poliovirus 2	301	271	239	70
Poliovirus 3	300	271	239	70
Rhinovirus 2	315	261	237	69
Rhinovirus 14	301	261	231	69
EMC virus	288	255	231	69
FMDV O1K	213	218	220	69
FMDV A10	212	218	221	82

structural proteins are synthesized and utilized as a unit to form a complex and well-defined quarternary structure. There is considerable evidence that the conformation of the individual coat proteins and their relationship to each other are major factors in the antigenicity of picornaviruses.

II. ANTIGENICITY OF VIRAL AND SUBVIRAL PARTICLES AND ISOLATED COAT PROTEINS

Poliovirus subjected to relatively mild denaturing conditions undergoes a major change in antigenic reactivity (Le Bouvier, 1955; Hummeler and Tumilowicz, 1960). Naturally occurring empty capsids express similar, but not necessarily identical, antigenic determinants to denatured virus (Mayer et al., 1957) and for some time it was considered that empty capsids or heated virions expressed a completely distinct set of determinants (C or H antigen) from those found on native infectious virions (D or N antigen). However, monoclonal antibodies exist that will react with both types of particle (Blondel et al., 1983; Ferguson et al., 1984) although most are specific for either C(H) antigen or D(N) antigen (Rombaut et al., 1983; Ferguson et al., 1984).

The existence of major antigenic differences between full and empty capsids has been reported for several picornaviruses, including poliovirus (Mayer et al., 1957), rhinovirus (Lonberg-Holm and Yin, 1973), Coxsackie B virus (Frommhagen, 1965) and foot-and-mouth disease virus (Rowlands et al., 1975). The properties of particles expressing C antigen suggested that it might be associated with VP4, the smallest virion protein (Joklik and Darnell, 1961; Lonberg-Holm et al., 1975; Breindl, 1971). This possibility is rendered unlikely by the inaccessibility of VP4 to surface labelling techniques (Lonberg-Holm and Butterworth, 1976) and the observation that empty capsids with D antigenic character can be isolated and converted to particles expressing C antigen without loss of any protein (Lonberg-Holm and Yin, 1973; Rombaut et al., 1983). The difference between D and C antigenic forms of picornaviruses is thus generally held to be a function of the conformation of the virion proteins. Subviral particles and isolated proteins may express specific antigenic characteristics in addition to the conventional D(N) or C(H) determinants (Rowlands et al., 1975; Rombaut et al., 1983; Wiegers and Dernick, 1983).

In view of the importance of protein conformation in the antigenic structure of picornaviruses it is unlikely that any virion protein will be very effective in the induction of neutralizing antibodies. The published data are summarized in Table 15.2. The most consistent success has been achieved with VP1 of foot-and-mouth disease virus both when isolated from virions by SDS-polyacrylamide gel electrophoresis (Meloen et al., 1979) or from a genetically engineered fusion protein (Kleid et al., 1981). There has been less success with proteins isolated from poliovirus (Meloen et al., 1979; Blondel et al., 1982; Chow and Baltimore, 1982). It has been reported that this is a function of the methods used in the isolation of the

Table 15.2. Picornaviral proteins raising neutralizing antibody

Virus	Protein	Reference
Coxsackie B3	VP2	[1]
Mengovirus	VP1	[41]
FMDV	VP1	[44]
		[36]
Polio	none	[44]
	VP1	[3]
		[9]
	VP1, VP2, VP3	[12]
		[61]
	VP4	[18]

proteins, as VP1, VP2 and VP3, but not VP4, could raise neutralizing antibodies (Dernick et al., 1983). Neutralizing antibodies have been raised with VP1, VP2 and VP3 isolated from formalin-inactivated virus (van der Marel et al., 1983). Emini et al. (1983a) reported that VP4 was able to induce neutralizing antibody, although the antibody in question reacted specifically with VP3.

In all cases, the isolated proteins from picornaviruses are far less effective in stimulating neutralizing antibody than whole virions even where consistent responses are readily demonstrable, as in the case of VP1 from foot-and-mouth disease virus.

The accurate delineation of antigenic sites has required more sophisticated methods of analysis. These include theoretical predictions from sequence data, the use of fragments of proteins and synthetic peptides to induce neutralizing antibody, and the identification of sites on the virion to which neutralizing antibodies bind by the isolation of antigenic variants resistant to monoclonal antibodies or the reaction of neutralizing antibodies with synthetic peptides.

III. ANALYSIS OF SEQUENCE DATA IN THE IDENTIFICATION OF ANTIGENIC SITES IN PICORNAVIRUSES

Until recently, the available data on the sequences of picornavirus genomes greatly exceeded the information on the structure of the virion and various attempts have thus been made to deduce the location of antigenic features from sequence data alone. Two approaches have generally been used in combination. Firstly, a theoretical analysis is made to predict which regions of the proteins are likely to be on the surface. The most generally used is the hydrophilicity analysis of Hopp and Woods (1981), although more sophisticated predictive methods are sometimes employed (Pfaff et *al.*, 1982). There are generally several peaks of hydrophilicity in a given sequence, of which only a few may correspond to antigenic sites. Secondly, when sequences of related but antigenically distinct strains are compared, at least some of the differences in the sequences would be expected to be related to differences in antigenic properties. The antigenic sites can thus be identified, as they are predicted to be regions of both high hydrophilicity and sequence heterogeneity (Makoff *et al.*, 1982; Emini *et al.*, 1983c, 1984).

There are several instances in which this simple combined approach has predicted antigenic sites that were later confirmed by other more empirical methods. Thus Pfaff *et al.* (1982) and Bittle *et al.* (1982) identified regions of the VP1 of the foot-and-mouth strain O1K, and subsequently demonstrated that synthetic peptides corresponding to the amino acid sequences of these areas were immunogenic, and Emini *et al.* (1983c) identified a region in the VP1 of poliovirus type 1. In the case of FMDV, however, the regions had already been identified by other means, as described below. A theoretically predicted site in VP2 of poliovirus type 1 was identified by Emini *et al.* (1984), and a region homologous to this was subsequently shown by other workers to be a major mutational locus in antigenic mutants derived from rhinovirus 14 and poliovirus types 1 and 3 by the use of monoclonal antibodies (Sherry and Rueckert, 1985; Rossman, Chapter 17, this volume; Minor *et al.*, 1986). There are many examples, however, in which the theoretical predictions proved incorrect or highly unlikely (Pfaff *et al.*, 1982; Rossman, Chapter 17).

While theoretical predictions are easily made, they are usually insufficient to identify antigenic sites reliably. The experimental methods used in conjunction with predictive methods will be illustrated by work with foot-and-mouth disease virus, poliovirus and rhinovirus 14.

IV. THE USE OF IMMUNOGENIC PROTEIN FRAGMENTS AND SYNTHETIC PEPTIDES IN STUDIES OF THE ANTIGENIC STRUCTURE OF FMDV

Trypsin treatment of intact FMDV virions destroys infectivity and reduces the ability of the virus to induce neutralizing antibody in animals, while apparently leaving the particles morphologically intact (Cavanagh *et al.*, 1977). Trypsin-treated virus can be shown to have undergone a selective cleavage of VP1 and the extent of VP1 cleavage parallels the biological changes in the virus. This phenomenon varies with the strain of virus examined so that, for example, strain C997 has been shown to be far less affected than O6 (Rowlands *et al.*, 1971). While it is possible that trypsin treatment results in a major conformational rearrangement of the capsid that alters antigenicity without affecting virion morphology, it is considered more likely that it destroys the major antigenic sites on VP1, consistent with the immunogenicity of this virion protein outlined above. These findings suggest that other sites are of far less significance in the neutralization of FMDV by antibodies.

In the light of these observations, Strohmaier et al. (1982) prepared fragments from the purified VP1 protein of strain O1K, using a variety of methods that included cyanogen bromide cleavage and digestion by specific enzymes such as trypsin and mouse submaxillary gland protease (MSGP). The fragments were identified by sequencing the termini and were tested for immunogenicity in mice. It was shown that two fragments resulting from cyanogen bromide cleavage (residues 55–180 and 181–213) and the C-terminal fragment generated by MSGP digestion (residues 146–213) induced neutralizing antibody, while no product of trypsin digestion would do so. As trypsin digestion removed the C-terminal 13 amino acids by multiple cleavages, and also a region from residue 138 to 154, it was concluded that there were at least two antigenic sites that were encompassed by residues 146–154 and 201–213 respectively. The data are summarized in Table 15.3.

These findings were exploited by Pfaff et al. (1982) and Bittle et al. (1982) who also used theoretical considerations of hydrophilicity, strain heterogeneity and predicted structure to identify regions of VP1 that were likely to be antigenic sites. The synthetic peptides having the amino acid sequences of these regions were then made and used to immunize animals. Pfaff et al. (1982) reported that a peptide including the sequence of the VP1 of the O1K strain of FMDV from residues 144 to 160 induced high levels of neutralizing antibody (end point titres of 3.5 $\log_{10}$) while peptides from other theoretically promising regions, including the terminal sequence from 205 to 213 did not induce a measurable neutralizing antibody response. Bittle et al. (1982), using the same strain of virus, reported that a peptide representing residues 141–160 induced antibody able to reduce the titre of virus by more than 3.7 $\log_{10}$ in an end point dilution assay in suckling mice, and that a slightly less effective peptide representing the C-terminal 13 amino acids (201–213) induced sera that reduced the virus titre by 3.5 $\log_{10}$. All peptides used were coupled to a carrier protein for immunization and Bittle et al. (1982) reported protection of guinea pigs from subsequent challenge. On the basis of the level of antibody induced per unit mass of material, the peptides were greatly superior to purified VP1 as immunogens but were less effective than whole virus. The

Table 15.3. Immunogenicity of fragments of FMDV VP1

Method of cleavage	Fragments raising neutralizing antibody
CNBr	55–180
	181–213
MSGP	146–213
Trypsin	none (138 to 154 and 200–213 totally digested)

molecular weight of the most effective peptide is about one-tenth that of VP1, so that it is possible that the two immunogens may be of similar potency on a molar basis.

Subsequent studies have shown that peptides including the amino acid sequence homologous to residues 141–160 will induce neutralizing antibodies for the A10, O1, C3 and SAT2 serotypes of FMDV, and that such antibodies mimic subtype specificity in that peptides deduced from the sequences of A10 or A12 virus would induce antibodies that neutralized the homologous but not the heterologous strains (Clarke et al., 1983). The peptides based on the A10 and A12 subtypes differed at eight residues.

Four strains (A, B, C, USA) were isolated from a pool of A12 virus that varied within the region between residues 141–160 of VP1, specifically at residues 148 and 153 (Rowlands et al., 1983). The amino acid sequences are shown in Table 15.4, where it can be seen that strain A differs from strain C by a single amino acid at residue 153, which is leucine in A and serine in C, and that strain B differs from strain USA by a single amino acid at residue 148, which is a leucine in B and phenylalanine in USA. Strains A and C differ from strains B and USA by substitutions at both 148 and 153, and it was found that sera raised by immunizing animals with virus particle vaccines reflected the differences in sequence. Thus a 100–200 times dilution of a serum raised with A virions would neutralize A virus most effectively, followed by C virus, while being relatively ineffective against either B or USA. More concentrated preparations of serum would neutralize all strains, but these findings implied that the most potent antibodies were directed at a region including 148 and 153 of VP1. Similar considerations applied to sera raised with virions of the other three strains. These findings mimic those encountered *in vivo*, where there is evidence that minor variation plays a significant part in the failure of vaccines to protect against FMDV.

Antisera raised with peptides based on the sequences of the four strains from residues 141–160 of VP1 showed the same spectrum of neutralization as the antiviral sera, such that strains A and C were closely related but antigenically distinct, as were B and USA. Thus anti-peptide sera show a similar range of specificity to antivirion sera, even for minimal variants isolated from a common pool, where a single amino acid change can alter

Table 15.4. Sequences of residues 146 to 155 of VP1 of strains of A12 virus obtained from a single pool

	146	147	148	149	150	151	152	153	154	155
A	Gly	Asp	Ser	Gly	Ser	Leu	Ala	Leu	Arg	Val
B			Leu					Pro		
C			Ser					Ser		
USA			Phe					Pro		

the antigenic reactions significantly. Collectively, the data lead to the conclusion that viral infectivity is neutralized when antibodies bind to a region of VP1 encompassed by residues 141–160 or 200–213, and that these represent the targets of the most potent antibodies, although other sites may be recognized by other antibodies.

These studies were based on the identification of proteins and peptides that will induce a neutralizing antibody response. An alternative approach is to produce neutralizing antibody by immunization with intact virus, and then to attempt to identify the site to which it binds on the virus. This method has been far more extensively applied to poliovirus and rhinovirus than to FMDV, but neutralizing monoclonal antibodies specific for the O1K strain of FMDV have been obtained. It has been shown that peptides corresponding to the two antigenic sites identified above are able to inhibit the binding of the antibodies to the virus in competitive ELISA tests, consistently with the view that the antibodies bind to the previously identified sites (Parry et al., 1985). However, antigenic variants selected for resistance to these antibodies appear to have mutations away from these antigenic sites. This is a surprising finding in view of the molecular basis of the strain specificity of anti-peptide sera for the A serotype described above.

While most neutralizing monoclonal antibodies do not react with separated virion proteins in Western blots, Duchesne et al. (1984) have reported studies with a monoclonal antibody which will do so. The reaction of the antibody was abolished if the VP1 protein was cleaved with trypsin or mouse submaxillary gland protease (MSGP) but was unaffected by extensive digestion with other proteases. By analogy with the work of Strohmaier summarized in Table 15.3, this suggests that this antibody recognizes an amino acid sequence including residues 146 and 147 in VP1.

A more radical approach to the identification of antigenic sites has been described by Geysen et al. (1984), based on the ability to synthesize small quantities of peptide immobilized on polythene rods rather than on column resins. This technique greatly simplified the process of synthesis and enabled large numbers of peptides to be made with relative ease. The 213 amino acid sequence of VP1 of O1K FMDV can be divided into a total of 208 contiguous hexapeptides (including residues 1–6, 2–7, 3–8, etc). Geysen et al. (1984) synthesized all 208 hexapeptides and then examined their reactivity in an ELISA test with antiviral sera raised in animals. This experimental approach is based on at least three assumptions: (a) that antigenic sites on VP1 are continuous elements of the sequence of six amino acid residues or less; (b) that sufficient quantities of a synthetic peptide containing the correct sequence will adopt the appropriate conformation to be recognized by the antibody; and (c) that binding of antibody to the peptide is absolutely specific in the sense that antibody will not bind detectably to peptide of the wrong amino acid sequence. While it is possible to question each of these assumptions, and there is evidence that assumption (b) is wrong in some cases (Emini et al., 1983b; see below), the method revealed sites including the sequences 146–152 and 206–212 in FMDV VP1, consistent with the other studies reviewed above. The value of this approach remains to be assessed as yet.

V. THE USE OF SYNTHETIC PEPTIDE IMMUNOGENS IN THE STUDY OF THE ANTIGENIC STRUCTURE OF POLIOVIRUS

The major antigenic sites on FMDV have been identified by the use of synthetic peptides able to induce neutralizing antibodies, and it was therefore reasonable to apply the same approach to identify sites on poliovirus. Chow and Baltimore (1982) reported that poliovirus VP1 was capable of inducing neutralizing antibody at a low level, and Chow et al. (1985) identified sequences in VP1 that were likely to be antigenically significant by the theoretical criteria of hydrophilicity and heterogeneity between different polio types (Chow et al., 1985). Synthetic peptides having the amino acid sequences of these regions were then used to immunize rats and rabbits and neutralizing antibody was induced by certain of the peptides listed in Table 15.5. Antibody titres were measured by assaying the dilution of serum required to reduce the titre of virus by 50% in a plaque formation assay, and the dilution required to protect cell sheets from infection by 1000 infectious units of virus for 24 hours. These methods are more sensitive than standard clinical assays (Domok and Magrath, 1979) and it is likely that the titres of serum recorded would have been undetectable by more routine methods.

Emini and co-workers have reported studies with various synthetic

Table 15.5. Immunologically active peptides deduced from the sequence of poliovirus structural proteins

Neutralizing response			Priming response		
Peptide		Reference	Peptide		Reference
VP1	93–103	[20]	VP1	11–17	[20]
VP2	162–173	[21]	VP1	70–80	[20]
VP1	141–147	[33, 34]	VP1	70–75	[20]
VP1	113–121	[34]	VP1	93–103	[20]
VP1	165–172	[34]	VP1	97–103	[20]
VP1	61–80	[10]	VP2	162–173	[21]
VP1	86–103	[10]	VP1	141–147	[33, 34]
VP1	91–109	[10]	VP1	113–121	[34]
VP1	100–109	[10]	VP1	165–172	[34]
VP1	161–181	[10]			
VP1	182–201	[10]			
VP1	222–241	[10]			

peptides that were likely to be antigenic sites on the basis of hydrophilicity and sequence heterogeneity (Emini et al., 1983c, 1984; Jameson et al., 1985a,b). Two methods for assessing immunogenic activity were described. In the first the peptides were coupled to a suitable carrier protein and used to immunize animals, which were then tested for the induction of neutralizing antibody. Antibody was assayed by the reduction in plaque titre when virus was mixed with undiluted serum. The reduction in titre reported were 2.0 $\log_{10}$ or less, which should be compared to the figure of greater than 3.7 $\log_{10}$ reported by Bittle et al. (1982) for the most potent FMDV peptide and the figures of 8 $\log_{10}$ or greater reported for sera raised with virions (Emini et al., 1983c). In one instance (Jameson et al., 1985a) a standard neutralizing assay was performed. The titre reported was 1 in 2; this is extremely low (Domok and Magrath, 1980).

The second method for assessing immunogenic potency was to immunize the animals with peptide coupled to a suitable carrier and then with a subimmunizing dose of virus. If the peptide represents an antigenic site, the administration of the virus should produce a high-titre secondary response. This method is presumably more sensitive than that described above in that it can identify peptides as antigenic sequences even when they do not themselves induce neutralizing antibody (Emini et al., 1983c). It is not invariably successful, in that a peptide deduced from the VP1 of type 3 poliovirus by methods described below will induce high-titre neutralizing antibody but fails to prime for a secondary response. It is also not clear whether the anti-viral antibodies induced by boosting with virus are directed against only that sequence on the virus that is represented in the peptide, or whether they bind to the entire virion. In the second case it could be argued that the sequence itself is not a target of neutralizing antibody but is an inducer of T-helper-cell function. Finally it has been reported that sequences specific for poliovirus type 1 can prime for a secondary response to a subimmunizing dose of hepatitis A virus, although there is no sequence homology between the peptides used and the structural proteins of hepatitis A (Emini et al., 1985). If this observation is correct, it implies that a peptide of an inappropriate sequence can assume a conformation like that of a viral antigenic site. The sequence of the antigenic site is thus not necessarily the same as that of the peptide with priming activity, which makes the use of peptides to identify antigenic sites in this way uncertain.

Table 15.5 lists the peptides deduced from poliovirus structural proteins that have been reported to induce neutralizing antibody. There is independent evidence based on a different approach that two of the listed sequences are involved in antigenic sites (see below). Nonetheless, these peptides are far less potent as inducers of antiviral antibodies than the FMDV peptides, and the approach of selecting immunogenic peptides has consistently failed to identify any sequence that is able to induce high titres of antibody against poliovirus. Alternative methods have thus been used in the study of the antigenic structure of poliovirus.

VI. THE USE OF ANTIGENIC MUTATION IN THE STUDY OF THE ANTIGENIC STRUCTURES OF POLIOVIRUS AND RHINOVIRUS

Monoclonal antibodies with neutralizing activity for viruses may be used as tools for the location of antigenic sites involved in virus neutralization. The identification of the sites to which they bind has involved either the isolation of non-neutralizable antigenic variants or an examination of the binding of the antibodies to synthetic peptides. The isolation of mutants has the advantage that it can be unequivocally demonstrated that a particular amino acid substitution has a well-defined effect on the antigenic reactivity of the virus. However, it is possible that the mutations that are selected will affect the conformation of the binding site for the antibody on the virus while being distant from it.

Studies involving mutant isolations from type 3 poliovirus were carried out using a panel of monoclonal antibodies with neutralizing activity (Minor et al., 1983; Evans et al., 1983; Minor et al., 1985). Mutants were isolated by two cycles of plaque formation by antibody-treated virus and 230 individual primary plaques were picked from two strains of type 3 virus, namely P3/Leon/USA/1937 and the closely related Sabin vaccine strain P3/Leon/12a_1b. The 230 isolates were characterized by their reactions with the panel of antibodies and 16 distinct patterns of reaction were identified from the mutants of P3/Leon/USA/1937. The reactions of the Leon-derived mutants suggested that all the antibodies were recognizing a single operationally defined site, designated site 1. Thus there existed a mutation probably representing a single amino acid substitution that would affect neutralization of the virus by any pair of antibodies of the panel (Table 15.6). In contrast, the isolates from the Sabin vaccine strain P3/Leon/12a_1b fell into two sites, one corresponding to site 1 of the P3/Leon/USA/1937 mutants, and the other being a Sabin-specific site detected by a single monoclonal antibody which reacted only with the Sabin vaccine or strains derived from it. All site 1 mutants were still neutralized by this antibody, and conversely all mutants resistant to this antibody were still sensitive to the site-1-specific antibodies.

The location of the mutations in the site 1 mutant group was accomplished by primer extension sequencing of the genomic RNA. The site had been approximately localized to a region 100 amino acids from the N-terminus of VP1 by the fortuitous observation that a TI RNAase resistant oligonucleotide from this region underwent characteristic changes in mobility on two-dimensional electrophoresis of digests of RNA from mutants. The substitutions were concentrated in codons 89–100 of VP1 and are shown in Fig. 15.1 as filled squares, each one representing a distinct amino acid substitution. Where a mutant of the Leon strain had the same substitution as a mutant of the Sabin strain the antigenic reactions with site 1 antibodies were identical, providing strong evidence that the observed substitutions were the cause of the modified antigenicity.

Table 15.6. Resistance of antigenic mutants derived from P3/Leon/USA/1937 to monoclonal antibodies

Mutant	1–14	4–12	4–4	199	194	134	208	175	204	197	165
1	R	R		R	R					R	
2	R	R		R	R					R	R
3	R	R	R	R	R		R	R	R	R	R
4	R	R	R	R	R		R		R	R	R
5				R	R	R	R	R		R	R
6				R	R	R	R				R
7				R	R	R	R	R	R		R
8		R			R		R				R
9	R	R	R	R	R	R	R	R		R	R
10	R		R	R	R	R	R	R	R	R	R
11				R	R		R	R		R	R
12		R		R	R	R	R	R			R
13	R	R		R		R	R	R			R
14	R	R					R			R	R
15	R	R	R	R	R			R	R	R	R
16	R	R		R	R		R		R		R

R indicates resistance to neutralization.

In all cases, failure of an antibody to neutralize the mutant was associated with failure to bind to virions in a modified single-radial-diffusion test. The Sabin-specific site was located at codons 286–288 of VP1 by similar methods.

It is possible that the region of amino acids 89–104 from the N-terminus of VP1 is a structurally highly important part of the virus, such that alterations within this sequence produce changes in distant parts of the virion, so affecting antigenicity. This is unlikely for three reasons. Firstly, mutations in this region have extremely subtle effects on antigenic character. Secondly, over half of the single-base mutations possible within codons 93–100 have been isolated, implying little or no constraint on the substitutions compatible with virus viability. And finally, all the mutants, including a deliberately selected triple mutant, grow normally *in vitro* compared with the parental types. These properties would not be expected for a region of great structural importance. It is more likely, therefore, that the region 89–104 represents the binding site of the monoclonal antibodies.

Peptides including the amino acid sequence 89–104 residues from the N-terminus of VP1 have been shown to induce high-titre neutralizing antibody that reacted with most type 3 poliovirus strains tested (Ferguson *et al.*, 1985). In contrast, the anti-peptide sera did not react at all with type 1 or type 2 poliovirus, which is consistent with the low degree of homology between the serotypes of poliovirus at residues 89–104 of VP1. Moreover, the mutants isolated from the Leon or Sabin strains of type 3 poliovirus were also often not neutralized by the anti-peptide sera, especially if the substitutions were in positions 96–98. A single amino acid substitution was thus sufficient to render the virus unrecognizable, a situation resembling that found for the monoclonal antibodies and very similar to the reports of strain specificity of anti-peptide sera with strain A foot-and-mouth disease virus described above. The mutational locus is thus identified as an antigenic site on type 3 poliovirus. In addition almost all of the monoclonal antibodies were affected by mutations within the site, and this suggests that it is a strongly immunodominant feature of the virus. This work has been extended to type 2 poliovirus, in which a similar concentration of mutations in a region encompassing amino acids 96–102 of VP1 has been identified, affecting the binding and neutralizing activity of five out of seven type-2-specific monoclonal antibodies (Minor et al., 1986).

Other workers have studied the type 1 polio strain Mahoney and have obtained strikingly different results from mutants isolated using a panel of 13 monoclonal antibodies (Emini et al., 1983b; Diamond et al., 1985a,b; Blondel et al., 1986). By the criteria used for the type 3 mutants, these antibodies fell into three separate groups. Sequencing studies have revealed mutational loci in VP3 (at residues 60 and 73 from the N-terminus), in VP1 (at residues 221, 222 and 223) and in VP2 (at residue 270). We have independently performed studies with the Sabin type 1 vaccine strain with a further set of monoclonal antibodies prepared in our own laboratory, and identified four different mutational groups. One group has mutations in VP3 (at residues 58, 59, 60 and 71) and a second group has mutations which are either in VP1 at residues 220 or 222, or in VP2 at residues 169 or 170 (Minor et al., 1986). The mutations in the remaining groups have not yet been located, but they are known to be outside VP1. These findings are very similar to those of Diamond et al. (1985a,b). No mutations within the region of VP1 corresponding to site 1 have therefore been identified in these studies with type 1 poliovirus; the difference in this respect between type 1 and type 2 and 3 is striking and surprising.

Recently, antigenic variants of rhinovirus 14 have been isolated and characterized and mutational loci have been identified in VP1 (residues 93 and 95), VP3 (residues 72, 79 and 78), VP2 (residues 158–162) and VP1 (at residues 83, 89, 138 and 139) (Rossman, Chapter 17, this volume). The first of these loci appears to have counterparts in poliovirus type 2 and 3, and the second in poliovirus 1. The antigenically significant mutational loci are summarized in Table 15.7 and their positions in the genome in Fig. 15.2.

We would argue that all or most antigenically significant mutational loci represent sites to which antibodies bind. This view was not shared by others because of results obtained from direct methods designed to identify the binding site of neutralizing monoclonal antibodies (Diamond et al., 1985a,b; Blondel et al., 1986).

P3/Leon/USA/37-VP1

```
  GIEDLISEVAQGALTSLPKQQDSLPDTKASGPAHSKEVPALTAVETGAT    50

  NPLAPSDTVQTRHVVQRRSRSESTIESFFARGACVAIIEVDNEQPTTRAQ   100

  KLFAMWRITYKDTVQLRRKLEFFTYSRFDMEFTFVVTANFTNANNGHALN   150

  QVYQIMYIPPGAPTPKSWDDYTWQTSSNPSIFYTYGAAPARISVPYVGLA   200

  NAYSHFYDGFAKVPLKTDANDQIGDSLYSAMTVDDFGVLAVRVVNDHNPT   250

  KVTSKVRIYMKPKHVRVWCPRPPRAVPYYGPGVDYKNNLDPLSEKGLTTY   300
```

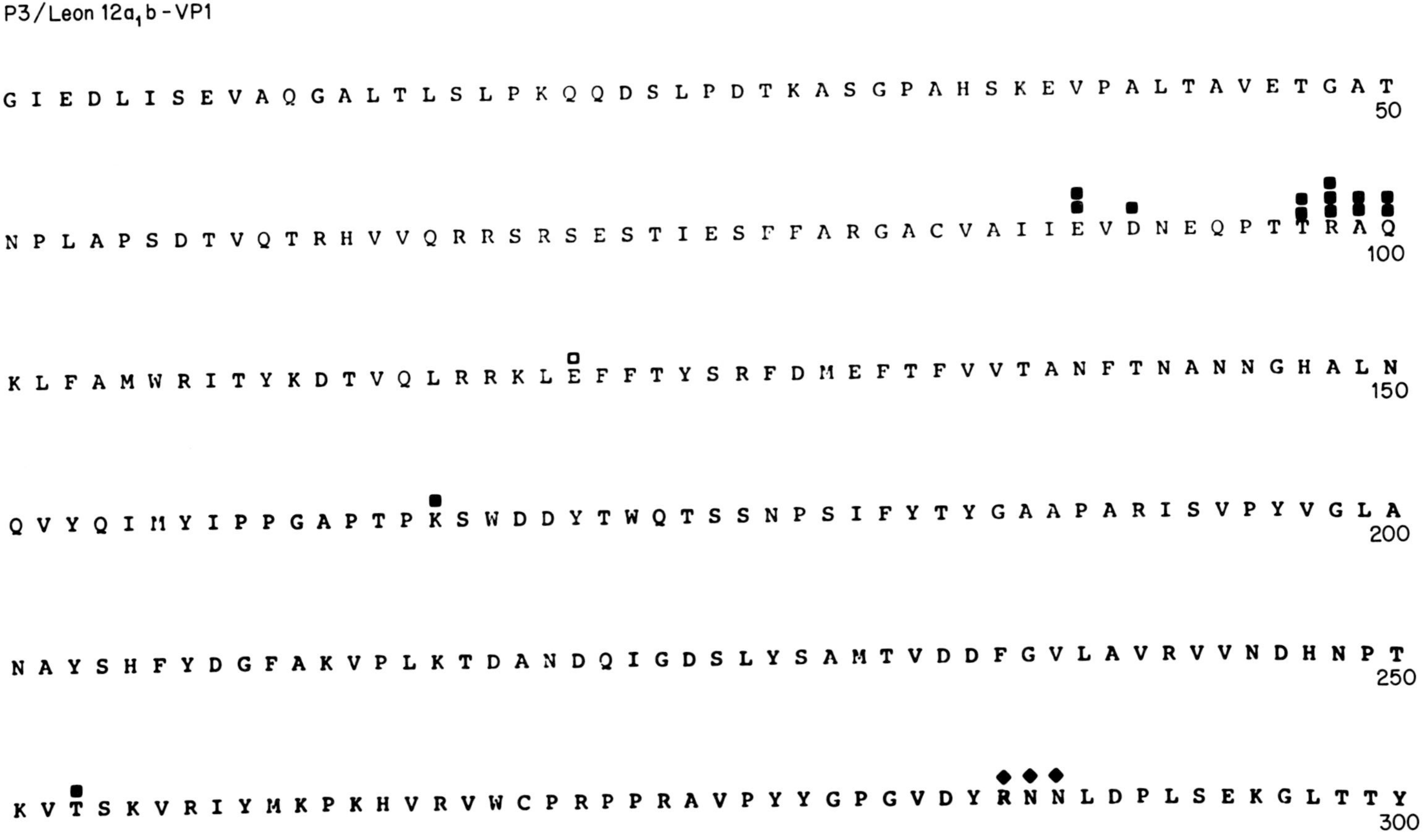

P3/Leon 12a₁b-VP1

```
              10        20        30        40        50
G I E D L I S E V A Q G A L T L S L P K Q Q D S L P D T K A S G P A H S K E V P A L T A V E T G A T

              60        70        80        90       100
                                                   ■ ■
                                                   ■ ■
                                                 ■ ■ ■
                                                 ■ ■ ■ ■
                                   □           ■ ■ ■ ■
N P L A P S D T V Q T R H V V Q R R S R S E S T I E S F F A R G A C V A I I E V D N E Q P T R A Q

             110       120       130       140       150
                        □
K L F A M W R I T Y K D T V Q L R R K L E F F T Y S R F D M E F T F V V T A N F T N A N N G H A L N

             160       170       180       190       200
                      ■ ■
Q V Y Q I M Y I P P G A P T P K S W D D Y T W Q T S S N P S I F Y T Y G A A P A R I S V P Y V G L A

             210       220       230       240       250
N A Y S H F Y D G F A K V P L K T D A N D Q I G D S L Y S A M T V D D F G V L A V R V V N D H N P T

             260       270       280       290       300
      ■                                     ◆
                                            ◆
                                            ◆
K V T S K V R I Y M K P K H V R V W C P R P P R A V P Y Y G P G V D Y R N N L D P L S E K G L T T Y
```

Figure 15.1. Location of amino acid substitutions in VP1 affecting the neutralization of type 3 poliovirus by monoclonal antibodies. Upper: P3/Leon/USA/37; lower: P3/Leon 12a₁b. Filled squares: mutations affecting site 1 antibodies. Filled diamonds: mutations affecting site 2 antibodies. Open square: silent mutation.

Table 15.7. Location of mutations in antigenic variants selected with monoclonal antibodies

Virus	Locus			Reference
Poliovirus 1	(1)	VP3	60, 71	[13, 14]
		VP3	58, 59, 60, 70	[47]
	(2)	VP1	221, 222, 223	[13, 14]
		VP1	220, 222	[47]
		VP2	169, 170	
	(3)	VP2	270	[93, 14]
Poliovirus 2	(1)	VP1	96–102	[47]
Poliovirus 3	(1)	VP1	89–100	[46]
	(2)	VP1	286–288	[46]
Rhinovirus 14	(1)	VP1	93, 95	Rossman†
	(2)	VP3	72, 75, 78	Rossman†
	(3)	VP2	158–162	Rossman†
	(4)	VP1	83, 85, 138, 139	Rossman†

†This volume, Chapter 17.

VII. IDENTIFICATION OF SITES TO WHICH ANTIBODIES BIND BY CROSS-LINKING AND PEPTIDE BINDING STUDIES

Emini *et al.* (1982) reported the use of the bifunctional reagent, toluene diisocyanate (TDI) to cross-link *Fab* fragments of neutralizing monoclonal antibodies to intact type 1 poliovirus. The antibody fragments were linked

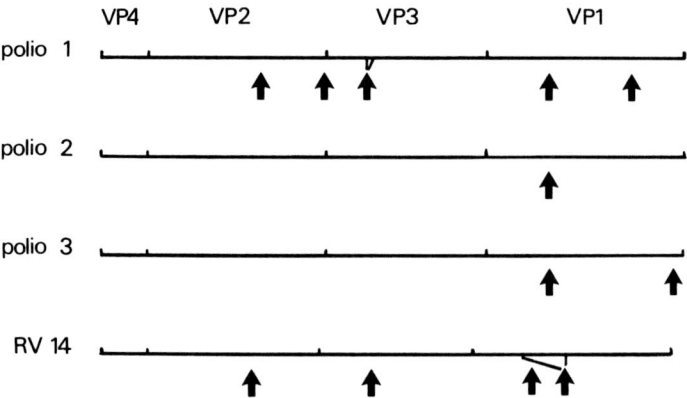

Figure 15.2. Location of antigenically significant mutational loci in polioviruses 1, 2 and 3 and rhinovirus 14.

to TDI at one pH, and then allowed to react with the virus before the second functional group was activated by exposure to a different pH to chemically link the antibody and virus. The treated preparation was analysed by gel electrophoresis, by which it was shown that VP1 was removed to large aggregates. This was not true to the same extent if the *Fab* fragments were reacted with EMC virus, or if ovalbumin was cross-linked to type 1 poliovirus, although there was some loss of VP1 in both cases. It was thus concluded that both antibodies were binding to VP1. Subsequent work showed that one of the antibodies (H3) selected for mutations within VP1 at residues 221–223; the other (D3) selected for mutation at codon 72 in VP2 (Diamond *et al.*, 1985a,b). An alternative method for the identification of binding sites involves the use of synthetic peptides in ELISA tests. Emini *et al.* (1983c) reported the identification of antigenic sites in the type 1 poliovirus strain Mahoney by examining the sequences of VP1 of all three serotypes for hydrophilic regions that were poorly conserved between the serotypes. The binding of monoclonal antibodies to peptides representing these putative antigenic sites was then examined by ELISA. One group of antibodies bound to a sequence representing residues 70–75 and another to a sequence representing residues 93–103. However, it was found that some antibodies which bound to the 70–75-residue peptide failed to bind to a longer peptide that included the same sequence. Conversely, some antibodies recognized the 93–103-residue peptide, but failed to react with a shorter peptide representing residues 97–103. The length of the peptide is therefore critical, and a sequence apparently reacting in one peptide could be masked in a slightly larger peptide. This finding has a bearing on the experimental approach of Geysen *et al.* (1984), outlined earlier.

Monoclonal antibodies that reacted in ELISA with peptide 93–103 were found to select for mutations within the VP3 site, while antibodies reacting with peptide 70–75 selected for mutations in VP1 at positions 221, 222 and 223 and in VP2 at position 270 (Diamond *et al.*, 1985a,b). It was thus concluded that the antigenically significant mutational loci in these instances were distinct from the antigenic sites. This implies that the region of VP1 of poliovirus type 1, which includes codons 93–103 from the *N*-terminus, cannot be altered to affect the binding of the antibodies considered without also affecting virus viability. Against this it can be argued that this region is known to vary between serotypes and between the Sabin type 1 vaccine strain and its progenitor Mahoney, and an antibody exists that will select for resistant mutants having mutations within it, as described.

The recent resolution of the atomic structures of type 1 Mahoney virus (Hogle *et al.*, 1985) and rhinovirus 14 (Rossman *et al.*, 1985; see Chapter 17, this volume) suggest strongly that mutations in general occur at the sites to which antibodies are directed. Thus residues 93–103 are prominently exposed, and distinct from residues 60, 72 and 73 of VP3, which form another exposed locus. It is thus unlikely that antibodies binding to residues 93–103 could select for mutations in the VP3 site. Additionally, residues 70–75, of VP1, which were reported to form a major antigenic site, were found to be internal and thus inaccessible to antibodies. Antibodies said to recognize this sequence selected for mutations in residues 220–222 of VP1, which form a prominent feature distinct from both the VP3 site and residues 93–103 of VP1, and thus a potentially distinct site.

It is likely that the identification of residues 93–103 and 70–75 of VP1 as the targets of monoclonal antibodies is a consequence of less specific reaction with the peptides than with the native antigen. Thus monoclonal antibodies neutralizing and binding to type 1 poliovirus and influenza haemagglutinin have been shown to bind to a peptide with a sequence corresponding to 16 amino acids of the VP1 of type 3 (Ferguson et al., 1986). The peptide induced type-3-specific antibody as described above. The monoclonal antibodies did not react with type 3 poliovirus or any other peptides tested. In this instance, therefore, the binding of the antibodies to the peptides was not related to their binding to the native antigen. This may have a bearing on the validity of the approach described by Geysen et al. (1984), which assumes complete specificity of reaction between peptide and antibody.

In one instance a direct method of establishing the region of binding of a monoclonal antibody to poliovirus was not in conflict with mutant studies. The type-1-specific monoclonal antibody C3 (Blondel et al., 1983) is unique among monoclonal antibodies with neutralizing activity for poliovirus so far reported in its ability to precipitate an isolated viral capsid protein, in this case VP1. A cDNA copy of the RNA coding for VP1 of poliovirus 1 was inserted into an expression vector, and produced a fusion protein that was still precipitable by antibody C3 (Wychowski et al., 1983). The inserted gene was then progressively deleted by Bal 31 nuclease treatment, and the products assessed for precipitability. It was found that only proteins containing amino acids 90–104 of VP1 reacted with C3, and that a synthetic peptide of this sequence coupled to a carrier protein was precipitable by C3. Moreover the synthetic peptide was able to reduce the efficiency of neutralization of virus by antibody C3, although a 10^{11}-fold molar excess of peptide over virus was required, implying a low affinity of antibody for peptide. Antibody C3 has been shown to select mutations in VP1 at residue 100 (Blondel et al., 1986; Diamond et al., 1985).

VIII. CONCLUSIONS

Four basic approaches have been taken to the identification of antigenic sites of picornaviruses, two involving consideration of the sequences of proteins to identify regions that will induce antibody, and two involving the identification of the binding sites of antibodies with known neutralizing activity.

(1) Theoretical treatments of the sequences, including predictions of

hydrophilicity and structure coupled with the identification of regions of low homology between antigenically distinct strains. This approach is based on the assumption that an antigenic site will be an area of low homology and high hydrophilicity.

(2) The identification of immunogenic protein fragments and synthetic peptides. This method has been used with great success for FMDV where one of the capsid proteins is strongly immunogenic in isolation from the virus. It has been less successful for poliovirus, in which none of the capsid proteins are potent inducers of neutralizing antibody.

(3) The identification of antigenically significant mutational loci by the selection of mutants with monoclonal antibodies, such loci probably being coincident with antigenic sites.

(4) The binding of monoclonal antibodies to synthetic peptides of known sequence in ELISA tests.

A fifth method is now available in a limited number of cases (Hogel et al., 1985; Rossman et al., 1985; see Chapter 17, this volume) in which the atomic structure of the virus is known, such that prominent features of the virion surface can be identified as potential antigenic sites.

The method of choice depends on the virus, with the most consistent results having been obtained by the use of immunogenic synthetic peptides of foot-and-mouth disease virus and the isolation of antigenic mutants from poliovirus and rhinovirus. Several sites have been identified by both methods, and it remains to be seen how the recognition of an antigenic site by an antibody effects neutralization of the virus.

REFERENCES

1. Beatrice, S. T., M. G. Katze, B. A. Zajac and R. L. Crowell (1980). *Virology* **104**, 426–438.
2. Bittle, J. L., R. A. Houghten, H. Alexander, T. M. Shinnick, J. G. Sutcliff, R. A. Lerner, D. J. Rowlands and F. Brown (1982). *Nature (London)* **298**, 30–33.
3. Blondel, B., R. Crainic and F. Horodniceanu (1982). *C. R. Acad. Sci. Paris* **294**, 91–94.
4. Blondel, B., O. Akachem, R. Crainic, P. Couillin and F. Horodniceanu (1983). *Virology* **126**, 707–710.
5. Blondel, B., R. Crainic, O. Fichot, G. Dufraisse, A. Cardrea, M. Girard and F. Horaud (1986). *J. Virol.* **57**, 81–90.
6. Breindl, M. (1971). *Virology* **46**, 962–964.
7. Carroll, A. R., D. J. Rowlands and B. E. Clarke (1984). *Nucl. Acids Res.* **12**, 2461–2472.
8. Cavanagh, D., D. V. Sangar, D. J. Rowlands and F. Brown (1977). *J. Gen. Virol.* **35**, 149–158.
9. Chow, M. and D. Baltimore (1982). *Proc. Natl Acad. Sci. USA* **79**, 7518–7521.
10. Chow, M., R. Yabrov, J. Bittle, J. Hogle and D. Baltimore (1985). *Proc. Natl Acad. Sci. USA* **82**, 910–914.
11. Clarke, B. E., A. R. Carroll, D. J. Rowlands, B. H. Nicholson, R. A. Houghten, R. A. Lerner and F. Brown (1983). *FEBS Lett.* **157**, 261–264.

12. Dernick, R., J. Heukeshoven and M. Hilbrig (1983). *Virology* **130**, 243–246.
13. Diamond, D. C., B. A. Jameson, E. A. Emini and E. Wimmer (1985). *In* "Vaccines '85: Molecular and Chemical Basis of Resistance to Parasitic, Bacterial and Viral Diseases", (Eds R. A. Lerner, R. M. Channock and F. Brown), pp. 241–247. Cold Spring Harbor Laboratory.
14. Diamond, D. C., B. A. Jameson, J. Brown, M. Kohara, S. Abe, H. Itoh, T. Komatsu, M. Arita, S. Kuge, A. D. M. E. Osterhaus, R. Crainic, A. Nomoto and E. Wimmer (1985b). *Science* **229**, 1090–1093.
15. Domok, I. and D. I. Magrath (1979). *WHO Offset Publication* No. 46.
16. Duchesne, M., T. Cartwright, A. Crespo, F. Boucher and I. Fallour (1984). *J. Gen. Virol.* **65**, 1559–1566.
17. Emini, E., B. A. Jameson, A. J. Lewis, G. R. Larsen and E. R. Wimmer (1982). *J. Virol.* **43**, 997–1005.
18. Emini, E. A., A. J. Dorner, L.-F. Dorner, B. A. Jameson and E. Wimmer (1983a). *Virology* **124**, 144–151.
19. Emini, E. A., Kao Shaw-Yi, A. J. Lewis, R. Crainic and E. Wimmer (1983b). *J. Virol.* **46**, 466–474.
20. Emini, E. A., B. A. Jameson and E. Wimmer (1983c). *Nature (London)* **304**, 699–703.
21. Emini, E. A., B. A. Jameson and E. Wimmer (1984). *J. Virol.* **52**, 719–721.
22. Emini, E. A., J. Berger, J. V. Hughes, S. W. Miba and D. L. Linemeyer (1985b). *In* "Vaccines 85. Molecular and Chemical Basis of Resistance to Parasitic Bacterial and Viral Diseases" (Eds R. A. Lerner, R. M. Channock and F. Brown), pp. 217–220. Cold Spring Harbor Laboratory.
23. Evans, D. M., P. D. Minor, G. S. Schild, J. W. Almond (1983). *Nature (London)* **304**, 459–462.
24. Ferguson, M., P. D. Minor, D. I. Magrath, Yi-Hua Qi, M. Spitz and G. C. Schild (1984). *J. Gen. Virol.* **65**, 197–201.
25. Ferguson, M., D. M. A. Evans, D. I. Magrath, P. D. Minor, J. W. Almond and G. C. Schild (1985). *Virology* **143**, 505–515.
26. Ferguson, M., S. E. Reed and P. D. Minor (1986). *J. Gen. Virol.* **67**, 2527–2531.
27. Forss, S., K. Strebel, E. Beck and H. Schaller (1984). *Nucl. Acids Res.* **12**, 6587–6601.
28. Frommhagen, L. L. (1965). *J. Immunol.* **95**, 818–822.
29. Geysen, H. M., R. H. Meloen and S. J. Barteling (1984). *Proc. Natl Acad. Sci. USA* **81**, 3998–4002.
30. Hogle, J. M., M. Chow and D. J. Filman (1985). *Science* **220**, 1358–1365.
31. Hopp, T. P. and K. R. Woods (1981). *Proc. Natl Acad. Sci. USA* **78**, 3824–3828.
32. Hummeler, K. and J. S. Tumilowicz (1960). *J. Immunol.* **84**, 630–634.
33. Jameson, B. A., J. Bonin, E. Wimmer and O. M. Kew (1985a). *Virology* **143**, 337–341.
34. Jameson, B. A., J. Bonin, M. G. Murray, E. Wimmer and O. Kew (1985b). *In* "Vaccines 85: Molecular and Chemical Basis of Resistance to Parasitic, Bacterial and Viral Diseases" (Eds R. A. Lerner, R. M. Channock and F. Brown), pp. 191–198. Cold Spring Harbor Laboratory.
35. Joklik, W. L. and J. E. Darnell (1961). *Virology* **13**, 439–447.
36. Kleid, D. G., D. Yansura, B. Small, D. Dowbenko, D. M. Moore, M. J.

Grubman, P. D. McKercher, D. O. Morgan, B. H. Robertson and H. L. Bachrach (1981). *Science* **214**, 1125–1129.
37. Lonberg-Holm, K. and B. E. Butterworth (1976). *Virology* **71**, 207–216.
38. Lonberg-Holm, K. and F. H. Yin (1973). *J. Virol.* **12**, 114–123.
39. Lonberg-Holm, K., L. B. Gosser and J. C. Kauer (1975). *J. Gen. Virol.* **27**, 329–342.
40. Le Bouvier, G. L. (1955). *Lancet* **ii**, 1013–1016.
41. Lund, G. A., B. R. Ziola, A. Salmi and D. G. Scraba (1977). *Virology* **78**, 35–44.
42. Makoff, A. J., C. A. Paynter, D. J. Rowlands and J. C. Boothroyd (1982). *Nucl. Acids Res.* **10**, 8285–8295.
43. Mayer, M. M., H. J. Rapp, B. Roizman, S. W. Klein, K. M. Cowan, D. Lukens, C. E. Schwerdt, F. L. Schaffer and J. Charney (1957). *J. Immunol.* **78**, 435–455.
44. Meloen, R. H., D. J. Rowlands and F. Brown (1979). *J. Gen. Virol.* **45**, 761–764.
45. Minor, P. D., G. C. Schild, J. Bootman, D. M. A. Evans, M. Ferguson, P. Reeve, M. Spitz, G. Stanway, A. J. Cann, R. Haptmann, L.-D. Clarke, R. C. Mountford and J. W. Almond (1983). *Nature (London)* **301**, 674–679.
46. Minor, P. D., D. M. A. Evans, M. Ferguson, G. C. Schild, G. Westrop and J. W. Almond (1985). *J. Gen. Virol.* **65**, 1159–1165.
47. Minor, P. D., M. Ferguson, D. M. A. Evans, J. W. Almond and J. P. Icenogle (1986). *J. Gen. Virol.* **67**, 1283–1291.
48. Palmenberg, A. C., E. M. Kirby, M. R. Janda, N. L. Drake, G. M. Duke, K. F. Potratz and M. S. Collett (1984). *Nucl. Acids Res.* **12**, 2969–2985.
49. Parry, N. R., E. J. Ouldridge, P. V. Barnett, D. J. Rowlands, F. Brown, J. L. Bittle, R. A. Houghten and R. A. Lerner (1985). *In* "Vaccines 85: Molecular and Chemical Basis of Resistance to Parasitic, Bacterial and Viral Diseases (Eds R. A. Lerner, R. M. Channock and F. Brown), pp. 211–216. Cold Spring Harbor Laboratory.
50. Pfaff, E., M. Mussgay, H. O. Bohm, G. E. Schultz and H. Schaller (1982). *EMBO J.* **1**, 869–874.
51. Rombaut, B., R. Vrijsen and A. Boeye (1983). *Arch. Virol.* **76**, 289–298.
52. Rossman, M. G., E. Arnold, J. W. Erickson, E. A. Frankenberger, J. P. Griffith, H. J. Hecht, J. E. Johnson, G. Kramer, Ming Luo, A. G. Mosser, R. R. Rueckert, B. Sherry and G. Vriend (1985). *Nature (London)* **317**, 145–153.
53. Rowlands, D. J., D. V. Sangar and F. Brown (1971). *J. Gen. Virol.* **13**, 85–93.
54. Rowlands, D. J., D. V. Sangar and F. Brown (1975). *J. Gen. Virol.* **26**, 227–238.
55. Rowlands, D. J., B. E. Clarke, A. R. Carroll, F. Brown, B. H. Nicholson, J. L. Bittle, R. A. Houghten and R. A. Lerner (1983). *Nature (London)* **306**, 694–697.
56. Sherry, B. and R. Rueckert (1985). *J. Virol.* **53**, 137–143.
57. Skern, T., W. Sommergruber, D. Blaas, P. Gruendler, F. Fraundorder, C. Pieler, I. Fogy and E. Kuechler (1985). *Nucl. Acids Res.* **13**, 2111–2126.
58. Stanway, G., P. J. Hughes, R. C. Mountford, P. D. Minor and J. W. Almond (1984). *Nucl. Acids Res.* **12**, 7859–7875.
59. Strohmaier, K., R. Franze and K. H. Adam (1982). *J. Gen. Virol.* **59**, 295–306.
60. Toyoda, H., M. Kohand, Y. Kataoka, T. Suganuma, T. Omata, N. Imuta and A. Nomoto (1984). *J. Mol. Biol.* **174**, 561–585.

61. van der Marel, P., T. G. Hazendonk, M. A. C. Henneke and A. L. Wezel (1983). *Vaccine* **1**, 17–22.
62. Wiegers, K. J. and R. Dernick (1983). *J. Gen. Virol.* **64**, 777–785.
63. Wychowski, C., S. van der Werf, O. Siffert, R. Crainic, P. Bruneau and M. Girard (1983). *EMBO J.* **2**, 2019–2024.

16. Enveloped Virus Entry

Mark Marsh

Institute of Cancer Research, Chester Beatty Laboratories, Fulham Road, London SW3 6JB, U.K.

I.	Introduction	281
II.	Structure of enveloped viruses	282
III.	The endocytic pathway	283
	A. Coated pits and coated vesicles	284
	B. Endosomes	285
IV.	Virus endocytosis	288
	A. Cell surface binding	288
	B. Receptor-mediated endocytosis	289
	C. Penetration	290
	D. Lysosomes	291
V.	Viral membrane fusion	292
	A. Influenza virus haemagglutinin(HA)	293
	B. SFV spike glycoprotein	295
VI.	The advantages of the endocytic pathway	297
VII.	Conclusions	297
	References	298

I. INTRODUCTION

Between 1981 and 1986 considerable progress has been made in understanding how animal viruses enter cells. Enveloped viruses must transfer their genome through two membranes, their own and that of the host cell, to gain access to the replicative apparatus of the cell. This task is accomplished by fusion between the viral membrane and a limiting membrane of the cell. For some viruses, such as the paramyxoviruses, the fusion reaction is independent of pH and may occur at the cell surface. With alpha-, orthomyxo-, bunya- and rhabdoviruses, membrane fusion is pH-dependent. The fusion reaction is triggered only at low pH and occurs, after endocytosis of intact virions, in acidic endocytic vesicles (for reviews see Poste and Pasternak, 1978; Dimmock, 1982; Lenard and Miller, 1982; Simons et al., 1982; White et al., 1983; Marsh, 1984; Kielian and Helenius, 1985 and Gonzalez-Scarano et al., 1984).

This chapter will consider our current understanding of the events in enveloped virus entry. Much of the discussion will concentrate on Semliki Forest virus (SFV, Alphaviridae), influenza virus (Orthomyxoviridae) and vesicular stomatitis virus (VSV, Rhabdoviridae). All of these viruses have been used extensively in entry studies and all exhibit pH-dependent fusion.

II. STRUCTURE OF ENVELOPED VIRUSES

In enveloped virus particles the genome is packaged, together with one or more proteins, into a nucleocapsid particle. The nucleocapsid is surrounded by a membrane, the envelope, that both protects the nucleic acid during the extracellular phase of the viral life cycle and mediates the receptor binding and fusion functions involved in entry. The membrane is a lipid bilayer, derived from the infected cell at the time of nucleocapsid budding. For SFV, influenza and VSV, budding occurs at the plasma membrane (see Wagner, 1975; Garoff et al., 1982; Compans and Choppin, 1975).

The lipids of the viral membrane are acquired from the host cell, but the membrane proteins are coded by the viral genome. The proteins are of two types: (a) the matrix, or M-proteins, and (b) the spike, or envelope, glycoproteins.

The matrix proteins are non-glycosylated polypeptides associated with the inner aspect of the membrane. M-proteins are only found in some enveloped viruses; they occur in the myxo- and rhabdoviruses, but not in the alphaviruses. They are thought to provide a link between the spike proteins and the nucleocapsids during budding (see Simons and Garoff, 1980).

The spike glycoproteins, in contrast, are ubiquitous in enveloped viruses and are amongst the best understood of all membrane proteins. The complete amino acid sequences of the spike glycoproteins of representatives of alpha-, orthomyxo-, retro- and rhabdoviruses, have been determined from complementary DNA (see White et al., 1983), and the three-dimensional structure of the influenza haemmagglutinin has been determined (Wilson et al., 1981). The spikes are transmembrane proteins that span the bilayer and have small internal domains. The majority of the protein, however, is presented on the surface of the virion as a spike that can be visualized in the electron microscope. The proteins are synthesized on membrane-bound polysomes and cotranslationally inserted into the rough endoplasmic reticulum where they are core glycosylated. The newly synthesized glycoproteins are transported to the Golgi apparatus, where they acquire fatty acids and complex oligosaccharides, and subsequently to the cell surface. For alphaviruses and orthomyxoviruses the proteins are proteolytically modified *en route* to the cell surface; e.g. influenza HA is cleaved to a disulphide bonded heterodimer (HA1-HA2) and the SFV spike glycoprotein P62 is cleaved to give the mature E2 and E3. At the cell surface the glycoproteins are incorporated into the membranes of budding virus particles (see White *et al.*, 1983; Garoff *et al.*, 1982).

Although significant differences exist in the organization and primary amino acid sequences of the various viral spike proteins, they all perform similar functions. (1) They specify the cellular site from which budding occurs and are involved in the mechanics of budding. (2) They are involved in receptor recognition, binding and receptor-dependent internalization of viruses. (3) They mediate membrane fusion.

III. THE ENDOCYTIC PATHWAY

Endocytosis is the process by which cells internalize their plasma membrane together with components of the extra-cellular environment. The mode of uptake for large particles (over 200 nm diameter) is an induced, receptor-dependent process termed phagocytosis, and is, in the main, a characteristic of cells termed phagocytes. Pinocytosis, the internalization of soluble ligands and small particles (less than 200 nm diameter) is, however, common to many cells and occurs through a constitutive endocytic pathway that is biochemically distinct from phagocytosis (see Steinman *et al.*, 1983).

Ligands are internalized through pinocytosis either non-specifically, as components of the medium (fluid-phase endocytosis), or specifically, through specialized receptors expressed on the cell surface (receptor-mediated or adsorptive endocytosis). In using receptors, ligands can be concentrated at the cell surface and taken up very efficiently (see Brown and Goldstein, 1979).

Pinocytosis normally functions to internalize many different ligands. Examples include nutrients and nutrient carriers, e.g. transferrin, low-density lipoprotein, intrinsic factor/transcobalamin; growth factors and effector molecules, e.g. polypeptide hormones, chemotactic peptides, immunoglobulins and lymphokines; modified serum proteins and components to be removed from the extracellular space, e.g. asialoglycoproteins, mannose and mannose-6-phosphate, labelled glycoproteins, alpha-2-macroglobulin/protease complexes. In addition to these physiological ligands, viruses and other pathogens (e.g. plant and bacterial toxins), which bind opportunistically to the cell surface, are internalized by the same endocytic mechanisms (see Steinman *et al.*, 1983).

A general understanding of the cellular pathways and compartments involved in fluid-phase/receptor-mediated endocytosis has come from studies with a number of different ligands. Receptors can be distributed randomly over the cell surface, in which case the ligand/receptor complexes relocate on the plasma membrane to the sites of internalization — the coated pits. Some specific receptors, e.g. the LDL receptor, have an affinity for coated pits, even in the absence of ligand (Brown and Goldstein, 1979; Hopkins, 1985). Bound ligands, receptors and fluid-phase components are internalized through coated pits in coated vesicles, delivered first to endosomes and, in some cases, to the lysosomes (see Fig. 16.1).

A. Coated Pits and Coated Vesicles

Coated pits are specialized cell surface domains. In fibroblasts they account for approximately 2% of the cell's surface area and appear to be the principal sites of pinocytic uptake. The characteristic coat structures

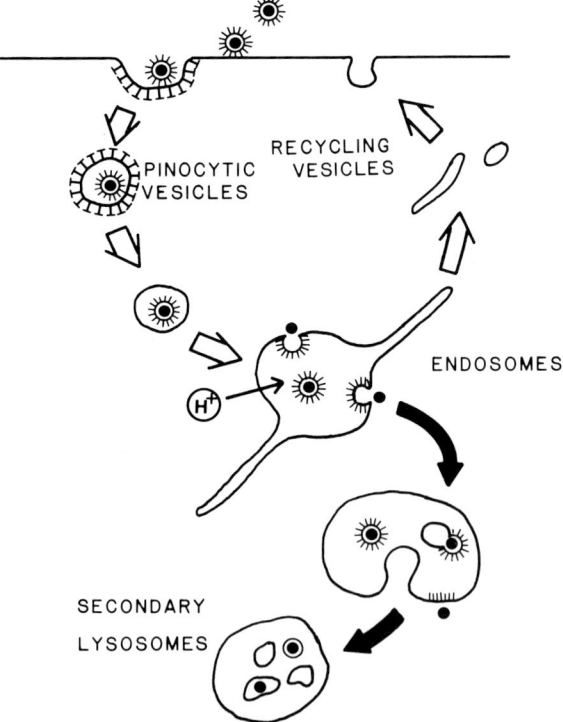

Figure 16.1. Schematic representation for SFV entry into BHK cells. SFV virions bind to the cell surface, are translocated to coated pits and coated vesicles, and internalized in coated vesicles. The internalized virions are delivered to endosomes where the low pH mediates fusion of the viral and endosomal membrane. Subsequently, non-fused virions and viral spike proteins are degraded in the lysosomes.

are located on the cytoplasmic aspect of the plasmalemma and are composed of a protein complex termed clathrin (Pearse, 1976). Coated pits invaginate to form coated vesicles and in the process internalize a portion of membrane together with ligands and medium contained in the pit. Following invagination, the coat is removed from the vesicle, potentially by an uncoating ATPase (Braell et al., 1984), and the clathrin is recycled to the plasma membrane to mediate subsequent rounds of internalization (Pearse, 1980).

Steinman et al. (1976) have estimated that L-cell fibroblasts and macrophages internalize the equivalent of 100% and 200% of their cell surface area respectively, and 4–8% and 25% of their volume respectively, each hour. In baby hamster kidney (BHK) fibroblasts about 2000 coated vesicles can form per cell per minute (Marsh and Helenius, 1980). Coated vesicle uptake is not induced by ligand binding and can account for most, if not all, of the measured membrane and fluid-phase uptake (Steinman et al., 1976; Marsh and Helenius, 1980). One consequence of this extensive internalization is that membrane must be recycled to the cell surface as cells do not have the capacity to replace the internalized membrane with newly synthesized material.

Coated pits show some selectivity for components to be internalized from the cell surface (see Bretscher, 1984). For example, the LDL and transferrin receptors appear to localize to coated pits whether occupied by ligand or not (Brown and Goldstein, 1979; Hopkins, 1985). It appears that most cell surface proteins and glycoproteins can be internalized (see Steinman et al., 1983). However, the internalization rates for different components, and the affects of specific and non-specific ligands (e.g. lectins or antibodies) on the internalization, remain unclear. Nevertheless, different ligands and receptors are internalized in the same coated pits and coated vesicles and delivered to the same endosomes (Dickson et al., 1981; Geuze et al., 1983).

B. Endosomes

Following internalization, coated vesicles deliver their contents within minutes to endosomes (see Helenius et al., 1983; Pastan and Willingham, 1983). Delivery occurs through fusion of decoated vesicles either with existing endosomes or with other uncoated vesicles. Endosomes are the first organelles of the endocytic, vacuolar apparatus. They are distinct from lysosomes by morphological criteria and, as their buoyant density is less than that of lysosomes, the two can be separated by density gradient centrifugation. The endosomes do not contain significant amounts of the acid hydrolases characteristic of lysosomes.

Within endosomes, internalized ligands and membrane are sorted to appropriate sites within the cell. The sorting process is dependent on the properties of individual ligand/receptor complexes and is influenced, at least in part, by the acidic environment in endosomes. At low pH many ligands dissociate from their receptors, e.g. LDL/LDL-receptor, asialoglycoproteins/ASGP-receptor. In these cases, the internalized receptors can be recycled to the cell surface and reutilized, while the ligands are directed to lysosomes and degraded. Not all receptors and ligands dissociate at low pH. If pH-insensitive ligand/receptor complexes are cross-linked by a multivalent ligand, e.g. immune complexes/Fc-receptor, both the ligand and receptor are delivered to lysosomes. However, with a monovalent ligand such as a Fab fragment of an antibody to Fc-receptor, the complex can recycle to the cell surface (Mellman and Plutner, 1984; Mellman et al., 1984; Ukkonen et al., 1986). Internalized transferrin/transferrin receptor complexes also recycle to the cell surface, though at low pH iron dissociates from transferrin and is retained by the cell (Dautry-Varsat et al., 1983; van Renswoude et al., 1983). In epithelial cells a further alternative exists. Ligands such as IgG, thyroglobin or IgA, after entering endosomes, are transported through the cell and released at another surface domain (Renston et al., 1980; Abrahamson and Rodewald, 1981; Herzog, 1983).

Low pH is an important component of the endosome sorting process. At least two compartments of the endocytic pathway, endosomes and lysosomes, have mildly acidic pH (pH 5.3–pH 6.0 and pH 4.8–pH 5.0 respectively), and these organelles, together with coated vesicles, contain membrane-associated proton-translocating ATPases (Tycko and Maxfield, 1982; Galloway et al., 1983; Merion et al., 1983; Forgac et al., 1983; Stone et al., 1983; Reeves, 1983). Carboxylic ionophores (e.g. monensin, nigericin) and weak bases (e.g. ammonium chloride, chloroquine) dissipate the proton gradients of acidic organelles (Ohkuma and Poole, 1978) and inhibit the dissociation of low-pH-sensitive receptor ligand complexes, the delivery of ligands to lysosomes and the recycling of receptors to the cell surface (Gonzalez-Noriega et al., 1980; Basu et al., 1981).

Most of the membrane components internalized during endocytosis are recycled to the cell surface. Recently we have studied the structural organization of endosomes using morphological techniques and these studies suggest how membrane may be recycled. Endosomes can be labelled by incubating cells for short times with markers internalized by fluid phase endocytosis. We have used horseradish peroxidase (HRP), which can be developed to give an electron-dense reaction product. In semi-thin section (0.25 µm) electron micrographs the endosomes are visualized as complex organelles comprising two components: a vacuolar body, 200–500 nm in diameter; and a number of associated 50–100 nm diameter tubules (Fig. 16.2). Morphometric estimates indicate that over 70% of the endosome membrane is associated with the tubules, while over 70% of the volume is contained within the vacuoles (Marsh et al., 1985). The tubules fulfil the requirement for recycling vehicles; a large surface-area-to-volume ratio for returning membrane, but not content, to the cell surface (Duncan and Pratten, 1977). Geuze et al. (1983) and Hopkins (1983) have also provided some evidence from labelling studies that these tubules are involved in recycling specific receptors.

16. Enveloped Virus Entry

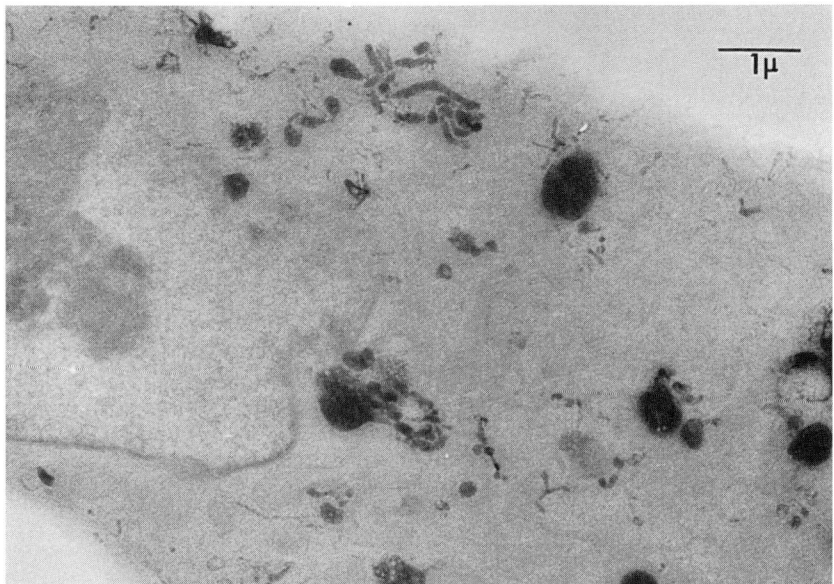

Figure 16.2. Semi-thin section of BHK-21 cell endosomes. BHK-21 cells were labelled with horseradish peroxidase for 15 min at 37°C, conditions under which the label is confined to the endosome compartment (see Marsh et al., 1986). The cells were washed, fixed, reacted with diaminobenzidine and processed for electron microscopy. Semi-thin (0.25 µm) sections were cut and viewed without staining. The HRP/DAB reaction product is seen in morphologically complex organelles which consist of a central vacuolar portion and a number of associated tubules.

Lysosomes can also be labelled with HRP. By incubating the pulsed labelled cells in HRP-free medium for 2 h, the label can be chased to the lysosomes. In the electron microscope the organelles appear as spherical, multivesicular organelles that do not have the extensive tubular components characteristic of endosomes (Marsh et al., 1985). If the membrane tubules associated with endosomes mediate membrane recycling, the absence of tubules on lysosomes suggests that recycling occurs prior to the delivery of membrane to lysosomes. The transport of ligands from endosomes to lysosomes is blocked at low temperatures (18–20°C) and internalized ligands are retained in the endosomes (Dunn et al., 1980; Marsh et al., 1983b).

The coated-vesicle-mediated endocytic pathway effects both constitutive fluid-phase and receptor-mediated (adsorptive) endocytosis. The considerable endocytic membrane traffic enables the cell continually to monitor the integrity and occupancy of cell surface components, to modulate cell surface receptor and antigen expression, and to redistribute cell surface components and associated ligands to different cellular compartments or surface domains.

IV. VIRUS ENDOCYTOSIS

Enveloped viruses deliver their genome to the cytosol following membrane fusion. For many of the viruses, the fusion reaction is triggered only at low pH. The viruses therefore require access to a suitable acidic compartment; by exploiting the properties of the endocytic pathway they can gain entry to the acidic endosome compartment.

A. Cell Surface Binding

Binding is believed to be the initial obligatory step in virus entry and it involves interaction between the viral spike glycoproteins and components of the cell surface. In general, the cell surface binding sites for viruses are not well characterized, but viruses clearly bind opportunistically to components expressed on uninfected cells. Specificity for different binding sites may exist and the restricted tissue distribution of these specific receptors can be the cause of viral tissue tropisms. An important feature of the receptors is that they should be able to mediate virus internalization (see Lonberg-Holm and Phillipson, 1979).

SFV can infect cells from such diverse sources as mammals and insects. The binding sites for this alphavirus are protease-sensitive glycoproteins (Helenius et al., 1980). With murine and human lymphoblastoid cells, SFV has specificity for major histocompatibility (MHC) antigens (Helenius et al., 1978). The virus can, however, infect cells that do not express MHC antigens (Oldstone et al., 1980), and clearly shows flexibility in its receptor specificity. Binding of SFV probably occurs through multiple-spike glycoprotein/receptor interactions; the avidity of viral binding to BHK cells is high (10^{11}–10^{10} mol^{-1}) whereas that of isolated spike glycoprotein complexes is known to be much lower (10^7 mol^{-1}: Fries and Helenius, 1979). On spatial grounds it seems that the intact virion can bring about 20 of its spike glycoproteins close to the cell surface (Marsh et al., 1983a). On BHK cells, SFV binds over the entire cell surface, preferentially on the microvilli, and occasionally in coated pits. Fusion figures are not observed at the cell surface unless the pH of the medium is reduced below pH 6.0 (Helenius et al., 1980; White et al., 1980).

The haemmagglutinin on myxoviruses binds to terminal sialic acid residues on cell surface glycoproteins and glycolipids. While binding is specific for sialic acid, the virus can clearly utilize many different oligosaccharides on cell surfaces, and consequently has a broad species and host-cell range. The disadvantage of using sialic acid as a binding site is that it is expressed on non-endocytic cells, for example on erythrocytes. To overcome this problem none of the myxoviruses have a spike glycoprotein with neuraminidase activity, to release virions from cells where they fail to be internalized, and give the opportunity to find a suitable host cell.

The broad binding specificity exhibited by SFV and influenza viruses will decrease the chances that mutations arising in specific receptors may render cells resistant to infection. However, not all enveloped viruses exhibit the broad binding range of the alpha- and myxoviruses. Genetic studies demonstrate that retroviruses, for example, show considerable receptor specificity and, as a consequence of restricted receptor expression, tissue tropisms (see Weiss, 1981). Recently, the specific receptors for several viruses have been identified; rabies virus, a rhabdovirus, binds to the nicotinic acetylcholine receptor (Lentz et al., 1982), and HTLV-3 appears to bind to the T4 antigen on T-cell helper subsets (Klatzmann et al., 1984; Dalgleish et al., 1984).

Antibodies can also facilitate viral binding. Thus antibodies against spike glycoproteins of alpha-, flavi-, bunya-, and myxoviruses can bridge virions to Fc-receptors on macrophages and lymphocytes and bring about their infection (see Millican and Porterfield, 1982; Peiris and Porterfield, 1982). Furthermore, viral spike glycoproteins may themselves function as viral receptors. VSV and SFV normally infect polarized MDCK-cells through the basolateral domain. However, MDCK-cells infected with influenza virus, and expressing the influenza HA on the apical domain, can be superinfected with VSV or SFV applied on the apical side. The HA binds sialic acid residues on the VSV and SFV spike glycoproteins and mediates the internalization of these other viruses. Similarly, influenza virus HA can function as a receptor for murine hepatitis virus, a coronavirus, in BHK cells; these cells do not normally express a receptor for murine hepatitis virus and are resistant to its infection (Fuller et al., 1985).

B. Receptor-mediated Endocytosis

Cell-surface-bound viruses are rapidly internalized through receptor-mediated endocytosis. Alpha-, rhabdo-, retro- and myxoviruses, have been observed in endocytic coated pits and coated vesicles (see Dales, 1973; Dimmock, 1982, Superti et al., 1984). As with the receptor-mediated endocytosis of physiological ligands, SFV endocytosis is inhibited at 4°C, partially inhibited by sodium azide and dinitrophenol but not by 2-deoxy-D-glucose. It is unaffected by colcemid and only partially inhibited by very high concentrations of cytochalasin B, both reagents that disrupt cytoskeletal function. In addition, it is not affected by weak bases and carboxylic ionophores (Marsh and Helenius, 1980).

The kinetics of virus internalization have been documented for SFV in BHK-cells, and for orthomyxoviruses and rhabdoviruses in MDCK-cells (Marsh and Helenius, 1980; Matlin et al., 1981, 1982; Yoshimura et al., 1982). It has been found that more than 50 000 SFV particles can be bound per BHK-cell at 0°C. On warming to 37°C, the bound viruses are internalized at a rate of 2000 particles per minute. The half-time for virions on the cell surface is in the order of 10–15 min. Influenza and VSV uptake on MDCK-cells is slower, with cell surface half-times of about 30 min (Matlin et al., 1981, 1982).

The internalization of viruses tends to be slower than that of physiological ligands. The rate of internalization may vary for different virus binding moieties, different cell surface components being internalized with different efficiencies (see above). Alternatively, if individual spike glycoprotein–receptor interactions have low affinity, the contacts may continually break and reform. Although the viruses would remain bound to the cell surface, the rate of translocation to coated pits could be slower (see Bretscher, 1984). The size of the virus particles may also limit the entry of virions into coated pits and reduce the rate of internalization. SFV virions are 75 nm in diameter and fit easily into 100 nm diameter coated vesicles (the average size of endocytic coated vesicles). However, larger viruses (influenza, 100 nm diameter, and VSV, 150 nm long) may not be accommodated with such ease in coated vesicles, and may be restricted to the larger vesicles.

Internalized virions are rapidly delivered to endosomes. The viruses can be seen within a few minutes in large, 200–500 nm, vesicles that, unlike multivesicular bodies or lysosomes, appear devoid of content. After longer incubation times (10–15 min), the endosomes containing viruses frequently have included vesicles (multivesicular bodies). The intact virions are generally restricted to the vesicular component of the endosomes (see Marsh et al., 1986).

Although internalized virus particles are within the confines of the cell membrane, they remain topologically outside the cytoplasm and the nucleocapsids are still separated from the cytosol by two membranes. However, the acidic pH in endosomes triggers the fusion activity of the pH-dependent viruses and the viral nucleic acid is delivered (penetration) to the cytosol for uncoating and replication.

C. Penetration

The kinetics of penetration for different enveloped viruses vary according to the pH required for fusion, the rate of delivery to endosomes and the rate of endosome acidification. SFV fuses at pH 6.0; penetration of this virus in BHK cells occurs rapidly after internalization with a lag of 1–2 min and a half-time for internalization and uncoating of 15 min (Marsh et al., 1983b). The kinetics of penetration for other enveloped viruses are not well documented. However, using a fusion mutant of SFV we have been able to study the uncoating of viruses that fuse at lower pHs. The fusion mutant, *Fus*-1, fuses at pH 5.5–5.3. This virus is bound and internalized with similar kinetics to the wild type, and penetration occurs from the endosomes (Kielian et al., 1984). However, the half-time for internalization and uncoating is extended from 15 to 45 min. The mutant is retained, unfused, within the endosome for a longer time until the endosome is acidified to the pH required for fusion. This observation is significant: it argues for heterogeneity in endosomes on the basis of pH and demonstrates that endosomes, and the endocytic pathway, are progressively acidified (Kielian et al., 1984). The pH-dependent enveloped viruses fuse at pH values between pH 6.5 and 5.3 and are thus able to gain access to the cytosol from the endosomes.

The proton gradient across the endosome membrane is dissipated by weak bases and carboxylic ionophores (see Ohkuma and Poole, 1978). As a consequence, the bases ammonium chloride, chloroquine, methylamine and amantadine inhibit infection by alpha-, myxo-, bunya-, flavi-, retro- and rhabdoviruses; and the ionophores, monensin and nigericin, inhibit infection by alpha- and rhabdoviruses (see Marsh, 1984). The reagents do not affect the binding, internalization or intracellular routing of SFV with BHK cells, nor do they affect the SFV fusion activity as such. Under experimental conditions SFV can be induced to fuse at the cell surface, or with liposomes, by lowering the pH of the medium. This fusion will still occur in the presence of the inhibitors providing the pH is reduced below the pH value for fusion (White and Helenius, 1980; Marsh et al., 1983b). The reagents do, however, inhibit the penetration event by increasing the pH in endosomes. The weak bases and ionophores both block infection at the same step in entry (Marsh et al., 1983), and the concentrations of the inhibitors required to inhibit different viruses reflect the fusion pH thresholds of these viruses (see Marsh and Helenius, 1984).

It is significant that infection by paramyxoviruses is also inhibited by the weak bases. The fusion activity of Sendai virus, for example, is pH-independent, and infection can potentially occur through the plasmamembrane. However, the kinetics of fusion are slow and cell-surface-bound viruses are rapidly internalized in coated vesicles (see Dales, 1973). Although the pH in endosomes should not affect fusion, the weak bases do inhibit infection. Lenard and Miller (1982) have suggested that low pH may be important for aspects of uncoating other than fusion.

Helenius et al. (1982) have used ammonium chloride to demonstrate that it is internalized viruses that infect cells. SFV can be internalized into BHK-cells in the presence of ammonium chloride. After non-internalized viruses are removed, the block can be reversed and infection will ensue. The infection can only result from those viruses internalized during the incubation with ammonium chloride. The acidification of endosomes, on reversing the ammonium chloride block, is relatively synchronous. In such cells, fusion profiles between internalized SFV and endosome membranes can be seen by electron microscopy (Helenius, 1984).

D. Lysosomes

Internalized viruses and viral components are delivered to lysosomes and degraded. The SFV glycoprotein E2 is highly sensitive to degradation at pH 7.0 or pH 5.0 (E1 is resistant to trypsin at pH 5.0: Kielian and Helenius, 1985). E2 is degraded rapidly after delivery to lysosomes (Marsh and Helenius, 1980). The low-molecular-weight products of degradation are released into the medium and serve as an indicator of delivery to lysosomes.

Lysosomes have an acidic pH and have been considered as a site for virus fusion (Helenius et al., 1980; Marsh et al., 1982a). The kinetics of SFV delivery to lysosomes has been studied in BHK-cells by cell fractionation and the release of degraded activity. Delivery of virus to lysosomes commences 20 min after internalization and occurs with a half-time of about 1 h. The time course suggests that viruses can remain in the endosome long enough to effect penetration. Thus the viral components delivered to lysosomes are either the spike glycoproteins inserted in the endosome membrane, their functions of binding and fusion having been completed, or viruses that have failed to fuse in endosomes.

At 20°C the transfer of SFV from endosomes to lysosomes is completely inhibited (Marsh et al., 1983b; Dunn et al., 1980). Inhibition is reversed on warming to 37°C, and degradation commences within minutes. The block occurs just prior to the delivery of ligands to lysosomes. Infection of cells can occur at these low temperatures, indicating that fusion occurs in the endosomes (Marsh et al., 1983b; Keilian et al., 1984).

The experiments reviewed above strongly support the notion that enveloped viruses infect cells through acidic endocytic vesicles. Although the evidence is compelling, there are objections to this interpretation (see Cassell et al., 1984). The major criticism is that the high multiplicities of viruses used in morphological and some biochemical experiments are not representative of the normal pathway leading to productive infection, However, SFV has a high pfu : particle ratio (approximately 1 : 3), and can be labelled to high specific activity and used to quantitate endocytosis over a range of multiplicities. With less than 1 particle per cell, or more than 50 000 particles per cell, the uptake, uncoating and intracellular processing of SFV occurs with similar efficiency and kinetics (Helenius et al., 1980; Marsh and Helenius, 1980; White and Helenius, 1980; Marsh et al., 1983). The biochemical and morphological studies have been paralleled with biological assays for infection and the inhibitor studies. The results all indicate that the cellular fate of individual virions is the same regardless of multiplicity, and that the pathway outlined by the biochemical and morphological studies does indeed reflect the infective pathway.

V. VIRAL MEMBRANE FUSION

The critical event in penetration is the fusion reaction. Helenius et al. (1980) first described the pH-dependence of this reaction with alphaviruses. Subsequently, pH-dependence has been demonstrated for orthomyxo-, rhabdo- and bunyaviruses and pH-independence for the paramyxoviruses (see White et al., 1983, Gonzalez-Scarano et al., 1984). The fusion event is mediated by the viral spike glycoprotein. Vesicles reconstituted with spike proteins or cells expressing cloned viral spike protein genes have been used to show that the spike glycoproteins are the only viral components required for membrane fusion (Marsh et al., 1983a; White et al., 1983; Florkiewiez and Rose, 1984). However, structural studies and amino acid sequences show the spike glycoproteins of different viruses to vary considerably. Three basic types of viral fusion protein have been distinguished (White et al., 1983).

(1) *Fusion proteins with N-terminal hydrophobic sequences.* The paramyxovirus fusion protein, the orthomyxovirus haemagglutinin HA2 portion and the envelope proteins of murine retroviruses all have a hydrophobic N-terminus. The hydrophobic terminus is derived by cleavage of a precursor protein that lacks the fusion activity. Although activating cleavages are required, these viruses show both pH-dependent (orthomyxo- and retroviruses) and pH-independent (paramyxoviruses) fusion.

(2) *Fusion proteins with internal hydrophobic sequences.* The three sequenced alpha-viruses (Ross River, SFV and Sindbis virus) have a highly conserved hydrophobic domain (non-transmembrane) between amino acids 80 and 96 of the E1 glycoprotein. These viruses contain two membrane-anchored, non-covalently associated glycoproteins, which have acid-dependent fusion activity but do not require activating cleavages.

(3) *Fusion proteins without apparent hydrophobic fusion sequences.* The rhabdoviruses, of which VSV and rabies virus have been sequenced, show low pH-dependent fusion activity. The spike glycoprotein, a single glycopolypeptide, does not require activating cleavages and there are no clear hydrophobic domains in the amino acid sequence other than the transmembrane domain.

For membrane fusion two opposing membranes must approach close enough to allow coalescence of the bilayers (White et al., 1983). Normally the hydration forces associated with these membranes keep the membranes separated by at least 30 nm. The function of the viral fusion proteins may be to overcome these hydration forces. For three groups of viruses (orthomyxo-, alpha- and bunyaviruses) it is now clear that acid conditions induce conformational changes in the fusion proteins (see White et al., 1983; Kielian and Helenius, 1985; Edwards et al., 1983; Gonzalez-Scarano, 1985). These changes will be discussed for influenza virus and SFV.

A. Influenza virus haemagglutinin (HA)

The influenza HA is the best characterized of the viral fusion proteins. It is synthesized as a precursor polypeptide (HA0) and assembled into non-covalently linked trimers. In the electron microscope the trimers are visible as 13 nm spikes projecting from the surface of virions, with about 1000 copies of HA per virion. During transport to the cell surface HA0 is proteolytically modified to give the mature disulphide-bonded heterodimers (HA1–HA2) found in the virus. This cleavage is essential for activating the fusion proteins and rendering the virus infective. Some tissue culture cell lines, e.g. MDCK, lack the proteolytic activity required for cleavage. These cells produce non-infective influenza virions lacking fusion activity. The fusion proteins in these virions can be activated and the virions rendered infective by treatment with trypsin. The new *N*-terminal of HA2 has a highly conserved sequence of 24 non-polar amino acids with only two or three polar residues; this sequence has homology with a similarly formed *N*-terminal sequence in the paramyxovirus fusion (F) glycoprotein. The properties of fusion have been studied with an avian influenza virus, fowl plague virus. The fusion with liposomes occurs only below pH 5.5, is completed within 1 min at 37°C, is more than 60% efficient, and is non-leaky (White et al., 1982).

The HA trimer can be isolated from the virus with bromelain. The enzyme cleaves very close to the membrane and releases the HA ectodomain as a water-soluble fragment that retains the trimeric organization (BHA). The three-dimensional structure of the BHA is roughly rectangular and comprises a globular head domain composed of HA1 polypeptide supported by a stalk composed of HA2 (Wilson et al., 1981). The cleavage point between HA1 and HA2 is in the stalk region of the molecule close to the viral membrane. The sialic acid binding site is on the distal globular head formed by HA1, and the *N*-terminal of HA2 is buried within the stalk.

The fusion activity for the influenza viruses varies between pH 6.0 and pH 5.3 depending on the virus strain (see White et al., 1983). At low pH an irreversible conformational change occurs in the BHA. Experimentally the conformational change is detected by alterations in (1) spectral and antigenic properties, (2) the sensitivity of the polypeptides to proteases and reducing agents, and (3) the increased hydrophobicity of the molecule. These latter amphiphilic properties of low-pH-treated BHA are indicated by its ability to bind non-ionic detergent and phospholipid vesicles and to form "rosettes" (Skehel et al., 1982; Doms et al., 1985; Graves et al., 1983). The changes are irreversible and reflect a reorganization of the trimer to expose a hydrophobic domain, presumably the *N*-terminal of HA2. From the three-dimensional structure of HA, the location of the trypsin cleavage sites and the disulphide bonds, it is apparent that these low-pH-induced rearrangements must involve extensive breakage of intra- and inter-subunit contacts. Such changes would be necessary to move the HA1 domains aside and permit the *N*-terminus of HA2 to interact directly with the target membrane (Doms et al., 1985).

Daniels et al. (1985) have sequenced the HA from a number of influenza virus mutants selected for growth in amantadine. The mutants have a pH threshold for fusion above that of the wild-type virus, and the DNA sequences of the HAs show that each mutant has base changes that result in one or two amino acid substitutions. The substitutions fall into two categories: those that destabilize the unexposed location of the hydrophobic *N*-terminal of HA2, and those that affect salt bridges and other contacts in the interfaces between the subunits of the HA trimer. The mutations appear to alter the interactions between subunits of the HA trimer, and to lower the energy barrier for the transition to the fusion-active state. In other words, the mutations reduce the concentration of protons required for the conformational change and raise the pH threshold for fusion.

Although acid-treated BHA will bind to membranes, it does not itself induce fusion. Only when anchored in a virus or cell membrane, through the transmembrane domain, can the N-terminal of HA2 form a bridge to the target membrane. Doms et al. (1985) have shown that the pH-induced changes characterized in BHA reflect low-pH-induced changes in the intact HA.

Despite the detailed studies with HA and BHA it remains unclear how the protein induces fusion of two opposing membranes. The HA may be required to overcome the hydration forces and enable the membranes to approach close enough to fuse spontaneously. Alternatively, HA may act directly on the target membrane to induce fusion by, for example, disruption of the lipid bilayers (Doms et al., 1985).

B. SFV spike glycoprotein

The sequence and organization of alphavirus spike proteins, and SFV in particular, are very different from that of HA and the three-dimensional structure of the proteins remains to be established. However, the membranes of these viruses have multiple copies (240) of a single spike glycoprotein complex that, like HA, mediates both binding and fusion. Each spike is a heterotrimer of three non-covalently associated glycoproteins E1, E2 and E3. E1 and E2 span the membrane. E2 and E3 are derived by proteolytic cleavage of a precursor protein P62. None of the proteins have hydrophobic N-terminal sequences, but E1 has a 16-amino-acid internal hydrophobic sequence between amino acids 80 and 96.

The properties of the SFV fusion reaction have been studied with both liposomes and cell membranes. The reaction occurs only at or below pH 6.2. It is rapid (90% fusion in 1 min), efficient (over 90% of viruses will fuse), non-leaky, only moderately reduced at 0°C, and is independent both of divalent cations and of the phospholipid composition of the membrane. However, with SFV the fusion is absolutely dependent on the presence of cholesterol in the target membrane (White and Helenius, 1980; White et al., 1980). Cholesterol analogues will substitute for cholesterol, but only those with a 3B-0H group (Kielian and Helenius, 1984).

As with HA, the spike glycoproteins of SFV undergo an irreversible conformational change at low pH. The change is reflected in the protease sensitivity of the spike glycoproteins. The neutral forms of E1 and E2 are resistant to trypsin treatment at 0°C; within 10 sec of incubation at pH 5.0 the E2 polypeptide is converted to a trypsin-sensitive form, whereas E1 remains trypsin-resistant (Fig. 16.3). At 37°C the neutral forms of both E1 and E2 are trypsin-sensitive; however, after treatment at pH 5.0, E1 becomes resistant. The changes are relevant to fusion because (1) the pH dependence and kinetics of the changes are similar to those of the fusion reaction, (2) with water-soluble forms of the spike, released from the membrane with proteinase K, the change to a trypsin resistant conformation for E1 is dependent on both low pH and cholesterol.

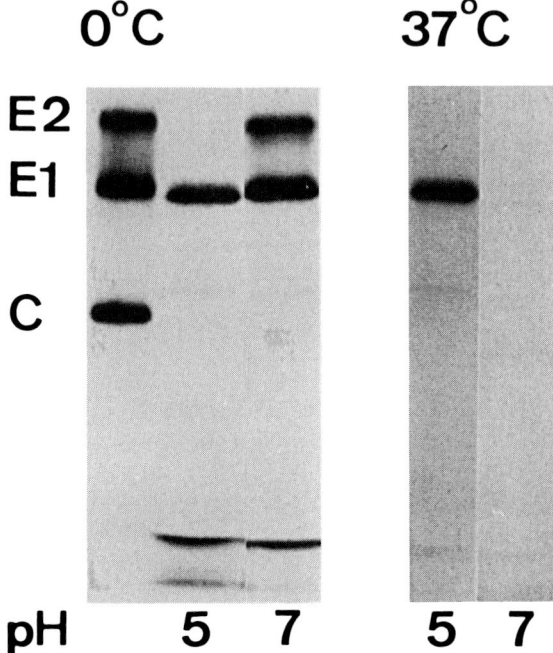

Figure 16.3. Trypsin sensitivity of SFV proteins after low-pH treatment. Intact SFV was incubated at pH 5 or pH 7 for 10 min at 37°C. The virus was neutralized and then digested with 100 μg ml^{-1} trypsin in 1% Triton X-100 for 10 min at 0°C or 37°C. The reaction was terminated with soy-bean trypsin inhibitor and the samples were analysed by gel electrophoresis. The control lane (extreme left) shows three of the viral proteins; E1 and E2 are the spike glycoproteins and C is the capsid protein. After low-pH treatment E2 becomes sensitive to protease treatment at 0°C, whereas E1 becomes resistant to the protease at 37°C. (Figure courtesy of Margaret Kielian, Department of Cell Biology, Yale University Medical School.)

Significantly, it is E1 that has the hydrophobic sequence between amino acids 80 and 96.

Similar changes occur in the proteins of a fusion mutant, *Fus* 1, but, in parallel with the shift in the pH for fusion, the pH for the conformational change is also altered. This mutant fuses at half a pH unit lower than the wild type, i.e. pH 5.3–5.5 (Kielian *et al.*, 1984; Kielian and Helenius, 1985).

In conclusion, although low pH is a common trigger for viral fusion it appears that the molecular mechanisms involved in the fusion reaction can vary. Variations in the structure of the fusion proteins and in the properties of the fusion reactions of various virus groups indicate that fusion induced by the three groups of spike proteins probably differs mechanistically. This notwithstanding, the different mechanisms achieve the same result—the viruses overcome the hydration forces that keep the membranes apart and the viral and target membranes fuse.

VI. THE ADVANTAGES OF THE ENDOCYTIC PATHWAY

The most direct route into a cell for a virus is penetration through the plasma membrane, but in most cases the viruses utilize the intracellular pathway. For the virus this clearly provides a number of advantages.

(1) The requirements for endocytosis and low-pH fusion ensure entry into metabolically active cells capable of constitutive endocytosis and ATP-dependent acidification and potentially able to support replication. Some of the enveloped viruses, the myxoviruses, have a neuraminidase glycoprotein to release themselves from cells when binding fails to result in internalization.

(2) If viruses fuse at the cell surface, the spike proteins will be inserted into the plasma membrane and can elicit an immune response. By fusing within the cell, viruses remain concealed until infection is established and the newly synthesized viral proteins reach the cell surface.

(3) Paramyxovirus fusion is pH-independent, but the infection is still inhibited by weak bases. Low pH may be required for uncoating events other than fusion (Lenard and Miller, 1982). Furthermore, an amantadine analogue, rimantadine, appears to block influenza virus infection by preventing the release of M-protein during viral uncoating (Bukrinskaya et al., 1982).

(4) Endocytosis may be required to deliver the viral capsids to specific regions of the cytoplasm for replication. Low-pH fusion of SFV at the surface of BHK cells can lead to infection (White et al., 1980). However, when VSV and fowl plague virus are fused into MDCK-cells infection does not occur, though these cells can be infected through the endocytic pathway (Matlin et al., 1981, 1982).

(5) The endocytic pathway may facilitate virus dispersal in the host. Epithelial and endothelial cells are able to transcytose media and ligands (see Simons and Fuller, 1985). By using receptors, such as secretory component, and insufficiently acidified transport pathways, viruses could be transported across epithelia without a cycle of replication.

VII. CONCLUSIONS

The mechanism by which enveloped viruses enter their host cells is now understood in some detail. The studies demonstrate that a variety of different enveloped viruses exploit the same cellular properties for entry. The hope is that by making detailed studies of the endocytic pathway and virus fusion we will not only understand the entry process, but be better able to design reagents to inhibit specific steps in the entry process. Furthermore, as the viruses exploit constitutive host-cell functions to replicate we can, in turn, exploit the viruses to study and understand the properties of these processes.

ACKNOWLEDGEMENTS

I wish to thank John Marbrook, Tony Davies and Sally Swift for critically reading the manuscript, Margaret Kielian for providing Fig. 16.3, Gareth Griffiths for assistance with Fig. 16.2 and the Cancer Research Campaign and Medical Research Council for financial support.

REFERENCES

1. Abrahamson, D. R. and R. Rodewald (1981). *J. Cell Biol.* **91**, 270–280.
2. Basu, S. K., J. L. Goldstein, R. G. W. Anderson and M. S. Brown (1981). *Cell* **24**, 493–502.
3. Braell, W. A., D. M. Schlossman, S. L. Schmid and J. E. Rothman (1984). *J. Cell Biol.* **99**, 734–7.
4. Bretscher, M. S. (1984). *Cell* **38**, 3–4.
5. Brown, M. S. and J. L. Goldstein (1979). *Proc. Natl Acad. Sci. USA* **76**, 3330–3337.
6. Bukrinskaya, A. G., N. K. Vorkunova, G. V. Kornilayeva, R. A. Narmanbetova and G. K. Vorkunova (1982). *J. Gen. Virol.* **60**, 49–59.
7. Cassell, S., J. Edwards and D. T. Brown (1984). *J. Virol.* **52**, 857–864.
8. Compans, R. and P. Choppin (1975). *Comp. Virol.* **4**, 179–252.
9. Dales, S. (1973). *Bacteriol. Rev.* **37**, 103–135.
10. Dalgleish, A. G., P. C. L. Beverley, P. R. Clapham, D. R. Crawford, M. F. Greaves and R. A. Weiss (1984). *Nature (London)* **312**, 763–766.
11. Daniels, R. S., J. C. Downie, A. J. Hay, M. Knossow, J. J. Skehel, M. L. Wang and D. C. Wiley (1985). *Cell* **40**, 431–439.
12. Dautry-Varsat, A., A. Ciechanover and H. F. Lodish (1983). *Proc. Natl Acad. Sci. USA* **80**, 2258–2262.
13. Dickson, R. B., M. C. Willingham and I. Pastan (1981). *J. Cell Biol.* **89**, 29–34.
14. Dimmock, N. J. (1982). *J. Gen. Virol.* **59**, 1–22.
15. Doms, R. W., A. Helenius and J. White (1985). *J. Biol. Chem.* **5**, 2973–2981.
16. Dunn, W. A., A. L. Hubbard and N. N. Aronson (1980). *J. Biol. Chem.* **255**, 5971–5978.
17. Duncan, R. and M. K. Pratten (1977). *J. Theor. Biol.* **66**, 727–735.
18. Edwards, J., E. Mann and D. T. Brown (1983). *J. Virol.* **45**, 1090–1097.
19. Florkiewiez, R. Z. and J. K. Rose (1984). *Science* **225**, 721–723.
20. Forgac, M., L. Cantley, B. Wiedenmann, L. Altsteill and D. Branton (1983). *Proc. Natl Acad. Sci. USA* **80**, 1300–1303.
21. Fries, E. and A. Helenius (1979). *Eur. J. Biochem.* **97**, 213–220.
22. Fuller, S. D., C.-H. von Bonsdorff and K. Simons (1985). *EMBO J.* **4**, 2475–2485.
23. Galloway, C. J., G. E. Dean, M. Marsh, G. Rudnick and I. Mellman (1983). *Proc. Natl Acad. Sci. USA* **80**, 3334–3338.
24. Garoff, H., C. Kondor-Koch and H. Reidel (1982). *Current Topics Microbiol. Immunol.* **99**, 1–50.
25. Geuze, H. J., J. M. Slot, G. A. M. Strous, H. F. Lodish and A. L. Schwartz (1983). *Cell* **32**, 277–287.
26. Gonzalez-Noriega, A., J. H. Grubb, V. Talkad and W. Sly (1980). *J. Cell Biol.* **85**, 839–852.
27. Gonzalez-Scarano, F. (1985). *Virology* **140**, 209–216.

28. Gonzalez-Scarano, F., N. Pobjecky and N. Nathanson (1984). *J. Gen. Virol.* **132**, 222–225.
29. Graves, P. N., J. L. Schulman, J. F. Young and P. Palese (1983). *Virology.* **126**, 106–116.
30. Helenius, A. (1984). *Biol. Cell* **51**, 181–186.
31. Helenius, A., B. Morein, E. Fries, K. Simons, P. Robinson, V. Schirrmacher, C. Terhorst and J. L. Strominger (1978). *Proc. Natl Acad. Sci. USA* **75**, 3846–3850.
31a. Helenius, A., M. Marsh and J. White (1982). *J. Gen. Virol.* **58**, 47–61.
32. Helenius, A., J. Kartenbeck., K. Simons and E. Fries (1980). *J. Cell Biol.* **84**, 404–420.
33. Helenius, A., I. Mellman, D. Wall and A. Hubbard (1983). *Trends Biochem. Sci.* **8**, 245–250.
34. Herzog, V. (1983). *J. Cell Biol.* **97**, 607–617.
35. Hopkins, C. R. (1983). *Cell* **35**, 321–330.
36. Hopkins, C. R. (1985). *Cell* **40**, 199–208.
37. Kielian, M. C. and A. Helenius (1984). *J. Virol.* **52**, 281–283.
38. Kielian, M. C. and A. Helenius (1986). *In* "The Togaviridae and Flaviviridae". (Eds S. Schlesinger and M. Schlesinger), pp. 91–119. Plenum Publishing Corporation.
39. Kielian, M. C., S. Keranen, L. Kaariainen and A. Helenius (1984). *J. Cell Biol.* **98**, 139–145.
40. Klatzmann, D., E. Champagne, S. Chamaret, J. Gruest, D. Guetard, T. Hercend, J.-C. Gluckman and L. Montagnier (1984). *Nature (London)*, **312**, 767–768.
41. Lenard, J. and D. K. Miller (1982). *Cell* **28**, 5–6.
42. Lentz, T. L., T. G. Burrage, A. L. Smith, J. Crick and G. H. Tignor (1982). *Science* **215**, 182–184.
43. Lonberg-Holm, K. and L. Philipson (1979). *In* "Cell Membranes and Viral Envelopes". (Eds H. A. Blough and J. M. Tiffany), pp. 789–848. Academic Press, London and Orlando.
44. Marsh, M. (1984). *Biochem. J.* **218**, 1–10.
45. Marsh, M. and A. Helenius (1980). *J. Mol. Biol.* **142**, 439–454.
46. Marsh, M. and A. Helenius (1984). *In* "Lysosomes in Biology and Pathology" (Eds J. T. Dingle, R. T. Dean and W. Sly), Vol. 7, pp. 297–313. Elsevier, Amsterdam.
47. Marsh, M., K. Matlin, K. Simons, H. Reggio, M. White, J. Kartenbeck and A. Helenius (1982). *Cold Spring Harbor Symp. Quant. Biol.* **46**, 835–843.
47a. Marsh, M., E. Bolzau, M. White and A. Helenius (1983a). *J. Cell Biol.* **96**, 455–461.
48. Marsh, M., E. Bolzau and A. Helenius (1983b). *Cell* **32**, 931–940.
49. Marsh, M., G. Griffiths, G. E. Dean, I. Mellman and A. Helenius (1986). *Proc. Natl Acad. Sci. USA* **83**, 2899–2903.
50. Matlin, K., H. Reggio, A. Helenius and K. Simons (1981). *J. Cell Biol.* **91**, 601–613.
51. Matlin, K., H. Reggio, A. Helenius and K. Simons (1982). *J. Mol. Biol.* **156**, 609–631.
52. Mellman, I. and H. Plutner (1984). *J. Cell Biol.* **98**, 1170–1177.
53. Mellman, I., H. Plutner and P. Ukkonen (1984). *J. Cell Biol.* **98**, 1163–1169.

54. Merion, M., P. Schlessinger, J. M. Brooks, J. M. Moehring, T. J. Moehring and W. Sly (1983). *Proc. Natl Acad. Sci. USA* **80**, 5315–5319.
55. Millican, D. and J. S. Porterfield (1982). *J. Gen. Virol.* **63**, 233–236.
56. Ohkuma, S. and B. Poole (1978). *Proc. Natl Acad. Sci. USA* **75**, 3327–3331.
57. Oldstone, M. B. A., A. Tishon, F. Dutko, S. I. T. Kennedy, J. J. Holland and P. W. Lampert (1980). *J. Virol.* **34**, 256–265.
58. Pastan, I. and M. C. Willingham (1983). *Trends Biochem Sci.* **8**, 250–254.
59. Pearse, B. M. F. (1976). *Proc. Natl Acad. Sci. USA* **73**, 1255–1259.
60. Pearse, B. M. F. (1980). *Trends Biochem. Sci.* **53**, 131–134.
61. Peiris, J. S. M. and J. S. Porterfield (1982). *J. Gen. Virol.* **58**, 291–296.
62. Poste, G. and C. A. Pasternak (1978). *Cell Surface Rev.* **5**, 305–367.
63. Reeves, J. P. (1983). In "Lysosomes in Biology and Pathology" (Eds J. Dingle, R. Dean and W. Sly), Vol. 7, pp. 175–199. Elsevier, Amsterdam.
64. Renston, R. H., D. G. Maloney, A. L. Jones, G. T. Hradek, K. Y. Wong and I. D. Goldfine (1980). *Gastroenterology.* **78**, 1373–1388.
65. Simons, K. and S. D. Fuller (1985). *Annu. Rev. Cell Biol.* **1**, 243–288.
66. Simons, K. and H. Garoff (1980). *J. Gen. Virol.* **50**, 1–21.
67. Simons, K., H. Garoff and A. Helenius (1982). *Sci. Am.* **246**, 58–66.
68. Skehel, J., P. Bayley, E. Brown, S. Martin, M. Waterfield, J. White, I. Wilson and D. Wiley (1982). *Proc. Natl Acad. Sci. USA* **79**, 968–972.
69. Steinman, R. M., S. E. Brodie and Z. A. Cohn (1976). *J. Cell Biol.* **68**, 665–687.
70. Steinman, R. M., I. S. Mellman, W. A. Muller and Z. A. Cohn (1983). *J. Cell Biol.* **96**, 1–27.
71. Stone, D. K., X. Xiao-Song and E. Racker (1983). *J. Biol. Chem.* **258**, 4059–4062.
72. Superti, F., M. Derer and H. Tsiang (1984). *J. Gen Virol.* **65**, 781–789.
73. Tycko, B. and F. R. Maxfield (1982). *Cell* **28**, 643–651.
74. Ukkonen, P., V. Lewis, M. Marsh, A. Helenius and I. Mellman (1986). *J. Exp. Med.* **163**, 952–971.
75. van Renswoude, J., K. R. Bridges, J. B. Harford and E. D. Klausner (1983). *Proc. Natl Acad. Sci. USA* **79**, 6186–6190.
76. Wagner, R. (1975). *Comp. Virol.* **4**, 1–94.
77. Weiss, R. A. (1980). In "Virus receptors Part 2" (Eds K. Lonberg-Holm and L. Philipson), pp. 187–202. Chapman and Hall, London.
78. White, J. and A. Helenius (1980). *Proc. Natl Acad. Sci. USA* **77**, 3273–3277.
79. White, J., J. Kartenbeck and A. Helenius (1980). *J. Cell Biol.* **87**, 264–272.
80. White, M., A. Helenius and M. J. Gething (1982). *Nature (London)* **300**, 658–659.
81. White, M., M. Kielian and A. Helenius (1983). *Q. Rev. Biophys.* **16**, 151–195.
82. Wilson, I., J. Skehel and D. Wiley (1981). *Nature (London)* **289**, 366–373.
83. Yoshimura, A., K. Kuroda, K. Kawasaki, S. Yamashina, T. Maeda and S.-I. Ohnishi (1982). *J. Virol.* **43**, 284–293.

17. The Structure of a Human Common Cold Virus (Rhinovirus 14) and its Structural and Functional Relations to Other Picornaviruses and Plant Viruses

Michael G. Rossmann, Edward Arnold, John W. Erickson[a], Elizabeth A. Frankenberger[b], James P. Griffith, Hans-Jürgen Hecht[c], John E. Johnson, Greg Kamer, Ming Luo, Anne G. Mosser*, Roland R. Rueckert*, Barbara Sherry* and Gerrit Vriend

Department of Biological Sciences, Purdue University, W. Lafayette, Indiana 47907, U.S.A. and *Biophysics Lab, University of Wisconsin, 1525 Linden Drive, Madison, Wisconsin 53706, U.S.A.

I.	Introduction	302
II.	The structure	303
III.	Immunogenic sites on HRV14	309
IV.	Antigenic sites on poliovirus and FMDV	312
V.	The canyon as receptor binding site	314
VI.	Neutralization	315
VII.	Assembly	316
VIII.	Evolution	316
	References	318

[a]*Present address*: Department of Physical Biochemistry, AP-9A D-47E, Abbott Laboratories, Abbott Park, North Chicago, Illinois 60064, U.S.A.
[b]*Present address*: Department of Agronomy, Purdue University, W. Lafayette, Indiana 47907, U.S.A.
[c]*Present address*: F. G. Roentgenstrukturanalyse, Universität Wuerzburg, Zentralbau Chemie, Am Hubland, D-8700 Wurzburg, Federal Republic of Germany.

I. INTRODUCTION

Picornavirions contain 60 protomers (Rueckert et al., 1969), each composed of four structural proteins VP1, VP2, VP3, and VP4 corresponding to genes *1D*, *1B*, *1C* and *1A*, respectively (for nomenclature see Rueckert and Wimmer, 1984). Their molecular weights in human rhinovirus 14 (HRV14) are 32 kD, 29 kD, 26 kD and 7 kD. The capsid protein VP0, corresponding to gene *1AB*, is cleaved into its components VP4 and VP2 only in the final stages of assembly (Rueckert et al., 1969; McGregor et al., 1975; Jacobson et al., 1970). The cleavage occurs at an Asn-Ser peptide in HRV14, and hence is not affected by the viral protease 3C whose specificity is for Gln-Gly peptides. The assembly of the capsid occurs in a series of steps culminating in the insertion of RNA into capsids to produce mature virions with the concomitant cleavage of VP0 (Rueckert, 1976; McGregor and Rueckert, 1977; Fernandez-Tomas et al., 1973; Jacobson and Baltimore, 1968; Fernandez-Tomas and Baltimore, 1973).

Considerable effort has been devoted to mapping topological relationships among VP1, VP2, VP3 and VP4 within the capsid (cf. Putnak and Phillips, 1981) using chemical labelling of the surface of intact particles (Carthew and Martin, 1974; Lonberg-Holm and Butterworth, 1976; Beneke et al., 1977), treatment with cross-linking reagents (Lund et al., 1977; Wetz and Habermehl, 1979; Hordern et al., 1979), reaction with specific antibodies and cross-linking with UV light (Wetz and Habermehl, 1982). The consensus is that VP1 is the most external and immunodominant protein, while VP4 is inaccessible from the outside but can be cross-linked with RNA on the interior. Heat treatment or mild denaturing agents cause a conformational change to the capsid, thus altering the response to antisera. Surprisingly, the internal capsid protein VP4 can dissociate and escape from the capsid during antigenic conversion (Rueckert, 1976).

The RNA, and hence by inference the polyprotein, has been sequenced for all three strains of poliovirus (Toyoda et al., 1984; Kitamura et al., 1981; Stanway et al., 1983; Racaniello and Baltimore, 1981), for various foot-and-mouth disease virus (FMDV) strains (Makoff et al., 1982; Carroll et al., 1984; Forss et al., 1984; Boothroyd et al., 1981), for two strains of rhinovirus (Stanway et al., 1984; Callahan et al., 1985; Skern et al., 1985), for encephalomyocarditis virus (EMCV) (Palmenberg et al., 1984) and for hepatitis A virus (Najarian et al., 1985; Linemeyer et al., 1985). Protein-to-protein comparisons between HRV, poliovirus, EMCV, FMDV and hepatitis A virus sequences show that HRV and poliovirus are closely related, whereas sequence homology of HRV to EMCV, FMDV or hepatitis A virus is not immediately obvious, particularly for the structural proteins.

Antibodies can bind to viruses but they do not necessarily neutralize infectivity (cf. Dimmock, 1984). In spite of the 60-fold equivalence of each potential binding site on the virus, as few as four neutralizing antibodies per virion can be sufficient to inhibit infectivity of poliovirus (Icenogle *et al.*, 1983). Neutralizing antibodies usually change the isoelectric point of the picornavirions (Mandel, 1976; Emini *et al.*, 1983a), indicating that a conformational change frequently accompanies neutralization. Antibodies may neutralize by interfering with cell attachment, membrane penetration or virus uncoating (Schrom *et al.*, 1982; Emini *et al.*, 1983b). Antibodies that bind to poliovirus may require bivalent attachment for neutralization of the virus (Icenogle *et al.*, 1983; Emini *et al.*, 1983c). Extensive studies have been reported (see below) on mapping the antigenic surfaces of HRV14 (Sherry and Rueckert, 1985; Sherry *et al.*, 1985), polio and FMDV. Four major immunogenic sites have been identified for HRV14. One of the HRV14 immunogenic sites on VP1 coincides with the dominant immunogen of polio.

In spite of the sequence and surface similarities of picornaviruses, they have different host and tissue specificity. Abraham and Colonno (1984) and Colonno (personal communication) have shown, using 24 rhinovirus serotypes in competition binding assays, that, while the majority recognize one receptor, a second smaller group recognizes a different receptor. HRV14 (sequenced by Callahan *et al.*, 1985 and by Stanway *et al.*, 1984) belongs to the larger receptor group while HRV2 (sequenced by Skern *et al.*, 1985) belongs to the smaller one. Minor *et al.* (1984) have been able to produce monoclonal antibodies that block cellular receptors of poliovirus as have Colonno and co-workers for the large rhinovirus receptor group (personal communication). Krah and Crowell (1985) have characterized some properties of HeLa cell receptors for group B Coxsackie viruses. They found that concanavalin A and other lectins absorbed to receptors and inhibited virus attachment, a finding similar to that of Lonberg-Holm (1975).

X-ray diffraction studies of crystalline picornaviruses have been limited. We describe here the first atomic resolution structure of any animal virus, namely that of HRV14.

II. THE STRUCTURE

The particle consists of an icosahedral protein shell (Fig. 17.1(*a*)) surrounding an RNA core. The lack of visible structure in the central cavity results from the random orientation of the asymmetric RNA molecule. Both the tertiary fold of the VP1, VP2 and VP3 polypeptide chains and their quaternary organization within the HRV14 capsid are closely similar to the two published high-resolution structures of $T=3$ (180 identical subunits per capsid) RNA plant viruses, tomato bushy stunt virus (TBSV) (Harrison *et al.*, 1978) and southern bean mosaic virus (SBMV) (Abad-Zapatero *et al.*, 1980). Although the subunit organization within the icosahedron is somewhat different (Rossmann *et al.*, 1983a), a similar tertiary structure has also been found for $T=1$ (60 subunits per capsid) satellite tobacco necrosis virus (STNV) (Liljas *et al.*, 1982). The radial position and orientation of structurally equivalent atoms of HRV14 and

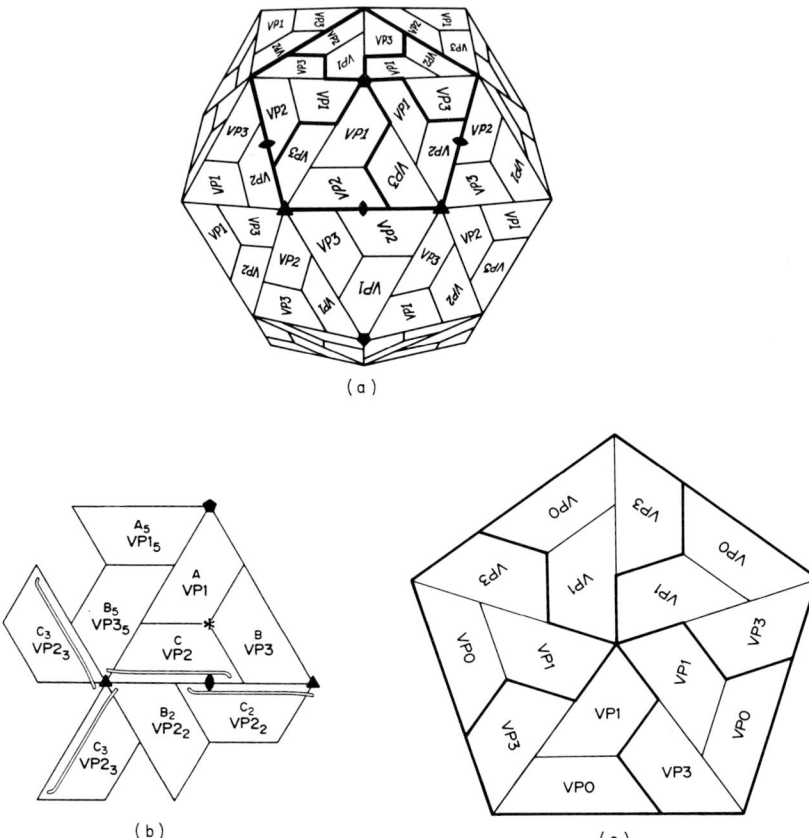

Figure 17.1. Relation of the pseudo-equivalent VP1, VP2 and VP3 subunits to the quasi-equivalent subunits A, C and B in TBSV and SBMV. (a) Icosahedral capsid. (b) Icosahedral asymmetric unit: (b) shows the ordered amino-terminal arm βA present only in the C subunit of the plant viruses and VP2 of HRV14. The amino end of the arm interacts with two other VP2 arms across the three-fold axis, while the carboxy end of the arm interacts with a VP2 across the two-fold axis. An ★ indicates the position of the quasi-three-fold axis in SBMV and TBSV analogous to the pseudo-three-fold axis in HRV14. Subscripts designate the symmetry operation required to obtain the given subunit from the basic triangle. (c) The thickly outlined VP1, VP3, VP0 unit corresponds to the 6S protomer, and the 15-mer cap to the 14S pentamer observed in assembly experiments.

SBMV generally agree to better than 0.3 nm relative to the icosahedral symmetry axes. In the plant viruses, the three quasi-equivalent subunits A, B and C have the same amino acid sequence but cannot have identical geometrical environments (Caspar and Klug, 1962). For SBMV, there are 260 amino acids per subunit, but the first 63 residues are associated with the RNA and therefore do not have icosahedral symmetry. These are said to be in the "random" domain. In the C subunit, unlike the A and B subunits, residues 38–63 are ordered with their ends forming a "β-annulus" about the icosahedral three-fold axes (Fig. 17.1(b)). Residues 64–260 of SBMV are referred to as the "shell" domain and form an eight-stranded anti-parallel β-barrel (Fig. 17.2(a)). The β-barrels in HRV14, like those in SBMV, are wedge-shaped with the thin end (left in Fig. 17.2) pointing toward the five- or three-fold (quasi-six-fold) axes. There are four excursions of the polypeptide chain toward the wedge-shaped end. Each excursion makes a sharp bend or "corner". The most exterior (top left in Fig. 17.2) corner is formed between the β-sheets βB and βC, the second corner down is formed between βH and βI, the third corner is between βD and βE1, while the most internal corner connects βF and βG. The βF-βG corner is the site of a 25-residue insertion in SBMV, including the α-helix αC, that is not present in any of the viral proteins of picornaviruses, TBSV or STNV.

The three larger capsid proteins VP1, VP2 and VP3 in HRV14 are oriented and situated at essentially the same radius and position as the A, C and B subunits in SBMV, respectively. The capsid proteins of HRV14 are related by a pseudo-three-fold axis, analogous to the quasi-three-fold axis in SBMV and TBSV, at the ★ in Fig. 17.1(b). However, VP1 and VP3 have additions at their amino and carboxy termini. The amino ends, as in plant viruses, are in contact with the RNA, but unlike plant viruses, they are more acidic than basic and have only a very few amino-terminal residues "disordered" (lacking icosahedral symmetry). The first 64 of the 73 amino-terminal residues of VP1 reside under VP3, while the first 42 of the 71 amino-terminal residues of VP3 are under VP1. Thus, the predominant positions of VP1 and VP3 at the RNA-protein interface are exchanged relative to their positions at the exterior surface.

The first 25 of the 69 residues of the internal structural protein VP4 are not seen in the electron density map, implying that they lack icosahedral symmetry. VP4 is positioned in part below VP1 and VP2 with its visible amino end surrounding the five-fold axis. The carboxy ends of VP1 and VP3 are external and function in part to associate proteins within a protomer (Fig. 17.1(c)).

Large sequence insertions relative to the typical shell domain form protrusions on VP1, VP2 and VP3 and create a deep cleft or "canyon wall" on the viral surface. The canyon separates the major part of five VP1 subunits (in the "North") clustered about a pentamer axis from the surrounding VP2 and VP3 subunits (Fig. 17.3) (in the "South"), thus forming a moat around the VP1 protrusions on the five-fold axis. The South canyon walls are lined with the carboxy-terminal ends of VP1 and a large sequence insertion in VP1 corresponding to helices αD and αE in the equivalent SBMV capsid protein. The North canyon wall is partially lined with the carboxy-terminus of VP3. VP2 is hardly associated at all with the canyon, while VP1 is the major contributor to the residues lining the canyon. The canyon is 2.5 nm deep and 1.2–3.0 nm wide.

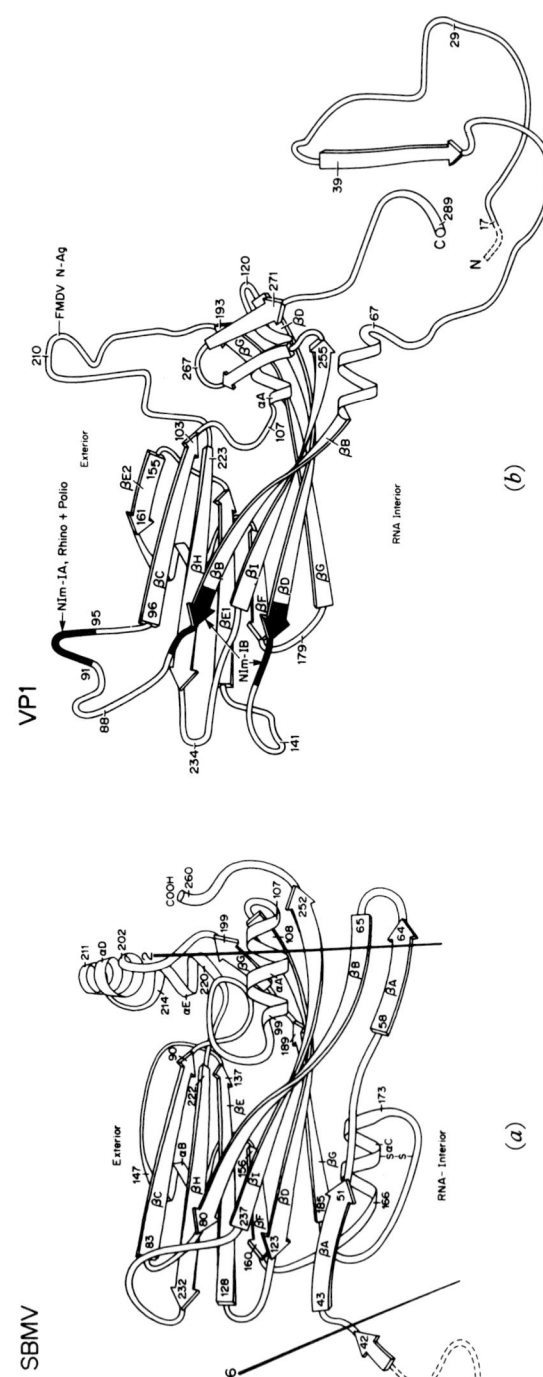

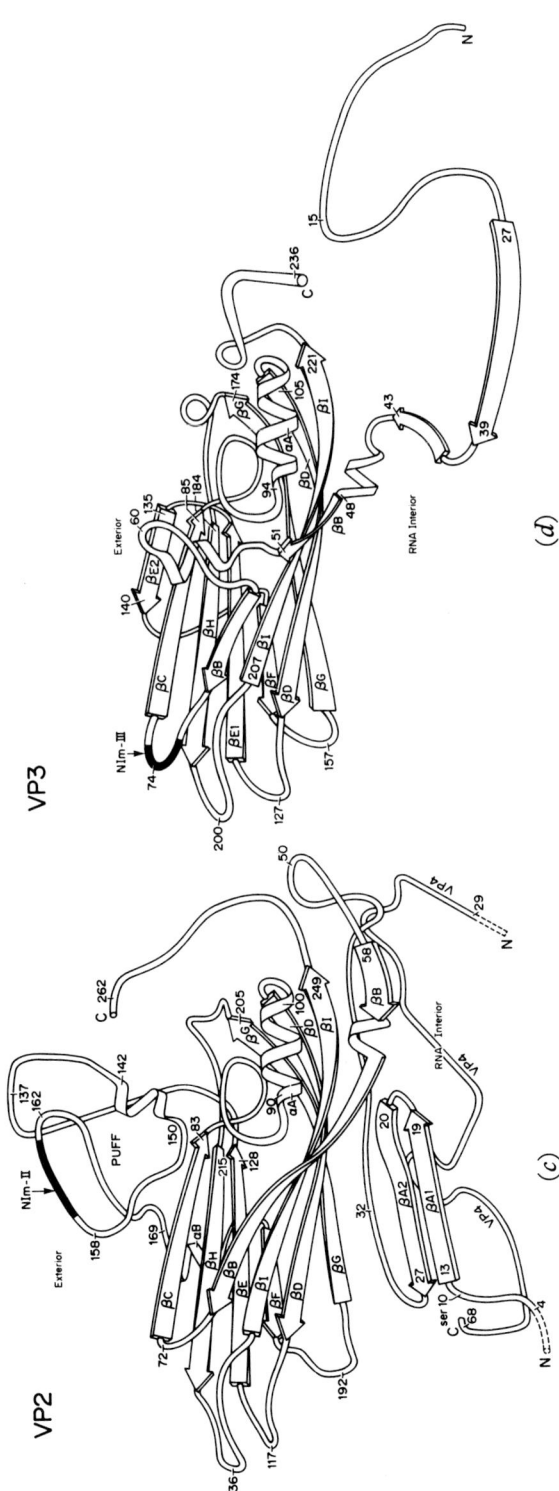

Figure 17.2. Diagrams showing the polypeptide fold of SBMV and of each of the three larger capsid proteins of HRV14. The nomenclature of the secondary structural elements is derived from that of SBMV (Rossmann et al., 1983b). Amino acid sequence numbers, appropriate for each protein, are also shown: (a) SBMV; (b) VP1 of HRV14; (c) VP2 and VP4 of HRV14 and (d) VP3 of HRV14. (Adapted from a drawing of SBMV by Jane Richardson.)

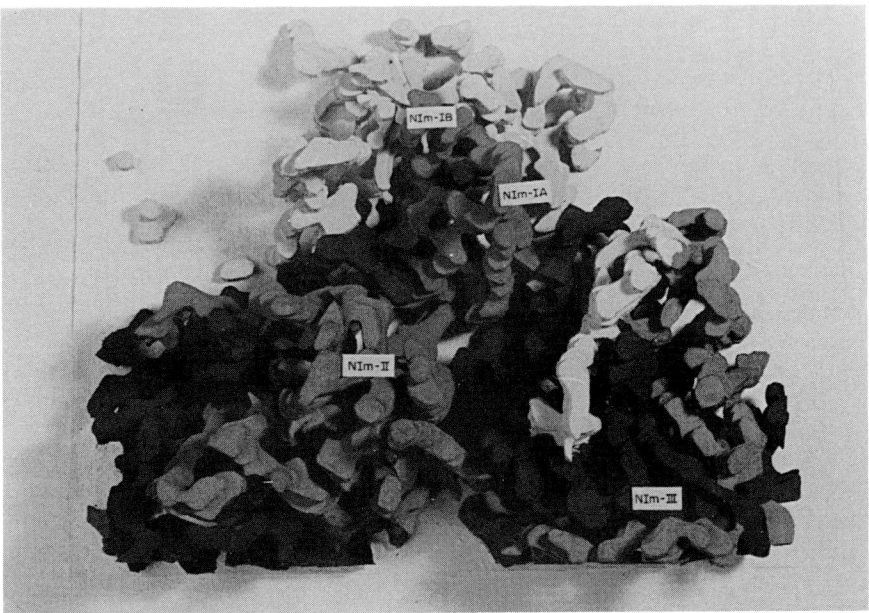

Figure 17.3. Balsawood model of the 0.5 nm electron density map showing the external portion of the virus. The balsawood sheets are equivalent to a thickness of 0.15 nm and are perpendicular to an icosahedral two-fold axis. The positions of the major HRV14 immunogens are shown. Note the large canyon that is the proposed binding site of the host-cell receptor. The base of the canyon is inaccessible to antibody binding, while the canyon rim is lined with antigenic sites.

Owing to the additional elaborations that VP1 has on the surface, relative to VP2 and VP3, its overall shape is that of a kidney, with the depression forming a large part of the canyon (Fig. 17.3). The first 16 residues of VP1 (Fig. 17.2(b)) are not seen in the electron density map. The shell domain of VP1 in HRV14 starts at residue 74. The small sequence insertion between βB and βC in rhinovirus and poliovirus is not found in FMDV. This loop forms a major immunogen in HRV14 (NIm-IA* to be discussed below) and poliovirus. The residues in VP1 of HRV14, which are analogous to αD and αE helices in SBMV, protrude to the surface and form part of the South rim of the canyon, but do not form helices in HRV14. There is an eight-residue insertion in polio and FMDV relative to HRV14 at the most external portion of this segment. These additional residues contain the major antigenic site of FMDV and have been predicted to form an α-helix (Pfaff et al., 1982). The carboxy-terminal 23

---

*The previously reported (Sherry and Rueckert, 1985) N-AgI and N-AgII have been renamed NIm-IA and NIm-III to identify them as neutralizing immunogens and to indicate their major coat protein locations on VP1 and VP3, respectively.

residues of VP1 line the South rim of the canyon in HRV14. There is little conservation of these residues among picornaviruses.

The first three residues of VP2 (Fig. 17.2(c)) are not seen in the electron density. The proximity of serine-10 in VP2 to the carboxy-terminus of VP4 suggests that the VP0 cleavage, which occurs during the virus maturation step, may be autoproteolytic (cf. serine proteases (Neurath, 1984; Steitz and Shulman, 1982)). The first nine residues of VP2 are likely to have been rearranged after cleavage. Portions of the region between residues 10 and 72 are involved in contacts between two-fold-related VP2 polypeptide chains (Fig. 17.1(c)). The αD and αE helices are absent in VP2 (Fig. 17.2(c)). There is a large 43-residue insertion, in the VP2 position corresponding to βE2 of VP1 and VP3, forming an external mushroom-shaped "puff". This is positioned adjacent to the VP1 elaborations, associated with the major antigenic site in FMDV, which line the South canyon wall (Fig. 17.3). The most external residues of this puff correspond to NIm-II of HRV14. In contrast to VP1 and VP3, the carboxy-terminus of VP2 has no extension beyond the shell domain.

All residues of VP3 (Fig. 17.2(d)) can be seen in the electron density. The 26 amino-terminal residues form a five-fold β-barrel about the pentamer axis analogous to the β-annulus (Harrison et al., 1978) about trimer axes in SBMV and TBSV. This five-fold annulus extends down into the RNA to a radius of 11.1 nm. The polypeptide emerges from the β-annulus, circles around the base of the VP1 shell domain while making extensive contact with the RNA, emerges on the viral surface near residue 61 and then enters the shell domain at residue 72. The top corner of the VP3 shell domain, between βB and βC, is the NIm-III site of HRV14, structurally equivalent to NIm-IA in VP1. Helix αB of SBMV is replaced by an extended chain in VP3 and the external helices αD and αE of SBMV are absent as in VP2.

III. IMMUNOGENIC SITES ON HRV14

Amino acid residues within the major neutralization immunogens of HRV14 have been identified by Sherry and Rueckert (1985) and by Sherry et al. (1986). They isolated mouse hybridoma lines that secreted monoclonal antibodies that neutralized HRV14. Each was then used to select several viral mutants resistant to neutralization by that antibody. Finally, every monoclonal antibody was assayed for its ability to neutralize the mutants. The results revealed four major immunogenic neutralization sites. Each immunogenic site was composed of overlapping epitopes where a given mutant was resistant to many or all of the antibodies directed against that site. These results were obtained without knowledge of the three-dimensional structure. However, when the electron density map became available, it was immediately clear that the substitutions that could confer resistance to neutralization, regardless of their location in the amino acid sequences, were localized into four distinct areas corresponding exactly to the proposed immunogens. Moreover, these residues invariably faced outward toward the viral exterior. The possible effects a given mutation might have on the capacity of the antibody to neutralize the virus have not yet been fully scrutinized in light of the structure. For a fuller understanding of these phenomena, it will be necessary to differentiate between mutants that resist neutralization while still maintaining their ability to bind antibodies and those mutants which entirely fail to bind antibodies.

The immunogenic site NIm-IA (Figs 17.3 and 17.4(a)) is an insertion between βB and βC at the topmost corner of the VP1 wedge. It is the most external portion of the complete virion being 16 nm from the centre.

(a)

(b)

17. *Structure of a Human Common Cold Virus* 311

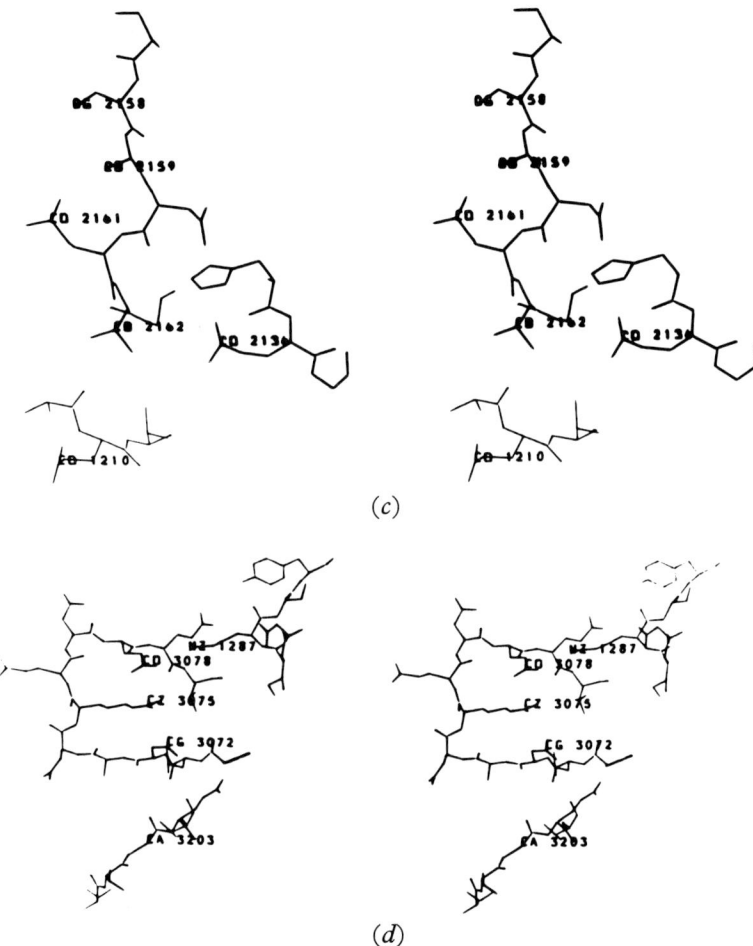

Figure 17.4. Stereo view of each of the four antigenic sites on the HRV14 surface. The models were built with respect to the 0.3 nm resolution electron density on an Evans and Sutherland PS300 system using the FRODO program package. (*a*) NIm-IA; (*b*) NIm-IB; (*c*) NIm-II; (*d*) NIm-III. The extremities of those residues associated with each neutralizing immunogenic site have been identified. 1000 has been added to residue numbers in VP1, 2000 for residues in VP2 and 3000 for residues in VP3. Identified atoms are C-α (CA), C-β (CB), C-γ (CG), O-γ (OG), C-δ (CD), C-ζ (CZ), and Nζ (NZ).

Residues 91 and 95, associated with NIm-IA, are on the extreme external portion of the loop.

Residues associated with site NIm-IB on VP1 (Figs 17.3 and 17.4(*b*)) are on the carboxy ends of βB and βD, which are situated on either side of the amino end of strand βI. No mutations conferring resistance have yet been found on the βI strand. The NIm-IB site is close to the five-fold axis, a little below the canyon rim.

NIm-II on VP2 (Figs 17.3 and 17.4(c)) is at the extreme outside of the puff, 15.5 nm from the viral centre. This immunogenic puff is adjacent to the external loop formed by the sequence insertion in VP1 corresponding to αD and αE of SBMV. One of the residues (210E on VP1) that is associated with NIm-II is, indeed, on VP1.

NIm-III on VP3 (Figs 17.3 and 17.4(d)) is 14.9 nm from the viral centre, to some extent in the shadow of a larger protrusion of VP3 (residues 58–63) and the carboxy end of VP1 (282–286). NIm-III is in a position on VP3 corresponding to NIm-IA on VP1, that is in the loop between βB and βC. Residue K287 on VP1, associated with this site, points directly toward and is adjacent to the other residues on VP3 associated with NIm-III. The large protrusion near NIm-III, consisting of residues 282–286 in VP1 and 58–63 in VP3, has not been shown to be antigenic for HRV14 but in FMDV residues associated with the carboxy-terminal end of VP1 have shown antigenicity (Table 17.1). It is possible that this structural component lacks antigenicity in HRV14 because of greater rigidity (Westhof et al., 1984; Tainer et al., 1984), and this hypothesis will be checked when the HRV14 structure is refined and temperature factors have been determined.

IV. ANTIGENIC SITES ON POLIOVIRUS AND FMDV

Results obtained in mapping the immunogenic and antigenic surfaces of other picornaviruses are summarized in Table 17.1. There is a dominant immunogen in poliovirus corresponding to NIm-IA of rhinovirus. Absence of a corresponding immunogen in FMDV is explained by deletion of this loop from its VP1.

The dominant antigenic site in FMDV resides in a region homologous to NIm-II in HRV14 (Table 17.1). The VP1 contribution to this antigen has also been shown for HRV14 (E210 of NIm-II) and for poliovirus (residues 222–241, Table 17.1). In FMDV, however, the immunogenic puff on VP2 is absent and eight amino acids are inserted at the extreme surface of the αD-αE region of VP1. The resulting protrusion of NIm-II in FMDV would be unsupported by the puff. This protrusion may, therefore, occupy the space which in HRV14 is occupied by the puff. The inability of FMDV VP1 to elicit neutralizing antibodies (Meloen et al., 1983) is consistent with the lack of structural support for this NIm-II protrusion in the absence of the neighboring VP2.

Poliovirus can also be neutralized by antibodies that bind to NIm-II or NIm-III (Table 17.1). Thus there is an overall consistency in identifying at least three of the major HRV14 neutralization sites in poliovirus.

Many of the methods used to determine antibody binding sites depended upon the use of synthetic peptides as antigens (Table 17.1). Peptides associated with neutralizing antigenic regions do, in general, elicit

Table 17.1. Evidence for antigenic regions in poliovirus and FMDV

	Virus	Coat protein	Amino acid number	HRV14[a] amino acid equivalent	Sero-type[b]	Ref.	Method[c]	Position in HRV14 structure
1.	Polio	VP1	11–17	deleted	1M	[19]	3	Within RNA
2.	Polio	VP1	24–40	14–30	1M	[16]	5	Faces RNA
3.	Polio	VP1	70–75	60–65	1M	[19]	3,6	Faces RNA
4.	Polio	VP1	61–80	51–70	1M	[16]	2	Faces RNA
5.	Polio	VP1	86–103	77–97	1M	[16]	5	NIm-IB + NIm-IA
6.	Polio	VP1	91–109	84–103	1M	[16]	2	NIm-IB + NIm-IA
7.	Polio	VP1	100–109	93–103	1M	[16]	2	Part of NIm-IA
8.	Polio	VP1	93–103	86–97	1M	[19]	2,3,6	NIm-IB(?) + NIm-IA[d]
9.	Polio	VP1	93–104	86–98	1	[84]	4	NIm-IB(?) + NIm-IA[d]
10.	Polio	VP1	93–100	86–93	3L	[24, 49]	1	NIm-IB(?) + part of NIm-IA[d]
11.	Polio	VP1	141–147	135–141	1M, S	[78]	3	NIm-IB
12.	Polio	VP1	161–181	154–174	1M	[16]	5	Partly exposed on canyon
13.	Polio	VP1	182–201	175–193	1M	[16]	2	Buried on 4th corner down
14.	Polio	VP1	222–241	212–225	1M	[16]	2	NIm-II
15.	FMDV	VP1	141–160	210–228	O_1K	[9]	2	NIm-II
16.	FMDV	VP1	144–159	211–227	O_1K	[54]	2	NIm-II
17.	FMDV	VP1	145–168	211–236	A12	[57]	8	NIm-II
18.	FMDV	VP1	146–154	211–223	O_1K	[73]	7	NIm-II
19.	FMDV	VP1	169–179	237–247	A12	[6, 57]	8	NIm-IB central strand
20.	Polio	VP1	270–287	254–272	1M	[16]	5	Partly exposed on canyon wall
21.	FMDV	VP1	200–213	268–294	O_1K	[9]	2	NIm-III(?)
22.	FMDV	VP1	200–213	268–294	O_1K	[73]	7	NIm-III(?)
23.	Polio	VP2	162–173	161–170	1M	[22]	2,3	NIm-II
24.	Polio	VP3	71–82	70–81	1M	[22]	6	NIm-III

[a] Aligned by eye and by fitting to the shell domain structure.
[b] 1M (type 1, Mahoney), S (Sabin), 3L (type 3, Leon), O_1K (type O_1, strain Kaufbeuern), A12 (type A, subtype 12).
[c] Method key:
1. Monoclonal antibodies raised against intact virus select for resistant mutations.
2. Synthetic peptides induce neutralizing antibodies.
3. Synthetic peptides prime for high-titre neutralizing response.
4. Synthetic peptide competes with monoclonal antibody to inhibit neutralization.
5. Synthetic peptides induce antibodies that bind virus but neutralize poorly.
6. Neutralizing antisera or monoclonal antibodies bind peptide in ELISA.
7. Deduced from ability or inability of protein fragments to induce neutralizing antibodies.
8. Deduced from ability or inability of neutralizing monoclonal antibodies to bind protein fragments.

[d] Uncertainty owing to somewhat arbitrary nature of computer alignments.

antibodies that can neutralize the intact virus. In a significant number of cases (lines 1, 2, 3, 4, 8, 13 of Table 17.1), however, the sequence in question lies far below the viral surface or even buried in the RNA. This suggests that some peptides can elicit antibodies that subsequently bind to totally unrelated portions of the native virus. Alternatively, some of the results might be accounted for by conformational changes occurring during the isoelectric transition of the virus (Mandel, 1979).

The correspondence of sequence variability with antigenic sites has frequently been pointed out in the study of picornaviruses. This is certainly apparent on comparing the two available rhinovirus sequences of HRV14 (Stanway et al., 1984; Callahan et al., 1985) and HRV2 (Skern et al., 1985), as well as in comparative studies of polio (Toyoda et al., 1984) and FMDV (Makoff et al., 1982; Beck et al., 1983) sequences. Nevertheless, the surface protrusion caused by HRV14 VP3 58–63 together with VP1 282–286 is equally variable, but has not yet been associated with an antigenic site in any picornaviruses.

V. THE CANYON AS RECEPTOR BINDING SITE

The 2.5 nm deep canyon, circulating around each of the 12 pentamer vertices (Fig. 17.3), suggests this to be the site for cell receptor binding. An antibody molecule would have difficulty in reaching the canyon floor, its entrance being blocked by the canyon rim. Thus, the residues in the deeper recesses of the canyon would not be under immune selection and could remain constant, permitting the virus to retain its ability to seek out the same cell receptors. A similar situation of a protected sialic acid binding site exists in the neuraminidase spike of influenza virus (Colman et al., 1983; Varghese et al., 1983).

While retention of the canyon structure for all picornaviruses is to be expected, variation in the residues lining the canyon should be anticipated between viruses that attach themselves to different host-cell receptors. That is, FMDV, polio, HRV14 and HRV2, all of which recognize different receptors, should exhibit some variation in the residues lining the canyon wall. It is thus noteworthy that not only are those parts of the carboxy-terminal ends of VP1 and VP3 which line the canyon walls some of the least conserved amino acids among picornaviruses, but also there is a deletion in HRV2 relative to HRV14 in βB at NIm-IB which is possibly associated with cell receptor carbohydrate recognition (Argos et al., 1980).

Residues lining the proposed cell receptor binding site of HRV14 include a number of charged residues. Of 48 residues on the floor and wall of the canyon in one icosahedral asymmetric unit, there are six aspartates, three glutamates, three lysines and one arginine. Thus, correct docking of the virus to cell receptors may in part be mediated by Coulombic forces.

FMDV can be treated with trypsin causing cleavage at residues between 138 and 154 of VP1. This causes the virus to lose its ability to attach to cells and its ability to stimulate neutralizing antibody (Cavanagh et al., 1977; Wild et al., 1969; Strohmaier et al., 1982; Baxt et al., 1984). The enzymic cleavages occur in NIm-II on the αD-αE protrusion of VP1, a large loop which also forms part of the presumed host receptor binding site.

The difference between attenuated and virulent virus has sometimes been attributed to alterations in cell specificity. Experiments involving recombination of RNA from virulent and attenuated strains suggest that the virulence of poliovirus lies in the 5′ end of the RNA (Agol et al., 1984). These authors concluded that the source of virulence probably lies in the coat proteins. However, the non-coding region, also contained in the recombinant RNA segment, has been implicated as well (Evans et al., 1985). Comparison of the amino acid sequence between type 1 Mahoney (virulent) and type 1 Sabin (attenuated) shows four amino acid changes in the position equivalent to a proposed carbohydrate site (Argos et al., 1980), while there are almost no changes anywhere else on any of the four structural proteins. This raises the possibility that the difference between the virulent and attenuated strains may be in an altered receptor specificity for carbohydrate recognition.

VI. NEUTRALIZATION

Some poliovirus neutralizing antibodies that bind to VP1 require bivalent attachment (Emini et al., 1983b, 1983c) associated with a change in isoelectric point from pI 6 to pI 4. Cleavage with papain of the bound antibody restores infectivity and the original isoelectric point. Icenogle et al. (1983) have shown that only about four copies per virion of a particular, bivalently bound, monoclonal antibody were needed to neutralize poliovirus, without altering the pI. Both these observations suggest that neutralization may involve conformational changes that interfere with cell attachment, penetration or uncoating.

Since an antibody itself has a two-fold axis, bivalent attachment could occur across icosahedral two-fold axes. The distances between the nearest two-fold-related immunogenic sites IA, IB, II and III are 12 nm, 12 nm, 5 nm and 6 nm respectively. All other symmetrical counterparts of these antigens related by two-fold axes are at least 17 nm apart. Lower and upper limits of the distances of the two antibody binding sites on an immunoglobulin molecule are not well known but probably lie in the range of 5–18 nm (Icenogle et al., 1983; Werner et al., 1972; Marquart et al., 1980). Hence symmetrical bivalent attachment of antibodies to the known immunogens may be possible, although NIm-II and III are close to the likely limit. Since Fab arms of antibodies are capable of some rotation, it is conceivable that bivalent asymmetric attachment might also occur across three-fold or five-fold axes. The resultant torque may be sufficient to interfere with the normal process of infection. Alternatively, these constraints may result in monovalent attachment, permitting cross-linking between viruses (Icenogle et al., 1983; Thomas et al., 1985).

VII. ASSEMBLY

Assembly of picornaviruses (Rueckert, 1976; McGregor and Rueckert, 1977; Fernandez-Tomas et al., 1973; Jacobson and Baltimore, 1968; Fernandez-Tomas and Baltimore, 1973; Putnak and Phillips, 1981; Sangar, 1979) proceeds from 6S protomers of VP1, VP3 and VP0, via 14S pentamers of five 6S protomers, to mature virions. The final step involves inclusion of the RNA into empty capsids or partially assembled shells with simultaneous cleavage of VP0 into VP2 and VP4. Conversely, disassembly *in vitro*, produced by mild denaturation, proceeds via the expulsion of VP4 followed by the RNA (Rueckert, 1976).

Both the amino and carboxy ends of VP1 and $VP3_5$ are intertwined with each other. Furthermore, if VP4 and VP2 are considered as VP0, then VP0 is also intertwined with VP1 and $VP3_5$. This strongly suggests that the 6S protomer is as shown in Figure 17.1(c). These protomers are themselves intertwined by virtue of the five-fold β-annulus formed by the amino ends of the VP3's and the proximity of the observed amino ends of VP4's to the five-fold axis. Thus, the 14S pentamers closely correlate with the observed structure, shown diagrammatically in Figure 17.1(c). Such an assembly sequence matches that observed in plant viruses, in particular that of SBMV, where the building blocks are dimers corresponding to VP1 and $VP3_5$ and where the formation of intermediates with five-fold symmetry is considered to be a critical stage in the formation of $T=1$ and $T=3$ capsids (Rossmann et al., 1983b). Indeed, it has now been shown (Erickson et al., 1985) that the 15-protein cluster, corresponding to the one shown in Figure 17.1(c), is conserved between $T=1$ and $T=3$ SBMV particles.

Once cleaved from VP4, VP2 is globular and does not contact the other proteins extensively. There are large solvent-accessible regions between VP2 and the surrounding proteins. This, as well as the extraordinarily internal heavy-atom sites on VP2 (used in the crystallographic structure determination), is consistent with the loose binding of VP2 to the capsid. Disruption of pentamer–pentamer contacts, mediated by a slight reorientation of VP2 or its complete removal, could provide a port by which the VP4 and RNA exit. Binding of a cell receptor in the canyon adjacent to VP3 could facilitate this process, possibly accompanied by an isoelectric change.

It is remarkable how, in both plant and animal RNA viruses, there is a β-annulus type structure between the amino ends of some subunits. It is equally remarkable how the amino ends of the capsid proteins are invariably associated with the RNA. In TBSV, SBMV and STNV they are basic, whereas in HRV14 they are acidic. These properties may be significant for the initial events of assembly.

VIII. EVOLUTION

Conservation of three-dimensional structure is almost invariably greater than conservation of amino acid sequence homology (Rossmann and

Argos, 1981; Rossmann et al., 1974; Bajaj and Blundell, 1984). Thus, structural comparison can be used to trace divergent evolution over longer time-spans than is possible by amino acid comparisons. Numerous structural comparisons have now been made and their probability for divergence assessed. These provide benchmarks to which other comparisons can be related. For example, the similarity among TBSV, SBMV and STNV is so great that their divergence from a common primordial ancestor is highly probable or, conversely, convergence is highly improbable (Rossmann et al., 1983a; Hopper et al., 1984). Although no quantitative structural comparisons are given here, the great similarity of the VP1, VP2 and VP3 shell domain structures to those of the plant viruses leaves little doubt as to their common ancestry.

ACKNOWLEDGEMENTS

We are most grateful to Sharon Wilder for her outstanding assistance in the organization of the Purdue structural laboratory over many years and in the preparation of this manuscript. We wish to thank Kathy Shuster, Bill Boyle and Jun Tsao for preparation of the figures. The crystallographic data used in this study were collected at the Cornell High Energy Synchrotron Source (CHESS). We are, indeed, most grateful for all the dedicated help we have received from Keith Moffat, Wilfred Schildkamp, Robert Hunt, Don Bilderbeck, Boris Batterman, Aggie Sirrine and all the CHESS staff and operators. We are also most grateful to the Purdue University Computer Center staff (in particular Saul Rosen, John Steele, Tom Putnam and Paul Townsend) for their generous help and encouragement of our use of the Cyber 205 supercomputer. We have benefited greatly by the helpful and stimulating discussions with our other colleagues in the Purdue structural groups including Jeffrey Bolin, Abelardo Silva, Ignacio Fita, Celerino Abad-Zapatero, R. Usha, M. V. Hosur, Cynthia Stauffacher, Patrick Argos and J. K. Mohana Rao. We thank Richard Colonno (Merck Sharp & Dohme Co.) for his interest in the work and providing the HRV14 RNA sequence prior to publication and also thank Ann Palmenberg (University of Wisconsin) for providing her sequence alignments of picornaviruses. We also thank Tim Schmidt for maintaining the x-ray equipment at Purdue University. The work was supported by an NIH grant to M.G.R., an NSF grant for supercomputer time to M.G.R., and an ACS grant to R.R.R. CHESS is supported by an NSF grant to Boris Batterman and the macromolecular diffraction facility at CHESS by an NIH grant to Keith Moffat. A postdoctoral Walter Winchell–Damon Runyon fellowship supported E.A. for some of the time, a predoctoral NIH training grant supported B.S. and travel support was provided by the Dutch Government to G.V. A Purdue University Showalter Foundation grant was awarded to M.G.R. to equip the cell culture laboratory and a recent grant from the Merck Sharp & Dohme Co. contributed to the salary of one technician involved in the virus propagation.

REFERENCES

1. Abad-Zapatero, C., S. S. Abdel-Meguid, J. E. Johnson, A. G. W. Leslie, I. Rayment, M. G. Rossmann, D. Suck and T. Tsukihara (1980). *Nature (London)* **286**, 33–39.
2. Abraham, G. and R. J. Colonno (1984). *J. Virol.* **51**, 340–345.
3. Agol, V. I., V. P. Grachev, S. G. Drozdov, M. S. Kolesnikova, V. G. Kozlov, N. M. Ralph, L. I. Romanova, E. A. Tolskaya, A. V. Tyufanov and E. G. Viktorova (1984). *Virology* **136**, 41–55.
4. Argos, P., T. Tsukihara and M. G. Rossmann (1980). *J. Mol. Evol.* **15**, 169–179.
5. Bajaj, M. and T. L. Blundell (1984). *Annu. Rev. Biophys. Bioeng.* **13**, 453–492.
6. Baxt, B., D. O. Morgan, B. H. Robertson and C. A. Timpone (1984). *J. Virol.* **51**, 298–305.
7. Beck, E., G. Feil and K. Strohmaier (1983). *EMBO J* **2**, 555–559.
8. Beneke, T. W., K. O. Habermehl, W. Diefenthal and M. Buchholz (1977). *J. Gen. Virol.* **34**, 387–390.
9. Bittle, J. L., R. A. Houghten, H. Alexander, T. M. Shinnick, J. G. Sutcliffe, R. A. Lerner, D. J. Rowlands and F. Brown (1982). *Nature (London)* **298**, 30–33.
10. Boothroyd, J. C., P. E. Highfield, G. A. M. Cross, D. J. Rowlands, P. A. Lowe, F. Brown and T. J. R. Harris (1981). *Nature (London)* **290**, 800–802.
11. Callahan, P. L., S. Mizutani and R. J. Colonno (1985). *Proc. Natl Acad. Sci. USA* **82**, 732–736.
12. Carroll, A. R., D. J. Rowlands and B. E. Clarke (1984). *Nucl. Acids Res.* **12**, 2461–2472.
13. Carthew, P. and S. J. Martin (1974). *J. Gen. Virol.* **24**, 525–534.
14. Caspar, D. L. D. and A. Klug (1962). *Cold Spring Harbor Symp. Quant. Biol.* **27**, 1–24.
15. Cavanagh, D., D. V. Sangar, D. J. Rowlands and F. Brown (1977). *J. Gen. Virol.* **35**, 149–158.
16. Chow, M., R. Yabrov, J. Bittle, J. Hogle and D. Baltimore (1985). *Proc. Natl Acad. Sci. USA* **82**, 910–914.
17. Colman, P. M., J. N. Varghese and W. G. Laver (1983). *Nature (London)* **303**, 41–44.
18. Dimmock, N. J. (1984). *J. Gen. Virol.* **65**, 1015–1022.
19. Emini, E. A., B. A. Jameson and E. Wimmer (1983a). *Nature (London)* **304**, 699–703.
20. Emini, E. A., S. Kao, A. J. Lewis, R. Crainic and E. Wimmer (1983b). *J. Virol.* **46**, 466–474.
21. Emini, E. A., P. Ostapchuk and E. Wimmer (1983c). *J. Virol.* **48**, 547–550.
22. Emini, E. A., B. A. Jameson and E. Wimmer (1984). In "Modern Approaches to Vaccines" (Eds R. M. Chanock and R. A. Lerner), pp. 65–75. Cold Spring Harbor Laboratory.
23. Erickson, J. W., A. M. Silva, M. R. N. Murthy, I. Fita and M. G. Rossmann (1985). *Science* **229**, 625–629.
24. Evans, D. M. A., P. D. Minor, G. S. Schild and J. W. Almond (1983). *Nature (London)* **304**, 459–462.
25. Evans, D. M. A., G. Dunn, P. D. Minor, G. C. Schild, A. J. Cann, G. Stanway, J. W. Almond, K. Currey and J. V. Maizel, Jr. (1985). *Nature (London)* **314**, 548–550.
26. Fernandez-Tomas, C. B. and D. Baltimore (1973). *J. Virol.* **12**, 1122–1130.

27. Fernandez-Tomas, C. B., N. Guttman and D. Baltimore (1973). *J. Virol.* **12**, 1181–1183.
28. Forss, S., K. Strebel, E. Beck and H. Schaller (1984). *Nucl. Acids Res.* **12**, 6587–6601.
29. Harrison, S. C., A. J. Olson, C. E. Schutt, F. K. Winkler and G. Bricogne (1978). *Nature (London)* **276**, 368–373.
30. Hopper, P., S. C. Harrison and R. T. Sauer (1984). *J. Mol. Biol.* **177**, 701–713.
31. Hordern, J. S., J. D. Leonard and D. G. Scraba (1979). *Virology* **97**, 131–140.
32. Icenogle, J., H. Shiwen, G. Duke, S. Gilbert, R. Rueckert and J. Anderegg (1983). *Virology* **127**, 412–425.
33. Jacobson, M. F. and D. Baltimore (1968). *J. Mol. Biol.* **33**, 369–378.
34. Jacobson, M. F., J. Asso and D. Baltimore (1970). *J. Mol. Biol.* **49**, 657–669.
35. Kitamura, N., B. L. Semler, P. G. Rothberg, G. R. Larsen, C. J. Adler, A. J. Dorner, E. A. Emini, R. Hanecak, J. T. Lee, S. van der Werf, C. W. Anderson and E. Wimmer (1981). *Nature (London)* **291**, 547–553.
36. Krah, D. L. and R. L. Crowell (1985). *J. Virol.* **53**, 867–870.
37. Liljas, L., T. Unge, T. A. Jones, K. Fridborg, S. Lövgren, U. Skoglund and B. Strandberg (1982). *J. Mol. Biol.* **159**, 93–108.
38. Linemeyer, D. L., J. G. Menke, A. Martin-Gallardo, J. V. Hughes, A. Young and S. W. Mitra (1985). *J. Virol.* **54**, 247–255.
39. Lonberg-Holm, K. (1975). *J. Gen. Virol.* **28**, 313–327.
40. Lonberg-Holm, K. and B. E. Butterworth (1976). *Virology* **71**, 207–216.
41. Lund, G. A., B. R. Ziola, A. Salmi and D. G. Scraba (1977). *Virology* **78**, 35–44.
42. Makoff, A. J., C. A. Paynter, D. J. Rowlands and J. C. Boothroyd (1982). *Nucl. Acids Res.* **10**, 8285–8295.
43. Mandel, B. (1976). *Virology* **69**, 500–510.
44. Mandel, B. (1979). *In* "Comprehensive Virology" (Eds H. Fraenkel-Conrat and R. R. Wagner), vol. 15, pp. 37–121. Plenum Press, New York.
45. Marquart, M., J. Deisenhofer, R. Huber and W. Palm (1980). *J. Mol. Biol.* **141**, 369–391.
46. McGregor, S. and R. R. Rueckert (1977). *J. Virol.* **21**, 548–553.
47. McGregor, S., L. Hall and R. R. Rueckert (1975). *J. Virol.* **15**, 1107–1120.
48. Meloen, R. H., J. Briaire, R. J. Woortmeyer and D. Van Zaane (1983). *J. Gen. Virol.* **64**, 1193–1198.
49. Minor, P. D., G. C. Schild, J. Bootman, D. M. A. Evans, M. Ferguson, P. Reeve, M. Spitz, G. Stanway, A. J. Cann, R. Hauptmann, L. D. Clarke, R. C. Mountford and J. W. Almond (1983). *Nature (London)* **301**, 674–679.
50. Minor, P. D., P. A. Pipkin, D. Hockley, G. C. Schild and J. W. Almond (1984). *Virus Res.* **1**, 203–212.
51. Najarian, R., D. Caput, W. Gee, S. J. Potter, A. Renard, J. Merryweather, G. Van Nest and D. Dina (1985). *Proc. Natl Acad. Sci. USA* **82**, 2627–2631.
52. Neurath, H. (1984). *Science* **224**, 350–357.
53. Palmenberg, A. C., E. M. Kirby, M. R. Janda, N. L. Drake, G. M. Duke, K. F. Potratz and M. S. Collett (1984). *Nucl. Acids Res.* **12**, 2969–2985.
54. Pfaff, E., M. Mussgay, H. O. Böhm, G. E. Schulz and H. Schaller (1982). *EMBO J.* **1**, 869–874.
55. Putnak, J. R. and B. A. Phillips (1981). *Microbiol. Rev.* **45**, 287–315.
56. Racaniello, V. R. and D. Baltimore (1981). *Proc. Natl Acad. Sci. USA* **78**, 4887–4891.

57. Robertson, B. H., D. O. Morgan and D. M. Moore (1984). *Virus Res.* **1**, 489–500.
58. Rossmann, M. G. and P. Argos (1981). *Annu. Rev. Biochem.* **50**, 497–532.
59. Rossmann, M. G., D. Moras and K. W. Olsen (1974). *Nature (London)* **250**, 194–199.
60. Rossmann, M. G., C. Abad-Zapatero, M. R. N. Murthy, L. Liljas, T. A. Jones and B. Strandberg (1983a). *J. Mol. Biol.* **165**, 711–736.
61. Rossmann, M. G., C. Abad-Zapatero, M. A. Hermodson and J. W. Erickson (1983b). *J. Mol. Biol.* **166**, 37–83.
62. Rueckert, R. R. (1976). *In* "Comprehensive Virology" (Eds H. Fraenkel-Conrat and R. R. Wagner), vol. 6, pp. 131–213. Plenum Press, New York.
63. Rueckert, R. R. and E. Wimmer (1984). *J. Virol.* **50**, 957–959.
64. Rueckert, R. R., A. K. Dunker and C. M. Stoltzfus (1969). *Proc. Natl Acad. Sci. USA* **62**, 912–919.
65. Sangar, D. V. (1979). *J. Gen. Virol.* **45**, 1–13.
66. Schrom, M., J. A. Laffin, B. Evans, J. J. McSharry and L. A. Caliguiri (1982). *Virology* **122**, 492–497.
67. Sherry, B. and R. Rueckert (1985). *J. Virol.* **53**, 137–143.
68. Sherry, B., A. G. Mosser, R. J. Colonno and R. R. Rueckert (1986). *J. Virol.* **57**, 246–257.
69. Skern, T., W. Sommergruber, D. Blaas, P. Gruendler, F. Fraundorfer, C. Pieler, I. Fogy and E. Kuechler (1985). *Nucl. Acids Res.* **13**, 2111–2126.
70. Stanway, G., A. J. Cann, R. Hauptmann, P. Hughes, L. D. Clarke, R. C. Mountford, P. D. Minor, G. C. Schild and J. W. Almond (1983). *Nucl. Acids Res.* **11**, 5629–5643.
71. Stanway, G., P. J. Hughes, R. C. Mountford, P. D. Minor, and J. W. Almond (1984). *Nucl. Acids Res.* **12**, 7859–7875.
72. Steitz, T. A. and R. G. Shulman (1982). *Annu. Rev. Biophys. Bioeng.* **11**, 419–444.
73. Strohmaier, K., R. Franze and K. H. Adam (1982). *J. Gen. Virol.* **59**, 295–306.
74. Tainer, J. A., E. D. Getzoff, H. Alexander, R. A. Houghten, A. J. Olson, R. A. Lerner and W. A. Hendrickson (1984). *Nature (London)* **312**, 127–134.
75. Thomas, A. A. M., P. Brioen and A. Boeyé (1985). *J. Virol.* **54**, 7–13.
76. Toyoda, H., M. Kohara, Y. Kataoka, T. Suganuma, T. Omata, N. Imura and A. Nomoto (1984). *J. Mol. Biol.* **174**, 561–585.
77. Varghese, J. N., W. G. Laver and P. M. Colman (1983). *Nature (London)* **303**, 35–40.
78. van der Werf, S., C. Wychowski, P. Bruneau, B. Blondel, R. Crainic, F. Horodniceanu and M. Girard (1983). *Proc. Natl Acad. Sci. USA* **80**, 5080–5084.
79. Werner, T. C., J. R. Bunting and R. E. Cathou (1972). *Proc. Natl Acad. Sci. USA* **69**, 795–799.
80. Westhof, E., D. Altschuh, D. Moras, A. C. Bloomer, A. Mondragon, A. Klug and M. H. V. Van Regenmortel (1984). *Nature (London)* **311**, 123–131.
81. Wetz, K. and K. O. Habermehl (1979). *J. Gen. Virol.* **44**, 525–534.
82. Wetz, K. and K. O. Habermehl (1982). *J. Gen. Virol.* **59**, 397–401.
83. Wild, T. F., J. N. Burroughs and F. Brown (1969). *J. Gen. Virol.* **4**, 313–320.
84. Wychowski, C., S. van der Werf, O. Siffert, R. Crainic, P. Bruneau and M. Girard (1983). *EMBO J.* **2**, 2019–2024.

Index

AAG binding to disk aggregates, 255
AAGAAG (and related oligonucleotide) binding to disk aggregates, 255–256
Adenine plus uracil content, 5
Alfalfa mosaic virus (ALMV), 221
 alphavirus relationship to, 91–95, 165
 genome structure, 227, 229
 homologies with other viruses, 91–95, 227–229, 230
 virion structure, 221–222
Alphaviruses, 7, 75–95
 fusion, 293, 295–297
 genome
  conserved sequences, 86–88, 230
  organization, 76–78
  plant viruses and their relationship to, 91–95, 165
  proteins, processing, 78–86
  replication, see Replication
  spike proteins, 295–297
  translation, 76–78
Amino acid sequence homologies, see Homologies
Antibodies
 to cricket paralysis virus in mammals, 68–69
 monoclonal, 266, 269–276
  C3, poliovirus type-1 specific, 276
 neutralizing, 315
  picornovirus, 267–271, 303, 309–310, 312–314, 315
  poliovirus, 267–271, 315
  rhinovirus, 309–310
  use or production, antigenic sites determined with, 263–276, 303, 309–314
 viral binding facilitated by, 289
Antigenicity, poliovirus, 55
 recombinant, 57
Antigenic/immunogenic sites
 foot-and-mouth disease virus, 263, 263–268, 312–314
 picornaviruses, 261–277, 303, 308, 309–314, 315
 polioviruses, 263, 267–274, 312–314, 315
 rhinoviruses, 303, 308, 309–312
Antigenic mutation, antigenic structures studied using, 269–274
Aphid-transmitted viruses, 197
A-protein (alkalischer protein), 240, 247
Apthoviruses, see Foot-and-mouth disease viruses
Assembly
 picornavirus, 316
 tobacco mosaic virus, 243–256
Assembly origin for encapsidation in tobacco mosaic virus, 165
Attenuation
 Sabin type 3 vaccine, 53–59
  mutations causing, 55–57, 57–58, 58, 59
  physiological basis, 59
  virulence versus, 315
AUG codon in picornaviruses, 4, 21
Autoproteases/autoproteolysis, 79–83, 168

Base composition of picornavirus genomes, 4, 5
Base-pairing, RNA production inhibited by, 213–215
Bean pod mosaic virus (BPMV), 222
Beet necrotic yellow vein virus (BNYVV), 193–196, 199
BHK cells, virus entry, 284, 285, 288, 290–291
Binding, viral, to cell surface, 288–289
Biological control agents, insects viruses as, 73
Bipartite genomes, 178–201
 advantages, 200

Bipartite genomes—*cont.*
  disadvantages, 201
  insect virus, 207–217
  plant virus, 177–201
  RNA species in, 179–201 *passim*
    characters assigned to, 200
Black beetle virus (BBV), 207, 208, 209–217
  genetic map, 213
  isolation and growth, 209
  molecular biology, 209–211
  plaque assay, 211
  RNA species, 209–211
    sequencing, 211–215
Broad bean mottle virus (BBMV), RNA secondary structures, 225–227
Bromelain, haemagglutinin isolated with, 295
Brome(grass) mosaic virus (BMV), 219–234
  alphavirus relationship to, 91–95 *passim*, 165
  early history, 220–221
  homologies with other viruses, 91–95 *passim*, 227–231 *passim*
  RNA replication, 231–232
  RNA species and sequences, 222–234
  translation, 223–224
  virion structure, 221–223
Bronchitis virus, infectious, *see* Infectious bronchitis virus

Canyons, picornavirus surface, 305–309
  as receptor binding sites, 314–315
Cap(s), tobravirus, 185, 187
Cap-binding protein (CBP) complex, picornavirus, 20
Capsid, full vs. empty, antigenic differences, 261
Capsid proteins
  alphavirus, 79–83
    cloned copies, expression, 83, 84
  picornavirus, *see* P1 poluypeptides, VP protein
  plant virus, 303–309
  rhinovirus, 303–309
Cardioviruses, physical properties, 2, 3

Carlavirus, 155
Carnation mottle virus, 155, 157
cDNA, *see* DNA, complementary
Cell
  components, replication-associated, 31–32, 34
  surface, virus binding to, 288–289
Cell-to-cell transport/spread/infection (in plant viruses), 192
  nepoviruses, 192
  tobacco mosaic virus, 166, 188
Cherry leafroll virus (CLRV), 190–192
Chromatin, host, viral proteins associating with, 166
Cistrons, silent, 224
Classification
  picornavirus, 1–2, 17, 259
  plant virus, 154, 155
  togavirus, 106, 107
Clathrin, 285
Cleavage sites/sequences, proteolytic, within polyproteins, 8, 12, 21–22, *see also* Proteolysis
Cleft, viral surface, *see* Canyons
Coated pits, 284–285
Coated vesicles, 284–285, 287
Coat protein
  picornavirus, 259–261
    antigenicity, 261–277, 303, 308, 309, 314
  synthesis
    black beetle virus, 210
    brome mosaic virus, 223–224
    dianthovirus, 196
    fungus-transmitted rod-shaped viruses, 193–194, 194
    plant viruses, 168, 186, 193–194, 194, 196, 223–224
    tobravirus, 186
    tymovirus, 168
    tobacco mosaic virus, 239–242
      aggregation, 239–242
Codon
  randomization, 89–91
  usage/selection
    alphavirus, 88–91
    picornavirus, 6, 7
Comoviruses 179–184, 199–201
  general properties, 179–180

proteases, 183–184
RNA species, 179–183
sequence comparisons with other viruses, 184
translation products, 180–183
Conserved sequences/domains
alphavirus, 86–88, 230
alphavirus-plant virus homologies, 93
bromovirus and cucumovirus, 225–227
plant virus, 225–227, 230
Control agents, biological, insect viruses as, 73
Cornell High Energy Synchrotron Source (CHESS), 317
Coronavirus, 117–125
recombination in, 140, 148, 149
sequence analysis, 120–121
subgenomic mRNA, 113–114, 118–119
translation products, 119–120
Cowpea chlorotic mottle virus (CCMV) RNA secondary structure, 225–227
Cowpea mosaic virus (CPMV), 179–183
nepovirus homologies, 192
Cowpea severe mosaic virus (CPSMV), 183
Cowpea strain of tobacco mosaic virus (CcTMV), genome organization, 164
Cricket paralysis virus (CrPV), 67
genome organization, 72
replication
in insect cell cultures, 69–71
in insect hosts, 68
in mammals, 68–69
Crossing viruses, procedure, 133, *see also* Recombinants; Recombination
Cross-linking of *Fab* fragments to poliovirus, 274–275
Cross-overs, *see* Recombination
Crude replication complex, 40–42
Cucumber green mottle mosaic virus (CGMMV), 159
genome organization, 164
Cucumber mosaic virus (CMV), RNA secondary structure, 225–227

Cyanogen bromide-treated VP1 fragments, antigenicity, 264
Cytoplasmic polyhedrosis viruses (CPVs), 73

Defective interfering (DI) RNA, 130–131, 131, 149
influenza virus, 130
Sindbis virus, 86–88, 130, 131
Deoxyribonucleic acid, *see* DNA
Dianthoviruses, 196–197, 199
Disks of tobacco mosaic virus
aggregation, 240–242
nucleation-related structure, 248, 249
oligonucleotide binding to, 255–256
role, 245–247
DNA, complementary (cDNA)
clones
infectious, 232–233
infectious RNA derived via, *see* Infectious RNA
recovery of parental poliovirus form, 54–55
RNA cross-linked to hybridized, TMV assembly employing, 253–255
DNA forms of RNA viruses, infectious, 232–233
Domains and sequences
conserved, *see* Conserved sequences/domains
functional, rhinovirus vs. poliovirus, 66
Double stranded RNA, *see* Replicative form RNA
Drosophila C virus (DCV), 67
replication
in insect cell cultures, 69–71
in insect hosts, 68
Drosophila melanogaster cells, culture in
black beetle virus, 209, 210, 211, 212, 215, 217
insect picornavirus, 69–71

E. Coli cell-free protein synthesizing systems, 223

Electrofocussing, gel, screening recombinants by, 133–135, 135, 145–146
Electron density map for TMV protein, 241
ELISA tests, antigenic sites determined with, 266, 276
Elongation of TMV protein on RNA, 248–255
  3' direction, 248
  5' direction, 248, 252
  bidirectional simultaneous, 251, 253, 255
Encapsidation
  black beetle virus, 213
  brome mosaic virus, 224, 239
  picornavirus, 25
  tobacco mosaic virus, 165
Encephalomyocarditis virus (EMCV)
  antigens common to cricket paralysis virus, 68–69
  comovirus homologies, 184
  homology with other viruses, 9–11, 184
  proteolytic processing, 11–12, 22, 23
Endocytic pathway, 283–287
  advantages, 297
Endocytosis (internalization), 283–292, 297
  fluid-phase, 283, 284
  receptor-mediated, 283, 284, 289–290
  virus entry via, 288–293
    kinetics of, 289–290
Endosomes, 285–287
  pH and virus fusion, 290–291
  virus delivery, 290
Enteroviruses, physical properties, 2
Entry/uptake
  enveloped virus, 281–298
  picornavirus, 18
Enveloped viruses
  entry/uptake, 281–298
  structure, 282–283
Envelope proteins, Sindbis, clones, expression from, 83, 84, 85
Enzyme-linked immunosorbent assays (ELISA), antigenic sites determined with, 266, 276

Equine arteritis virus, 105–114
  proteins, 107–108
  replication, 112–114
  RNA, 108–112
Evolution of RNA viruses, 90–91, 93, 94–95
  alphavirus, 90–91, 93, 94–95
  bipartite genome, 199
  convergent vs. divergent, 231
  picornavirus/rhinovirus, 316–317
  plant virus, 93, 94–95, 199, 230–231, 317
  structure related, 316–317
Expression, gene, *see* Gene expression
Expression systems, 86
  in vaccinia, 83

Fab fragment, cross-linking to poliovirus, 274–275
Fingerprinting with RNase, *see* RNase T_1
Flacherie virus, infectious, 72
Flavivirus
  genome organization and translation, 96–102
  replication, 91, 93, 106
Foot-and-mouth disease viruses (= aphtoviruses)
  antigenic structures, identification, 263, 263–268, 312–314
  canyon walls as receptor binding sites, 314–315
  genome general properties, 4, 26
  homology with other viruses, 9–11
  physical properties, 2, 2–3
polyprotein processing, 12
recombinants
  biochemical characterization, 135–138
  O/A serotypes, 144
  O/SAT2 serotypes, 144, 145, 146
  Rec B971, 146–147
recombination
  interstrain, 143–144
  sites, 140–143, 146–147
  recombination map, 133–135, 140, 141
  rhinovirus compared with, 308
  VPg protein, 30

Free energy of possible RNA secondary structures in yellow fever virus, 101
Functional domains, rhinovirus vs. poliovirus, 66
Fungus-transmitted rod-shaped viruses, 193–196, 199
Furoviruses, 193–196, 199
Fusion, 290, 291, 292, 292–297
  mutants
    influenza virus, 294
    Semliki Forest virus, 290, 296
  proteins, 293–297
    types, 293

Gene expression, see also Processing; Translation
  plant virus, 156–160, 160–172 passim
  regulation, coronavirus, 121–125
Genetic exchange, see Recombination
Genetic recombination, see Recombination
Genome
  base composition in picornaviruses, 4, 5
  size
    picornavirus, 4
    plant virus, 155, 179
  structure and properties, 129–151 passim
    alphavirus, 76–78
    alphavirus-plant virus similarities, 91, 92
    bipartite, 177–217
    coronavirus, 117–118
    flaviviruses, 96, 99, 100–101
    insect virus, 72
    mammalian viruses, 3–5, 8, 19, 108–114
    picornaviruses, 3–5, 8, 19, 72
    plant virus, 91, 92, 153–172, 177–201
    poliovirus, 19
    rhinovirus, 62–63
    tripartite, 223–234
Genome-linked protein, virus coded, see VPg

Glycoprotein
  cell, virus binding to, 288
  virus, 282–283
    membrane, 282–283
    Sindbis, vaccinia-encoded, expression, 83–84, 85
    yellow fever virus, 98
Glycosylation, protein function and the importance of, 99
Grapevine fanleaf virus (GFLV), 190
Guanidine resistant mutants, 30–31
  recombinant selection/mapping using, 133, 134, 135, 140

Haemagglutinin (HA) molecule, 260, 293–295
  cell surface binding, 288
  trimeric (BHA), 294, 295
  as a virus receptor, 289
Heat shocking of insect virus infected cultures, 70
Helix assembly in tobacco mosaic virus, 245–247
Homologies, 131
  amino acid/(poly)protein, 9–11, 227–231
    alphavirus, 79–81, 81–83
    alphavirus-plant virus, 91, 92, 93–94, 165
    flavivirus, 99, 100
    nepovirus-comovirus, 192
    picornavirus, 9–11, 31, 260, 302
    picornavirus-comovirus, 184
    rhinovirus-poliovirus, 63–64
    tobravirus-other plant virus, 188
  RNA
    bromovirus-cucumber mosaic virus, 225–227
    coronavirus, 124
    at recombination sites, 146–147, 148
    rhinovirus/poliovirus, 63, 64
Horse radish peroxidase, membrane recycling studies using, 286–287
Host factor, 32, 36
  function, 32, 36, 37, 42
Hydrophilicity analysis, antigenic sites predicted by, 262–263

Hydrophobic bonding, TMV protein aggregation effected by, 240
Hydrophobicity profiles, yellow fever and Murray Valley encephalitis proteins, 99–101
Hydrophobic sequences, fusion proteins with, 293

Immunofluorescence microscopy of Sindbis-vaccinia recombinant infected cells, 83, 85
Immunogens, see also Antigenicity; Antigenic sites
  picornavirus, 263–268
    protein fragment, antigenic studies using, 263–267
    rhinovirus, 303, 308, 309–312
    synthetic peptides, antigenic structures studied using, 263–268, 312–314
Immunoprecipitation of plant viral proteins, 160, 161
Inclusion bodies, potyvirus associated, 160
Inclusion proteins, potyvirus associated, 161
Infection, cell-to-cell see Cell-to-cell transport/spread/infection
Infectious bronchitis virus (IBV), 117–125
  sequence analysis, 120
  subgenomic RNA, 118
    translation products, 119–120
Infectious flacherie virus, 72
Infectious RNA
  cDNA derived, 215
    black beetle virus, 215–217
    brome mosaic virus, 232–234
  picornavirus replicative form (RF), 27
Influenza virus
  defective interfering RNA, 130, 131
  entry, 289, 290, 293–295
  fusion, 293–295
  haemagglutinin, see Haemagglutinin
  recombination, 130, 131, 149
  transcription, priming, 131
Initiation of helix assembly in tobacco mosaic virus, 245, 248

Initiation/start sites
  replication/transcription
    coronavirus, 124–125
    internal, 159
    picornavirus, 28
  translation, picornavirus, 4, 20, 21
Insect viruses
  bipartite genomic, 207–217
  picornavirus, 67–73
  viruses evolving from, 95
Internalization, plasma membrane and virus, see Endocytosis
Ionic strength, TMV coat protein aggregation effects, 239, 240
Isoelectric point of proteins, changes, mutations and recombinants mapped via, 133–135, 135, 145–146

Lactic dehydrogenase virus (LDV), 107
Leader RNA sequences, see Messenger RNA
Life cycle, virus, role of 5′ untranslated region in, 63
Ligand/receptor complexes, 283–284, 285, 286
L protein, 7
Lucerne transient streak virus (LTSV), 170, 171
Lysosomes
  membrane delivery to, 284, 287
  virus delivery to, 291–292

Map, genetic
  black beetle virus, 213
  tobacco mosaic virus, 163
  turnip yellow mosaic virus, 167
  tymovirus, 166, 167
Map(s), recombination, picornavirus, 133–135
Mapping, recombinant RNA, 139–140
Membrane
  cell
    internalization, see Endocytosis
    recycling, 286–287
  virus, 282–283
Membrane-associated polypeptides, poliovirus, 29

Membrane replication complexes, 32, 38–40
Messenger RNA
  equine arteritis virus, 108–112, 113–114
  leader-primed synthesis, 121, 124
  recombination involving, 149
  mouse hepatitis virus, 113–114, 118–120
  picornavirus, 20, 28
  splicing, 114
  subgenomic, 108–112
    transcription and translation, 112–114
Middleburg virus
  codon usage, 89, 90
  Sindbis virus homology with, 89, 90
Morphology, plant virus, 155
  alphavirus morphology compared with, 93
Mouse (murine) hepatitis virus (MHV)
  recombination, 148
  sequence analysis, 120, 121
  subgenomic mRNA, 113–114, 118–119
  translation products, 119–120
Mouse submaxillary gland protease-treated VP1 fragments, antigenicity, 264
Multiplicity of infection, cell entry patterns related to, 292
Murine hepatitis virus, see Mouse hepatitis virus
Murray Valley encephalitis (MVE) virus, 99–101
Mutations
  antigenic, antigenic structures studied using, 269, 274
  deleterious, recombination-eliminated, 150–151
  recombinants detected and selected using, 132–135
Myxoviruses, cell surface binding, 288, see also Orthomyxoviruses; Paramyxoviruses

N-AgI/II, see NIm
Nematode, plant virus transmission by, 188, 189

Nepoviruses, 188–193, 200
  general properties, 188–189
  RNA species, 188–192
  sequence comparisons with other viruses, 192
  sequence determinants, 192
  translation products, 190–192
Neurovirulence, poliovirus, 55
  recombinants, 57–58
NIm, 315
  -IA (N-AgIA), 308, 311
  -IB (N-AgIB), 311, 311–312
  -II (N-AgII), 309, 311, 312
  -III (N-AgIII), 309, 310–311, 312
Nodamura virus, 208
Nodaviridae, 207–208
NS (=NV) proteins of yellow fever virus, 97, 98–99
nsP proteins, alphavirus, homologies with plant viruses, 91–93, 93–94
nsP4, alphavirus, 76, 78
  codon randomization, 89–90
  homologies, 89
Nuclear polyhedrosis viruses (NPVs), 73
Nucleation of assembly in TMV, 247–248, 250
Nucleocapsid particle, 282
Nucleoproteins, brome mosaic virus, 222
Nucleotide composition of picornavirus genomes, 4, 5
NV (=NS) proteins of yellow fever virus, 97, 98–99

Oligonucleotide binding to disk aggregates, 255–256
Oligonucleotide fingerprinting, see RNase T_1
Open reading frames, see Reading frames
Orthomyxoviruses, 282, see also Myxoviruses
  fusion, 293

P1 (polypeptides), 7–9
  proteolytic processing, 13

P2 (polypeptides), 9
　C (=2C), 30–31
P3 (polypeptides), 9
　AB (=3AB), 30
　C (=3AC), 11–12
　D^{pol} (=$3D^{pol}$), see RNA polymerase
　proteolytic processing, 11–13
Packaging, alphavirus, sequences affecting, 86–88
Papaya ringspot virus (PRSV), 160
　post-translational cleavage in, 161–162
Paramyxoviruses, fusion, 291, 293, see also Myxoviruses
Pea-early browning virus (PEBV), 185
Pea ennation mosaic virus (PENV), 197–198
Peanut clump virus (PCV), 193, 194
Penetration, 290–291, see also Entry
Peptides
　binding studies, antibody binding sites identified by, 274–276
　synthetic, antigenic structures studied using, 263–268, 312–314
Pepper ringspot virus (PRV), 185, 186, 187–188
Pestivirus replication, 106
pH
　endosome sorting process and, 286
　TMV coat protein aggregation effects, 239, 240
　TMV disk structure effects, 239, 245, 246
　virus fusion requirements, 290–291, 292–293, 294, 295–296, 297
Phagocytosis, 283
Picornaviruses, 1–13, 67–73, 259–277, 302–303
　assembly, 316
　classification, 1–2, 17, 259
　coat proteins and their antigenicity, 259–277
　comovirus homologies, 184
　genome structure and properties, 3–5, 8, 19, 72, 302
　insect, 67–73
　Nodamura virus similarities and differences, 208
　physical/general properties, 2, 18
　plant virus homologies, 184

recombinants, biochemical characterization, 135–138
recombination, 133–148
recombination maps, 133–135, 139–140
replication, see Replication
rhinovirus homologies, 64–66
translation and processing, 5–13
Pinocytosis, 283, 284–285
Plant viruses, 153–201
　alphavirus relationship to, 91–95
　classification, 154, 155
　general features, 154–156
　genomes, 91, 92, 153, 172
　　bipartite, 177–201
　　tripartite, 219–234
　picornavirus/rhinovirus relationship to, 303–309, 316
Plaque assay
　black beetle virus, 211
　poliovirus, antibody titre measured by, 267
Polio 1 Sabin virus, homology with other viruses, 9–11
Poliovirus(es)
　antigenic structures, 216–277, 303, 308, 309–314, 315
　genome organization, 19
　plant virus homologies, 184
　poliovirus P3/Leon/12a$_1$b, 54, 55, 269, 273
　poliovirus P3/Leon/37, 54, 55
　poliovirus P3/Leon/USA/1937, 269, 270, 272
　proteolytic processing, 19
　recombinants
　　progenitor/vaccine, construction, 55–57
　　vaccine attenuation investigated with, 53–59
　recombination
　　interstrain, 143
　　nucleotide sequence homology at sites of, 146, 147
　relationship with other picornaviruses, 9–11, 63–66
　rhinovirus relationship to, 10, 11, 64–66, 308
　RNA, 4, 54
　　intracellular structures, 23, 24

type 1, 9–11, 26, 268, 270, 271, 274–276
 open reading frames, 62, 63
 Mahoney, 271, 275
 rhinovirus homologies, 10, 11, 63–64
poliovirus type 2, 270, 271
poliovirus type 3, 53–59, 268, 269–274, 276
 virulence vs. attenuation, 315
Poly A (polyadenylate) tracts in picornavirus, 25–26
 transcription initiated at, 28–29
Poly C (polycytidylate) tract in picornavirus, 3, 18, 26
Polyprotein
 definition, 5
 homologies, see Homologies
 organization, 7–9, 19, 21
 processing by proteolysis, see Processing
 translation, see Translation
Poly U
 synethesis, 37
 tracts, 25, 26
potato viruses, see Potexvirus; Potyvirus
Potexvirus (potato virus X), 155, 157
 expression strategies, 157, 159
Potyvirus (potato virus Y), 155, 157, 160–162
 expression strategies, 157, 160–162
Processing, proteolytic (=protein processing), see also Cleavage sites/sequences
 alphaviruses, 76, 78, 78–86
 enzymes, see Prote(in)ases, protein processing
 flaviviruses (yellow fever virus), 97, 98
 insect picornaviruses, 69, 71
  cellular site, 78
 mammalian picornaviruses, 7, 8, 11–13, 19, 21–23
Processing, RNA
 comoviruses, 181–183, 183–184
 equine arteritis virus, 113–114
 plant viruses, 161–162, 167–168, 181–183, 183–184
 polyprotein, 7, 8, 11–13, 19, 21–23, 76, 78, 78–86, 98, 161–162, 182, 198, 198–199, 229

comovirus-picornavirus comparisons, 184
Prote(in)ases, protein processing
 alphavirus, 79–83
 comovirus, 181–183, 183–184
 comovirus-picornavirus homologies, 184
 insect picornavirus, 71
 mammalian picornavirus, 11–13
Prote(in)ase 3C^{pro}, picornavirus
 cleavage by, 21–22, 22
 comovirus homologies, 184
Proteins
 comovirus, 181–183
 equine arteritis virus, 107–108
 insect virus induced, 69–71
 non-structural, role in plant viruses, 165–166, 168
 picornavirus, 28–33, 69–71
 poly-, see Polyprotein
 recombinant, 145–146
 RNA replication-involved, 28–33
 synthesis, endogenous, in plant virus infected cells, 171
Proteolysis (cleavage)
 haemagglutinin, 293
 polyprotein, see Processing
Pseudorecombination, 179, 189, 196, 201

Qβ replication, 34
 codon usage in, 90

Rasberry ringspot virus (RRV), 189, 201
rct 40^0 marker test, 57, 58, 59
Reading frames
 alternative/multiple, 91, 158
 open
  black beetle virus, 212
  coronavirus, 120
  flavivirus, 96–98
  fungus-transmitted rod-shaped viruses, 194–196
  potyvirus, 172
  rhinovirus vs poliovirus, 62–63
  tobamovirus, 164–165
  tobravirus, 187, 188
 overlapping, 91

Read-through, *see* Suppression
Receptor(s)
  binding sites, canyon walls as, 314–315
  picornavirus, 18, 62, 303, 314–315
  rhinovirus, 62, 314–315
  specificity, 289, 303
    virulence vs. attenuation and, 315
Receptor-mediated endocytosis, 314–315
Recombinants
  biochemical characterization, 135–138
  detection and isolation, 132–133
Recombination, genetic, 129–151
  brome mosaic virus, 234
  frequency of genomes undergoing, 135
  homologous, 130, 132–148, 148–149
  interstrain, 143–145
  maps/mapping, of picornaviruses, 133–135, 139–140, 149
  mechanisms, 131, 148–149
  non-homologous, 130–131
  proof, 138–139
  pseudo-, 179, 189, 196, 201
  reason/advantages, 149–151
  sites, 140–143
    nucleotide sequence homologies at, 146–147
  virus functions mapped by, 140
Red clover necrotic mosaic virus (RCNMV), 196–197
Repair, RNA, recombination-mediated, 150
Replicases
  alphavirus-plant virus homologies, 93–94, 95, 96
  brome mosaic virus, 231–232
  plant virus, 168, 231–232
  turnip yellow mosaic virus, 168
Replication/synthesis/transcription, RNA
  alphavirus, 106
  plant virus with similar replication patterns, 91–93
    sequences affecting, 86–88
  brome mosaic virus, 231–232
  flavivirus, 106
  functions, homologous, among plant virus, 230

pestivirus, 106
picornavirus, 23–44
  hair pin, snap-back/self-priming, 40–41
  in insect infecting forms, 68–72, 72
  initiation, 28
  in vitro, 34–40
  in vivo, 33–34
  minus strand, 33, 37
  models, 40–44
  phases, 33
  plus strand, 33
  proteins/polypeptides associated with, 9, 28–33
  soluble systems, 35–38
priming, in influenza virus, 131
recombination as a side-effect of, 149–50
rubivirus, 106
taxonomic comparisons, 106
of viruses with subgenomic mRNAs. 112–114
Replication complexes (RC), 28, 31
  crude (CRC), 38–40
  membrane associated, 31, 32, 33, 38–40
  partially purified (RCI), 39, 40
Replicative form (double stranded) RNA in picornaviruses, 23, 26–27, 40–41
  hetero-linked, 41
  homo-linked, 41, 43
  open vs. closed, 24
  synthesis, 33–34
Replicative intermediate RNA in picornaviruses, 23, 24, 27–28
Rhabdoviruses, fusion in, 293
Rhinoviruses, 61–66, 301–317
  antigenic structures, 269–274
  genome general properties, 4
  human, 61–66
  physical properties, 2
  poliovirus relationship to, 64–66
  polyprotein processing, 12
Rhinovirus 2, human (HRV2), 61–66
Rhinovirus 14, human, (HRV14), 61–66, 310–317
  antigenic sites, 303, 308, 309–312
  antigenic variants, 271, 274

homology withother viruses, 9–11
structure, 303–309
Ribonuclease, see RNase T$_1$
Ribonucleic acid, see RNA
RNA
  circular, 170
  conserved sequences, see Conserved sequences
  defective interfering, see Defective interfering RNA
  equine arteritis virus, 108–112
  genome, see Genome
  infectious, see Infectious RNA
  messenger, see messenger RNA
  monocistonic, 199
  repair, recombination-mediated, 150
  sequence homologies, see Homologies
  structure, secondary
    black beetle virus, 213, 214
    bromoviruses, 225–227
    cucumber mosaic virus, 225–227
    yellow fever virus, 101–102
  subgenomic, 227
    black beetle virus, 211
    brome mosaic virus, 224
    coronavirus, 113–114, 118–119
    dianthovirus, 196
    fungus-transmitted rod-shaped viruses, 196
    plant virus, 158–159, 162–163, 168, 186, 187, 196, 227, 229
    sequence relationships, 110–112
    tobravirus, 186, 187
    togavirus, 108–114
    transcription and translation strategies of viruses with, 112–114
  synthesis, see Replication
  tobacco mosaic virus assembly dependent on, 247–255
  transfer, suppressor, see Suppressor tRNA
RNA polymerase
  picornavirus, 28–29, 35–36, 36–37, 37
  picornavirus-comovirus homologies, 184
  3D^{pol}, 28–37
    priming, 28–29, 35–36, 36–37

tobravirus homologies with other plant viruses, 188
RNase T$_1$, oligonucleotide fingerprinting with
  equine arteritis virus RNA, 110–112
  recombinants analysed by, 135–138, 139, 140, 144
Ross River virus
  codon usage, 89, 90
  serine protease homologies with, 82
  Sindbis homologies with, 79–81, 82
Rubivirus replication, 106

Sabin vaccine strains
  antigenic structures, 269–270
  type 3, attenuation in, 53–59
Sabin virus, polio 1, homology with other viruses, 9–11
Satellite viruses and RNAs, 168–171
  bipartite genomes associated with, 178
Semliki Forest virus, 295–297
  codon usage, 89, 90
  entry, 284, 288, 289, 290, 291, 291–292, 295–297
  fusion reaction, 295–297
  serine protease homologies, 82
  spike glycoproteins, 295–297
Sequencing of RNA viruses, 224
Serine proteases, 81–83
Sialic acid residues, cell surface, virus binding to, 288
Silent cistrons, 224
Sindbis virus
  defective interfering (DI) RNA, 86–88, 130, 131
  genome organization and translation, 75–95, 229
  proteins, cloned copies, expression, 83–86
  recombination, 130, 131, 148
  relationships with other viruses, 91–95, 229–231
Single-stranded RNA of polioviruses, 23, 24
Size
  genome, see Genome
  virus, entry related to, 290

Sobemovirus, 155, 157
    satellites, 170–171
Soil-borne wheat mosaic virus (SBWMV), 193–194, 199
Solanum nodiflorum mottle virus (SNMV), 170–171
Southern bean mosaic virus (SBMV), picornavirus/rhinovirus relationship to, 303–309, 316
Spike proteins, 282–283, 293–297
    types, 293
Splicing, mRNA, in equine arteritis virus, possibility of, 114
Squash mosaic virus (SqMV), 183
Start sites for transcription and translation, see Initiation/start sites
Stop codons, UAG, suppression at, see also Suppression and read-through
    in fungus-transmitted rod-shaped viruses, 193–194
    in tobacco mosaic virus, 163
Strains, recombination between, 143–145
Suppression and read-through, see also Stop codons
    alphavirus, 76–78
    plant virus, 158, 163, 165, 170, 186, 193–194
    tobravirus, 180
Suppressor tRNAs
    alphavirus, 76
    plant virus, 163
Synthesis, see also Assembly
    protein, see Coat proteins; Proteins; Translation
    RNA, see Replication

Taxonomy, see Classification
Temperature sesitive mutants
    recombinant selection using, 132, 133–135, 140, 143
    Sindbis virus, polyprotein cleavage in, 79
Temperature sensitivity of poliovirus, 55
    recombinants, 58–59

Terminal blocking groups, 3′-, 212, see also Caps
Terminal uridyl transferase (TUT), 32, 37
    replication models involving, 40, 43
Termination, premature, 156–158
Tobacco etch virus (TEV), 160, 161
Tobacco mosaic virus, 162–166, 237–256
    alphavirus relationship to, 91–95
    assembly, 243–256
    cell-to-cell migration, mutations affecting, 166, 188
    evolution, 231
    homologies with other viruses, 91–95, 188, 227–231
    L strain, 164–165
    molecular architecture, 239–243
    RNA species, 162–163, 227, 229
    tobravirus homologies, 188
    translation strategies, 163–164
    vulgare strain, 164–165
Tobacco necrosis virus (TNV), 155, 157, 169
    satellite, 169
    in vitro translation, 223
Tobacco rattle virus (TRV), 185–188
Tobacco ringspot virus (TobRV), 189, 190
Tobacco vein mottling virus (TVMV), 160, 161, 172
    translational strategies, 162
Tobamoviruses, 155, 157, 162–166
    expression strategies, 157, 159
Tobraviruses, 185–188, 199–201
    general properties, 185
    RNA species, 185–188
    sequence comparisons with other viruses, 188
    sequence determination, 186–188
    translation products, 185–186
Togaviruses (togaviridae), 105, 106, 107
Tomato black ring virus (TBRV), 189, 190, 192
Tomato bushy stunt virus (TBSV)
    rhinovirus relationship, 303, 304, 305, 309
    satellites, 170
Tombusvirus, 157, 159
    satellites, 170

Transcription, *see* Replication
Transcription vectors for cDNA, 233
Transfer RNA(s), suppressor, *see* Suppressor tRNAs
Transfer RNA-like structures, 130–131
 in brome mosaic virus, 232
Translation of RNA
 alphavirus, 76–78
 bipartite genome viruses, 178–201, 209–211
 coronavirus, 119–120, 121
 flavivirus, 98
 in vitro
  insect viruses, 71
  plant viruses, 181, 190, 193
 picornavirus, 5–7, 20–21, 33, 71
 plant virus strategies, 157–159, 160–171, 178–201 *passim*
 polyprotein, 5–7, 20–21, 98
  initiation, 20, 21
  rate, 5–6
 rhinovirus, 63
 switch from, to transcription, 33
 tripartite genome viruses, 223–224
 untranslated regions and their role in, 63
 in viruses with subgenomic mRNA, 112–113
Transmission of plant viruses, 154–155
Tripartite plant virus genomes, organization, 219–234
Trypsin sensitivity of Semliki Forest virus spike proteins, 295, 296
Trypsin-treated foot-and-mouth disease virus/proteins, antigenicity, 263, 264
Turnip crinkle virus (TCV) and its satellites, 170
Turnip yellow mosaic virus, 166–168, 220
Tymoviruses, 155, 157, 166–168

UAG stop codons, *see* Stop codons
Uncoating, 290
Untranslated regions, 5'-, of rhinovirus, 62–63
Uptake, virus, *see* Entry

UV-target sizes of subgenomic mRNA template, 113–114

Vaccine, Sabin type 3, attenuation, 53–59
Vaccinia virus, Sindbis genes cloned in, 83–86
Variants, recombination generated, 150
Velvet tobacco mottle virus (VTMoV), 170
Vesicular stomatitis virus (VSV) entry, 289, 290
Viroids, 170, 171
Virulence vs. attenuation, 315
Virusoids, 170–171
VP (capsid) proteins
 equine arteritis virus, 107, 107–108
 picornavirus, 7–9, 31, 302, 303–309, 316
  antigenicity, 261–262, 263–276, 310–312
  assembly, 316
  gene region, absence of recombination in, 143
  plant virus equivalents, 303–309 *passim*
  protomers, 316
 rhinovirus, 302, 303–309, 310–312
VP0 protein, picornavirus, 316
VP1 protein, picornavirus, 303–309, 316
 antigenicity, 263–276, 310–312
 precipitation, monoclonal antibody, 276
VP2 protein, picornavirus, 303–309, 316
 antigenicity, 263, 267, 271, 274, 275, 312
VP3 protein, picornavirus, 303–309, 316
 antigenicity, 271, 274, 275, 312
 mutations in, attenuation-related, 58
VP4 protein
 picornavirus, 305, 316
 rhinovirus, 305
VPg (virus-coded genome-linked) proteins
 comovirus, 180, 181, 184

nepovirus, 189
picornavirus, 3, 9, 18, 20, 28, 29–30, 31, 184
  detection, 29
  function, 25
  linking to nascent RNA, 37, 38
  precursor, 29, 30
  -related proteins, 36
  removal/release, 25
  structure, 29
plant virus, 161, 170, 180, 181, 184, 189
poliovirus/rhinovirus sequence comparisons, 65, 66
sobemovirus, 170
uridylated
  RNA synthesis primed by, 41–43, 44
  synthesis, 38, 39–40, 42

Wheat cell-free translation system, 223
Wheat mosaic virus, soil-borne (SBWMV), 193, 194, 199

X-ray crystallography of TMV protein, 240

Yellow fever virus, genome organization and translation, 96–102
Yield test, 133